Generis

PUBLISHING

Apis Mellifera mtDNA: Diversity, Evolution, and Conservation Applications

Jaber Amin

Title: **Apis Mellifera mtDNA: Diversity, Evolution, and Conservation Applications**

ISBN: 979-8-89966-248-5

Author: Jaber Amin

Cover image: www.pixabay.com

Publisher: Generis Publishing
Online orders: www.generis-publishing.com
Contact email: info@generis-publishing.com

Contents

ACKNOWLEDGEMENTS

First: Supervisors and Professors: I would first like to express my deepest gratitude and appreciation to my supervisor, Prof. Dr. B. Devrim Özcan, for his invaluable guidance and continuous academic support throughout my research journey. His scholarly mentorship was the cornerstone of completing this work.

Secondly: Colleagues and Researchers: I thank my colleague researcher Elif Dikaya, and all laboratory members, for their valuable collaboration and stimulating discussions. I also acknowledge Prof. Dr.Ulviye Komova for her essential assistance in collecting critical research samples necessary for conducting the required experiments.

Thirdly: Family: The Source of Giving and Support; I dedicate this book to: The soul of my beloved father, Mr. Eshag Amin AbrahimThe inspirer, who cherished knowledge and culture), may Allah have mercy on him and grant him paradise. May this effort immortalize his passion for knowledge and serve future generations. My dear mother Ms. Kulthum Hasan Khamis, for her endless moral support, tireless dedication to my well-being, and daily reassurance about my circumstances. My debt of gratitude to her is immeasurable. All members of my honorable family, who never hesitated to offer unwavering support and encouragement throughout my doctoral journey.

Fifth: Friends and Journey Companions: Lastly, I extend my thanks to all friends and colleagues who supported me morally and materially. They were true pillars of strength during difficult times; with their support, the challenges of this journey were eased and enriched.

Chapter 1

1. Introduction to *Apis mellifera* and mtDNA

The western honeybee (*Apis mellifera)* is vital to the global economy and environment. It is a highly eusocial species with a complicated social structure. It is also very important for pollination to be able to live in a wide range of climates. Makes it essential for both natural ecosystems and modern farming. This species has a lot of genetic diversity within it. There are about 30 to 31 recognized subspecies spread out over continents and grouped into different evolutionary lineages, such as African (A), North Mediterranean (C), Western and Northern European (M), Oriental (O), and Ethiopian (Y) lineages (Ajao et al., 2021; Alburaki et al., 2023).

Mitochondrial DNA (mtDNA) has become an important molecular marker in honey bee genetics research because it has several advantages over nuclear markers. The mtDNA is a small, circular genome that is about 16–17 kilobases long. It comes from the mother and doesn't recombine, so it gives a clear picture of maternal lineage and evolutionary history (Cleary, 2022; Meixner et al., 2013).

Researchers have been able to map out fine-scale population structures, find historical introgression events, and track the movement of lineages across geographical regions because it mutates quickly compared to nuclear DNA and is found in large numbers in cells (Maria Cristina Arias et al., 2006).

There have been many natural and human-made events that have affected the evolution of *Apis mellifera.* Early biogeographic events, like the Pleistocene glaciations and the later recolonizations of Europe, have had a big impact on the genetic diversity of native bee populations (Bertrand, 2013; Ellis et al., 2018). At the same time, modern beekeeping practices, like the international trade of queens and colonies, have mixed even more genes into local populations, making the genetic landscape even more complex (Alburaki et al., 2023; Knoll et al., 2024).

We will go over the basics of *A. mellifera's* mitochondrial genomics in this chapter. We will also talk about how mtDNA diversity varies between and within subspecies and what these differences mean for evolution and conservation.

1.1. Overview of *Apis mellifera* Biology and Global Distribution

Apis mellifera can spread out on its own and adapt to many different types of environments. Over the past few hundred years, people have brought the species to almost every continent except Antarctica. It was originally found in Africa and parts of Asia (Alburaki et al., 2023; Ellis et al., 2018). Today, its genetic diversity is kept up by a mix of subspecies that have adapted to different climates, biological organisms, and beekeeping techniques. For example, *A. m. adansonii* in West Africa, A. m. ligustica in the Mediterranean basin, and *A. m. carnica* and *A. m. mellifera* in Europe all have different physical and behavioral traits that show how they evolved (Alburaki et al., 2023; Bertrand et al., 2015).

Apis mellifera has been domesticated and spread around the world because it is so important for pollinating crops, making honey, and other bee products. But this growth around the world hasn't been easy. Genetic bottlenecks, hybridization events, and the unintentional introduction of non-native lineages have often changed the genetic makeup of local populations (Ellis et al., 2018; Knoll et al., 2024). Mitochondrial DNA markers, like molecular archives that help us study how the composition of maternal lineages has changed over time (Cleary, 2022; Rortais et al., 2011).

1.2. Structure and Characteristics of Mitochondrial DNA in *Apis mellifera*

In comparison to other metazoans, the mitochondrial genome of *A. mellifera* has a highly consistent gene order and organization. It typically contains 13 protein-coding genes essential for oxidative phosphorylation, 22 tRNA molecules, and 2 rRNA molecules. This configuration results in a compact genome, around 16–17 kb in length (Meixner et al., 2013). The intergenic spacer between the cytochrome oxidase subunit I (COI) and II (COII) genes is among the most extensively researched regions of honey bee mitochondrial DNA (mtDNA). The tandem repetition of the P and Q components renders the tRNAleu-COX2 intergenic region highly polymorphic. Such distinctions are crucial in delineating several evolutionary lineages of honeybees and in identifying novel haplotypes exclusive to specific areas (D. Henriques, Chávez-Galarza, et al., 2019; Maria Cristina Arias et al., 2006).

The maternal passes down the same copy of mtDNA to all her offspring. This characteristic makes mtDNA an effective method for tracking maternal lineage histories. This trait significantly impacts conservation genetics and phylogeographic studies. Researchers can create family trees and understand specific family relationships because mtDNA doesn't mix with other DNA, and the changes that happen over generations act like a record of the maternal's evolutionary history (Sarhaddi Avval et al., 2024).

Furthermore, mtDNA mutates more quickly than nuclear DNA, which has made it possible to figure out recent diversification events and observe small genetic differences between populations that might look the same when looked at with nuclear markers. For instance, studies that use mtDNA markers have been advantageous in finding many different haplotypes in native subspecies. One example is the Nigerian *A. m. adansonii*, which has a lot of genetic variation that is linked to how it adapts to its environment (Ajao et al., 2021; Alsharhi et al., 2025).

Using mtDNA analyses to help protect honeybees is also critical in areas where new exotic genetic lineages have recently been introduced. For instance, African mtDNA lineages have spread to parts of Europe due to the importation of Africanized bees. These kinds of introductions could have detrimental effects on local populations by messing up established gene complexes that are perfectly suited to the local environment (Oleksa et al., 2021; Rortais et al., 2011). Conservationists can swiftly identify and address these new introgression events by continuously monitoring mtDNA haplotypes. This protects the evolutionary potential and long-term health of native honeybee stocks (Ellis et al., 2018; Knoll et al., 2024).

Furthermore, advances in next-generation sequencing technologies have simplified and reduced the cost of analyzing entire mitochondrial genomes. These investigations provide us with a wealth of information that will aid in the development of molecular diagnostic tools necessary for promptly determining a colony's genetic condition. This, in turn, promotes regulations and conservation initiatives aimed at preserving the distinct genetic heritage of native honeybee subspecies.

1.5. Integrative Approaches Combining mtDNA with Other Genetic Markers

While mtDNA offers unique benefits, its uniparental inheritance and lack of recombination make it difficult to fully capture the genetic diversity of honeybee populations. As a result, a lot of recent research uses integrative methods that look at both mtDNA and nuclear genetic markers. Nuclear markers, such as microsatellites and SNP panels, provide us more information about how genes are passed down from both parents. Such information helps us get a more complete picture of gene flow and population structure (D. Henriques, Parejo, et al., 2019).

Researchers can uncover cases of asymmetric introgression by comparing mtDNA and nuclear DNA data. This occurs when the mitochondrial genome remains largely intact despite significant nuclear admixture, or vice versa. These differences are crucial for understanding how evolution helps populations adapt to their surroundings and for creating conservation strategies that protect not just maternal lineages but also the overall genetic health of populations (Behura, 2007; Momeni et al., 2021).

Recent studies using various data types have found that nuclear markers can uncover hidden gene exchanges that might have occurred due to past or current interactions between humans, even if mtDNA shows a strong lineage purity in native populations. These results show how important it is to do full genetic monitoring that uses the best parts of both mitochondrial and nuclear markers to make strong conservation plans (Ellis et al., 2018).

1.6. Future Directions in Honey Bee mtDNA Research and Conservation

New research trends suggest that the next step in honeybee genetic studies will involve using complete genome sequencing, advanced genotyping methods, and sophisticated data analysis to better understand mtDNA diversity and its implications for conservation. By sequencing the entire mitochondrial genome, we can build better family trees for different groups and spot subtle signs of natural selection or changes in population size that earlier studies missed by only examining parts of the mtDNA (Eimanifar, Kimball, et al., 2018; D. Henriques, Chávez-Galarza, et al., 2019).

These kinds of detailed genetic studies will not only help us learn more about how *A. mellifera* evolved, but they will also help us create conservation programs that are more effective. For instance, combining mtDNA data with environmental and ecological factors could help identify genetic markers linked to local adaptation. Such data could help selective breeding programs that focus on natural resilience to stressors like pathogens, parasites, and climate change (Alsharhi et al., 2025; Szalanski et al., 2016).

Additionally, using consistent molecular tests like the DraI COI-COII test and mtSNP panels will simplify comparing results from various labs and help create worldwide genetic databases of honey bee mitochondrial haplotypes. Collaborative efforts are crucial in monitoring global changes in honeybee diversity, identifying new introgression events, and ensuring that conservation policies are grounded in the latest genetic data (Momeni et al., 2021; Rortais et al., 2011).

Due to rapid environmental changes and human pressure, integrative genetic tools are becoming increasingly important. Honeybee populations all over the world are facing problems like habitat loss, pesticide exposure, and new diseases that have never happened before. Protecting the genetic diversity of honey bees is crucial for maintaining their ability to adapt. Studies of mitochondrial DNA will continue to be one of the mainstays of honeybee conservation genetics because they help us understand both long-term evolutionary processes and recent changes in population size (Sarhaddi Avval et al., 2024; Taurisano et al., 2024).

1.7. In conclusion

To sum up, the mitochondrial DNA of *A. mellifera* is a useful and flexible molecular resource for figuring out the species' evolutionary history, population structure, and adaptive diversity. The conserved gene order and small size of mtDNA have changed the way we view honey bee genetics. For example, the tRNAleu-COX2 intergenic spacer is a region that varies a lot. Research using mtDNA markers has demonstrated the existence of distinct maternal lineages associated with historical biogeographical events. They have also shown how natural selection, changes in population size, and human activities have affected the genetics of this species (Cleary, 2022)

To protect *A. mellifera*, we need to protect not only its many ecological roles but also its genetic diversity, which is important for its ability to adapt and survive in the face of environmental and human-made problems. Mitochondrial DNA analysis remains a crucial tool for monitoring genetic integrity, guiding conservation breeding programs, and shaping regulations to safeguard native subspecies (Ellis et al., 2018; Taurisano et al., 2024).

In the future, combining mtDNA data with whole genome sequencing and nuclear marker analyses should reveal us even more information about how honeybees evolve. These efforts will help create better conservation strategies that protect genetic diversity, stop the mixing of non-native stocks, and make sure that *A. mellifera* populations are managed in a way that lasts around the world (Momeni et al., 2021).

This chapter has given an overview of the biological and molecular research on the mtDNA of *A. mellifera*. It has also set the stage for more research into the evolutionary and conservation uses of mitochondrial genetics. As threats to honey bee survival grow, it will be important to keep studying their mitochondrial diversity and evolutionary history in order to protect this important pollinator and keep ecosystems in balance in both agricultural and natural settings (Ajao et al., 2021; Sarhaddi Avval et al., 2024).

In conclusion, using traditional mtDNA tests, high-throughput sequencing technologies, and additional nuclear markers will help us learn more about the biodiversity of *A. mellifera*. This combined approach is necessary to develop effective strategies for protecting and managing honeybee populations worldwide, ensuring they remain strong, productive, and genetically sound (Bertrand, 2013; D. Henriques, Chávez-Galarza, et al., 2019). Molecular tools can greatly assist applied conservation biology, as demonstrated by the work in mitochondrial genomics. Such efforts. This will help ensure the long-term health of one of the most important pollinators on Earth.

Chapter 2

2. Mitochondrial DNA of *Apis mellifera*

2.1. Genome Structure and Functional Organization

The structure of the *A. mellifera* mitochondrial genome is similar to that of other metazoan mtDNA. The genome is circular and about 16 kilobases long. It has its 37 genes arranged in a tight space, with little space between them except for the non-coding control region. It is thought that this area, which has a lot of adenine and thymine bases, contains regulatory elements that control the start of replication and transcription (Alburaki et al., 2023). The fact that different honeybee subspecies have the same genes in the same order shows how limited the mitochondrial genome is in terms of function while still allowing for enough genetic variation for phylogenetic analysis (Alburaki et al., 2023).

2.2. Non-Coding Regions and Lineage Diagnostics

Besides the regular protein-coding and RNA genes, non-coding parts like the intergenic spacer regions display much of the mitochondrial variation in *A. mellifera*. The tRNAleu-COII intergenic region, also called the "COI-COII" region, is a significant marker because it has many different forms. The P and Q elements in this area consist of repeated patterns and sequence elements that vary in length. They are important diagnostic markers for telling apart major maternal lineages like the African (A), European (M and C), and Middle Eastern (O) lineages (Alburaki et al., 2023). The different lengths of the P element and its different forms (P0, P, P1, and P2) make up a molecular signature that has been used successfully for DNA barcoding and lineage tracing in honeybee populations all over the world (Alsharhi et al., 2025).

2.3. Phylogeography and Geographic Variation

To fully understand the diversity of *Apis mellifera* mtDNA, we need to do detailed population surveys and samples across its natural range. In places like Europe, Africa, and the Middle East, in-depth phylogeographic studies have shown that local populations are very different from each other. For example, mitochondrial markers have found different haplogroups that show historical barriers to gene flow and secondary contact zones, like the ones in the Mediterranean basin and across the Arabian Peninsula (Alsharhi et al., 2025). Molecular studies have shown that apicultural practices and repeated introductions have led to limited mtDNA haplotype diversity in populations like those in the USA. Most of this diversity comes from a few ancestral European lineages (Alburaki et al., 2023).

2.4. Molecular Techniques and Genomic Advances

There are many different molecular methods used to study *A. mellifera* mtDNA, and they have changed a lot over the past few decades. Researchers used a method called restriction fragment length polymorphism (RFLP) analysis on mtDNA regions that were copied using PCR, especially with enzymes like DraI, to quickly determine an individual's ancestral lineage. These PCR-RFLP tests are still popular because they are cheap and easy to understand for routine identifications in breeding and conservation programs (Abou-Shaara, Al-Ghamdi, et al., 2021). Recent improvements in next-generation sequencing (NGS) technologies have made it possible to fully sequence honeybee mitochondrial genomes. The process has created a more complete dataset that allows for more accurate phylogenetic reconstructions and evolutionary periods (Carr, 2023).

Sequencing the complete mitochondrial genome of various honeybee subspecies has demonstrated that certain lineages, especially those from Europe, have minimal genetic variety, whilst others, particularly from Africa and the Middle East, display significant variance (Carr, 2023). These studies demonstrate how historical demographic events, such as population bottlenecks and founder effects, have altered the genetic composition of contemporary honey bee populations. Research utilizing molecular clocks to analyze mtDNA mutation rates has provided estimations for the divergence of main lineages. Evidence indicates that the evolutionary history of *A. mellifera* extends back several hundred thousand years (Carr, 2023).

2.5. Annotation, Bioinformatics, and Functional Insight

For both functional and evolutionary studies, it is critical to accurately annotate the mitochondrial genome. Many bioinformatics tools, such as Geneious, MEGA, and GetOrganelle, which is specifically designed for assembling mitochondrial genomes, have been used to help accurately identify the 13 protein-coding genes, 22 tRNAs, and 2 rRNAs in the *A. mellifera* mtDNA (D. Henriques, Chávez-Galarza, et al., 2019). Comparing mtDNA sequences often shows small differences in the order of genes or the makeup of nucleotides that can be linked to changes in how species adapt or changes in population size. These kinds of differences are crucial for figuring out how species are related to each other and for figuring out the molecular basis of traits like disease resistance. This knowledge has real-world effects on honeybee breeding and the management of beekeeping (D. Henriques, Chávez-Galarza, et al., 2019).

2.7. Maternal Lineage and Evolutionary Insights

One of the main evolutionary benefits of studying mtDNA is that it can help us learn about maternal lineage history because it is passed down from only one parent. The mtDNA is a direct record of matrilineal evolution because only the queen passes on her mitochondrial genome to all of her off springs. This arrangement is different from nuclear DNA, which usually has genetic recombination (Abdulla et al., 2023). This special feature makes mtDNA a very useful tool for finding out where different subspecies of *A. mellifera* came from and how they moved around. Research using mtDNA markers has demonstrated that geographic and ecological factors frequently influence the distribution of haplotypes. This information provides us clues about how changes in climate in the past may have affected the spread and genetic differences between honeybee populations (Abou-Shaara, Al-Ghamdi, et al., 2021).

2.8. Phylogeographic Patterns and Conservation

Phylogeographic studies of honey bee mtDNA have found intriguing patterns of diversification that are closely related to the species' ecological history. For instance, mitochondrial studies have shown that the western honeybee is made up of several evolutionary lineages that were once separated by mountains, seas, and other geographic barriers, as well as by changes in climate and environment. In turn, these barriers have caused the development of unique haplotypes in certain areas that show both ancient divergence and more recent adaptations to the area's environmental conditions (Alsharhi et al., 2025). This detailed resolution of haplotype diversity has been very advantageous for conservation genetics because it lets scientists recognize and protect genetically unique populations, which is essential because of the many changes in the environment and the pressures that people put on it.

2.9. Applied Significance in Apiculture and Conservation

mtDNA is vital for species evolution, beekeeping, and honey bee population management. Breeders have started using mtDNA screening in their programs to cultivate queen bees (Sarhaddi Avval et al., 2024), because they know that certain mitochondrial haplotypes are linked to beneficial traits, like better resistance to pathogens or better foraging efficiency. Additionally, conservation efforts to safeguard the genetic integrity of native honey bee subspecies have increasingly relied on mtDNA-based identification to monitor the introduction of non-native lineages. This type of introgression often occurs when businesses bring in exotic queens. It can weaken the adaptations that native populations have made to their environment (Alburaki et al., 2023).

2.10. Hybridization, introgression, and reevaluation of subspecies

Recent studies that sequenced the entire mitochondrial genome have shown strong evidence for the complicated history of hybridization and admixture in *A. mellifera*. For example, research that looks at the mtDNA of multiple individuals from different subspecies has shown that some subspecies that have been traditionally defined as such might actually be polyphyletic. This goes against traditional morphology-based taxonomies. Some populations of honeybees, previously believed to be genetically distinct due to their physical traits, exhibit significant mitochondrial overlap. The evidence indicates that in the past, people brought queens from different regions, allowing genes to freely flow between them (Carr, 2023). These results show how natural selection, human intervention, and historical biogeography all work together to shape the current diversity of honey bee populations.

2.12. Regulatory Regions and Functional Adaptation

The *A. mellifera* mtDNA has a lot of variation in both coding and non-coding regions. The non-coding regions are less studied, but they may play a role in regulating genes. The control region, which is also called the D-loop in many vertebrate systems, is just as important in honeybees because it has parts that are necessary for starting mitochondrial transcription and replication. Changes in this area have been linked to differences in mitochondrial copy number and possible changes to how cells respond to changes in metabolic demand or environmental stress (Alburaki et al., 2023). Detailed annotation work has shown that even small changes in nucleotides in these non-coding segments can change how the organism works, which can affect its overall fitness (Alsharhi et al., 2025).

2.13. Sequencing Technologies and Evolutionary Timescales

Researchers have transitioned from classical PCR-RFLP methods to more advanced whole-genome sequencing platforms in their study of *A. mellifera* mtDNA. Early studies that used restriction enzyme digestion and Sanger sequencing helped us understand mtDNA variability by listing the most common haplotypes and setting baseline levels of nucleotide diversity (Bebane 2025 Research Paper, Ecology, pages 1–2). Illumina paired-end sequencing and long-read technologies are two examples of high-throughput sequencing methods that have made it possible to put together complete mitochondrial genomes with a high level of accuracy in recent years. These kinds of big datasets make it possible to identify not only single nucleotide polymorphisms (SNPs) but also structural changes like small insertions and deletions that have big effects on the phylogenetic reconstruction of honey bee lineages (Bebane 2025 Research Paper, Ecology, pages 7–8).

Accurate mtDNA data is essential for calibrating molecular clocks and determining the duration since the divergence of various lineages. Research of full mitochondrial genomes indicates that *A. mellifera* may have initially emerged approximately 780,000 years ago. Subsequently, other events resulted in the establishment of distinct African, European, and Middle Eastern lineages observed today (Carr, 2023). The estimations of evolutionary time are significant for both academic and practical applications in conservation genetics and apiculture, since understanding the historical lifestyles of subspecies can aid in management (Tihelka et al., 2020).

2.14. Computational Tools and Phylogenomic Modeling

Using bioinformatics tools to study bee mitochondrial genomes has helped us learn more about how *A. mellifera* mtDNA has changed over time. MEGA and Geneious are two examples of software packages that can do multiple sequence alignments, build phylogenetic trees, and analyze divergence. These features make it easier to compare mitochondrial genomes from different parts of the world (D. Henriques, Chávez-Galarza, et al., 2019). For phylogenetic inference, other specialized software like MACSE for codon-aware alignment and MAFFT with Q-INS-i for RNA structural considerations must ensure that the functional constraints of mitochondrial gene structure are modeled correctly (Y.-R. Li et al., 2024).

2.15. Subspecies Phylogeny and tRNA Gene Rearrangement

Comparative mitochondrial phylogenomics has helped us learn more about how different *Apis subspecies* are related to each other. For instance, studies of complete mitochondrial genomes have shown that some subspecies belong to a single group, while others show signs of historical mixing and introgression. Data indicate that European subspecies like *A. m. ligustica* and *A. m. carnica* typically cluster together, while African subspecies such as *A. m. scutellata* and *A. m. capensis* display a more complex pattern with various genetic differences found in different regions (Carr, 2023). These results show that natural dispersal, changes in the weather, and movements caused by people still have an effect on the genetic makeup of honeybee populations todays (Oleksa et al., 2021).

The mitochondrial genome has demonstrated the significance of variability and mobility in tRNA genes, despite the generally fixed gene order in *A. mellifera* subspecies. Minor variations in the configuration or copy quantity of tRNA genes may correlate with environmental changes and can facilitate a more precise delineation of clades (Ilyasov 2021, Phylogenetic Relationships, Pages 1–2). A detailed comparison of tRNA gene sequences and secondary structures among subspecies has enhanced our understanding of evolutionary relationships that remain ambiguous when solely examining protein-coding genes (Ilyasov 2021, Phylogenetic Relationships Among, pp. 3–4).

2.16. Emerging Epigenetic Dimensions in Mitochondrial Genomics

There is more and more interest in how epigenetics controls mitochondrial genomes, in addition to traditional sequence-based analyses. Researchers have started to look into the role of DNA methylation in the mtDNA of insects, but this area is still very new in *A. mellifera*. Early research on related bee species suggests that changes in mitochondrial DNA methylation patterns may affect gene expression, aging, and how organisms react to changes in their environment (Françoso et al., 2020). These new findings indicate that using mito-omics methods, which bring together genetic, gene activity, and chemical modification data, could improve our understanding of how mitochondria function and how they have evolved in honeybees (Françoso et al., 2020).

2.17. Human-Mediated Influences on Mitochondrial Diversity

Human actions have also had a big impact on the mtDNA landscape of *A. mellifera*. The large-scale trade and movement of queens for commercial beekeeping has caused a loss of mitochondrial diversity in some areas, especially in countries where apiculture relies heavily on a small number of European-derived subspecies (Alburaki et al., 2023). On the other hand, areas that have not had much human activity tend to have a wider range of mtDNA haplotypes, which is a sign of natural evolutionary processes and being cut off from the rest of the world (Modaber et al., 2019). Additionally, mtDNA analyses have shown that African mitochondrial lineages have been brought into areas where they were not native because of queen importation. This study shows how quickly honeybee population genetics can change in the modern era (Oleksa et al., 2021).

2.18. Uses for conservation and management

These genetic changes have real-world effects on how we manage beekeeping and conservation. Conservation genetics programs are increasingly utilizing mtDNA screening to identify pure native lineages and monitor the extent of introgression from non-native subspecies. This information is very useful for creating breeding programs that will help keep the genetic integrity and adaptive potential of local bee populations, which will make them more resistant to disease and environmental stressors (Alattal & Algamdi, 2022).

2.19. Forensic and diagnostic uses of mtDNA

Forensic and diagnostic uses of mtDNA research in *A. mellifera* are another area of study. Quick molecular tests can use mitochondrial markers to confirm bee subspecies and determine the origin of honey. These tests are fairly cheap and reliable. These diagnostic tools are especially advantageous in areas where mislabeling or false claims about where honeybees come from can have big effects on the economy. They are also advantageous in efforts to protect rare or endangered local lineages (Chávez-Galarza et al., 2021).

2.20. High-Resolution Sequencing and Phylogenomic Insights

Researchers can now make high-resolution datasets that include not only the full mitochondrial genome but also deep coverage of structural variants and rare polymorphisms thanks to improvements in sequencing technology. These kinds of datasets are making it possible to build detailed haplotype networks and evolutionary trees that provide us new information about the microevolutionary processes that are happening within and between *A. mellifera* populations. These in-depth phylogenomic studies have cleared up a lot of previously unclear connections between subspecies and given us a better idea of the genetic diversity that makes the species so ecologically successful (Tihelka et al., 2020).

2.21. Comparative Genomics and Long-Term Evolution

Furthermore, comparing the mitochondrial genomes of *A. mellifera* with those of closely related species, like *A. cerana,* has helped us understand the evolutionary pressures that have changed the mtDNA over millions of years. These kinds of studies have shown how important purifying selection is for keeping important mitochondrial genes intact, even though mtDNA has a high rate of mutation (Chávez-Galarza et al., 2017; Y.-R. Li et al., 2024).

2.22. Phenotypic Correlates of Mitochondrial Variation

The large amount of mitochondrial genomic data now available for *A. mellifera* has made it possible to study the evolution of life history traits, physiology, and even behavior in new ways. For instance, researchers have looked into the links between certain mitochondrial haplotypes and changes in traits related to metabolism, thermal tolerance, and immune function. These studies have shown that mtDNA variation can have real effects on phenotypes (Sarhaddi Avval et al., 2024). These studies could help us learn more about bee biology and improve breeding methods so bees thrive in different environments (Chávez-Galarza et al., 2021).

2.23. Integrative Approaches with Nuclear Genomic Data

Finally, combining mtDNA data with nuclear genomic data is a promising direction for the future. Mitochondrial DNA (mtDNA*)* presents a clear picture of matrilineage, but nuclear markers can show how both parents contributed to the genetic makeup of a population and provide us more information about its overall structure. When you look at both mitochondrial and nuclear data together, it can help us figure out complicated admixture histories that have been hidden in analyses that only look at one genomic compartment. This integrated approach is especially important because there is more and more evidence that the evolutionary history of *A. mellifera* is best understood as a mix of natural selection, demographic history, and gene flow caused by people (Gonçalves et al., 2023).

2.24. A Synopsis of Mitochondrial Contributions to Evolutionary Studies

The mitochondrial DNA of *A. mellifera,* constitutes a compact yet information-dense genome that has been highly beneficial in elucidating the evolutionary, ecological, and biogeographical history of the western honeybee. Researchers have elucidated intricate patterns of divergence, migration, and hybridization across the species' extensive geographic distribution, facilitated by its conserved genetic content and organization, alongside significant variations in non-coding areas and mitochondrial markers. Our proficiency in annotating and studying mtDNA is improving due to advanced sequencing and bioinformatics techniques. This research elucidates how evolutionary pressures have altered the genetic composition of this significant pollinator throughout time. The examination of *A. mellifera* mtDNA remains a pivotal aspect of molecular entomology (Abou-Shaara, Al-Ghamdi, et al., 2021; Tihelka et al., 2020), due to its significant implications for conservation, apiculture, and the comprehension of eusocial insect evolution.

2.25. Future Directions and Emerging Challenges

The study of mitochondrial genomics in *A. mellifera* will help us learn even more about the relationships between subspecies and their evolutionary history. Ongoing research that combines deep sequencing with careful phylogeographic analyses will almost certainly uncover more layers of genetic complexity. These new ideas will help us come up with better ways to protect the genetic diversity of this species, which is important for both the environment and the economy, in the face of global problems like climate change, habitat loss, and new diseases (Alattal & Algamdi, 2022; Alburaki et al., 2023)

2.26. Holistic Genomic Integration for Comprehensive Insight

As we look to the future, combining different genomic fields, such as mitochondrial genomics, nuclear genome analyses, transcriptomics, and even epigenetics, will present us with a better overall picture of honeybee biology. This all-encompassing approach will be essential for answering remaining questions about the adaptive significance of mtDNA variation and how mitochondrial and nuclear genomes work together to create phenotypic diversity (Françoso et al., 2020; Oleksa et al., 2021). This kind of research is important not just for academics, but also for making smart management choices that will protect honeybees, which are essential for agriculture and the health of ecosystems around the world (Modaber et al., 2019).

2.27. Conclusion: The Long-Term Effects of Mitochondrial Research

In conclusion, the mitochondrial DNA of *A. mellifera* is a powerful molecular record of the species' population history, evolutionary divergence, and how it has adapted to environmental challenges. The large body of research based on mtDNA studies, from early PCR-RFLP analyses to modern mitogenomic methods, shows how useful and deep the information that can be learned from this small genome can be. Researchers are still working on sequencing, annotating, and integrated phylogenomic analyses to learn more about the evolution of honeybees. This information is crucial for conservation efforts and sustainable beekeeping practices around the world (Françoso et al., 2020; D. Henriques, Chávez-Galarza, et al., 2019; Oleksa et al., 2021; Tihelka et al., 2020).

2.28. The Need for Ongoing High-Resolution Surveys

Future studies must continue to focus on high-resolution, comprehensive surveys of mitochondrial diversity because mtDNA is crucial for understanding the evolutionary story of *A. mellifera*. These kinds of studies will not only help us understand the fine-scale genetic structure of populations and between populations, but they will also help us find regionally specific haplotypes and adaptive traits that are important for the survival of honeybees in a world that is changing quickly. As new genomic technologies and bioinformatics pipelines improve, they will help us learn more about how the mitochondrial genome has changed over time. These discoveries will help us learn more about how insects have evolved and how many different kinds of insects there are (Gonçalves et al., 2023; H. Zhang et al., 2021).

2.29. The Central Role of mtDNA in Sustainable Beekeeping

Overall, the fascinating story of mitochondrial DNA studies in *A. mellifera* shows how important it is as a molecular tool for figuring out how species are related, helping conservation efforts, and keeping ecosystems that depend on these pollinators in balance. As research goes on, the mtDNA of *A. mellifera* will continue to be an important tool in the search for the genetic basis of adaptation, survival, and reproductive success in one of the most important insect species on Earth (Oleksa et al., 2021; Sarhaddi Avval et al., 2024).

2.30. Mitochondria: Engines of Discovery and Conservation

The large amount of research on *A. mellifera* mtDNA not only affords us a look into the past, but it also provides us with the molecular tools we need to face future problems. By combining traditional molecular techniques with next-generation sequencing and integrative omics, we are getting closer to learning even more about how mitochondrial genetics, phenotypic traits, and environmental adaptation interact in complex ways. This ongoing effort will continue to be crucial for making sure that conservation and breeding programs are based on a profound understanding of genetic diversity and the potential for evolution (Chávez-Galarza et al., 2021; D. Henriques, Chávez-Galarza, et al., 2019).

2.31. Final Remarks: Mitochondria as Pillars of Honey Bee Research

In short, the mitochondria of *A. mellifera* are more than just power sources; they are records of evolutionary history, markers of adaptation, and tools that help scientists and beekeepers alike. Studying these genomes over and over again will not only confirm the dynamic history written in their sequences, but it will also lead to new ways to protect honeybees around the world and manage them in a way that is beneficial for the environment. As we gather and improve mtDNA data from different geographic areas and subspecies, the story of honey bee evolution becomes more complex and informative. This means that the mitochondrial genome of *A. mellifera* will continue to be a key part of evolutionary biology and conservation science (Abou-Shaara, Al-Ghamdi, et al., 2021; Alattal & Algamdi, 2022).

In the end, researchers and apiculturists can better understand, protect, and use the genetic wealth of this wonderful species by using all of the possibilities of mitochondrial genomics. Studies of mitochondrial DNA (mtDNA) are helping us protect global biodiversity, promote sustainable farming practices, and uncover the genetic narratives of one of the most important pollinators on Earth (Alburaki et al., 2023; Oleksa et al., 2021).

Thereby, the mitochondrial genome of *A. mellifera* is a model for studying molecular diversity, evolutionary history, and how insects adapt. We will definitely learn more about this small but powerful genome as research continues. This will keep it at the forefront of both scientific discovery and practical conservation efforts for years to come (Abou-Shaara, Al-Ghamdi, et al., 2021; Tihelka et al., 2020).

Chapter 3

3. Structure and Function of Honey Bee Mitochondrial Genome

The mitochondrial genome of *Apis mellifera* is a small, circular structure about 16–17 kilobases long. It has a very stable set of 37 genes, including 13 protein-coding genes, 22 transfer RNA (tRNA) genes, and two ribosomal RNA (rRNA) genes. It also has a non-coding control region that is crucial for starting replication and transcription (Hu et al., 2016). This well-organized genome is both a powerhouse for making energy in cells and a molecular archive that shows how honeybees have changed over time.

The honeybee mitogenome has 13 genes that code for proteins that are important parts of oxidative phosphorylation and the electron transport chain. These include cytochrome c oxidase subunits 1, 2, and 3 (cox1, cox2, and cox3), ATP synthase subunits (atp6 and atp8), cytochrome b (cytb), and seven NADH dehydrogenase subunits (nd1, nd2, nd3, nd4, nd4l, nd5, and nd6). These genes are important for making ATP quickly and are very similar across Metazoa, showing how essential they are for keeping cellular metabolism going (Hu et al., 2016; Kahnt et al., 2015). Usually, each gene starts with one of the standard insect start codons (ATG, ATA, ATT, or ATC) and ends with the standard stop codon (TAA). However, variations in the use of stop codons demonstrate the evolution of genes over time (Hu et al., 2016).

Twenty-two tRNA genes adopt cloverleaf secondary structures, facilitating the precise and swift translation of mitochondrial mRNAs. The protein-coding regions intermingle with these genes. The majority of tRNA genes adhere to this conventional structure; however, a few do not. For instance, certain tRNAs may lack the dihydrouridine (DHU) arm, resulting in a more simplified loop shape (Karuppaiah et al., 2023). The structural and functional core of mitochondrial ribosomes comprises the two ribosomal RNA genes, 12S rRNA and 16S rRNA, in conjunction with tRNAs. They ensure that translation occurs accurately within the distinct milieu of the mitochondrial matrix (Hu et al., 2016; Kahnt et al., 2015).

The non-coding control region, also known as the D-loop or A+T-rich region, is a well-known part of the *Apis mellifera* mitochondrial genome. This area is about 821 base pairs long and has important regulatory elements that control the start of both replication and transcription (Hu et al., 2016; Knoll et al., 2024). The control region has a lot of adenine and thymine (A+T), which is a common trait in insect mitochondrial genomes. It also contains important sequence patterns like (TATA) boxes, PolyT stretches, and repeated elements that assist transcription factors and the mitochondrial replication machinery in attaching to them (Kahnt et al., 2015; Knoll et al., 2024).

The arrangement of the tRNA, rRNA, and protein-coding genes in the honeybee mitogenome is strikingly similar to that of other insects. There are notable variations, though, particularly in where some tRNA genes are located. Only in tRNA clusters do these gene order changes occur, demonstrating the evolution of the mitochondrial genome. They have also been used by researchers to differentiate between *Apis mellifera's* many subspecies and evolutionary lineages (Ilyasov et al., 2019; Knoll et al., 2024). Basic mitochondrial processes are aided by the constant gene order of rRNA and protein-coding genes. Nonetheless, localized alterations, primarily in tRNA regions, enhance the genetic diversity utilized in evolutionary and phylogeographic research (Ilyasov et al., 2019).

Another thing that affects the structure and function of the mitochondrial genome is its high A+T bias. This nucleotide bias is clear in both the coding regions and the control region, where the A+T content can be more than 80%. This section shows a strong skew that changes how codons are used and the secondary structures of RNA transcripts (Hu et al., 2016; Kahnt et al., 2015). Researchers believe that high A+T regions accelerate evolution because they are more prone to change and may also influence DNA replication by promoting strand asymmetry during the copying process (Karuppaiah et al., 2023).

The process of replicating mitochondrial DNA in *A. mellifera* is similar to that in other metazoans. It involves the coordinated action of nuclear-encoded proteins and mitochondrial-specific DNA polymerases. The control region typically initiates replication, copying the heavy (H) and light (L) strands from various locations. The transcription of the mitochondrial genome works like a polycistronic model, where long pieces of RNA are created from both strands and then split into individual mRNAs, tRNAs, and rRNAs (Hu et al., 2016;

Kahnt et al., 2015). The need for tRNA punctuation to process polycistronic transcripts demonstrates the diverse functions of tRNA genes. They not only help with translation, but they also help with cutting and maturing primary RNA transcripts (Meixner et al., 2013).

Research on the honeybee mitochondrial genome has revealed that it contains genes for proteins needed for the electron transport chain (ETC) to function. The ETC is responsible for most of the cell's ATP production through oxidative phosphorylation. The NADH dehydrogenase subunits, cytochrome c oxidase complexes, cytochrome b, and ATP synthase subunits all work together to move electrons through the ETC. At the same time, protons move across the inner mitochondrial membrane to create an electrochemical gradient (Hu et al., 2016; Kahnt et al., 2015). ATP synthase uses this gradient to make ATP from ADP and inorganic phosphate. This process is a key part of how cells use energy (Hu et al., 2016; Karuppaiah et al., 2023).

The mitochondrial genome also plays a big part in controlling apoptosis and the production of reactive oxygen species (ROS). When mitochondrial genes mutate or the control region changes, it can cause mitochondrial dysfunction, which can stop ATP production and make more ROS, which can then start cell death pathways. Even though these processes haven't been studied in honeybees in great detail in this context, the fact that the ETC components are conserved suggests that *Apis mellifera* probably has similar ways of controlling them (Hu et al., 2016; Kahnt et al., 2015).

The mitochondrial genome's own properties and outside regulatory signals both affect how *Apis mellifera* genes are expressed. Transcription factors made by the nucleus, along with mitochondrial transcription factor (ATFAM) and other helper proteins, attach to parts of the control region to change how transcription starts and how quickly it happens. This communication between the nuclear and mitochondrial genomes is important for keeping the mitochondria working properly when they need to respond to changes in development or stress in the environment. It could also affect how adaptable and resilient honeybee populations are (Hu et al., 2016; Kahnt et al., 2015).

Researchers have used the structural and functional properties of the *A. mellifera* mitochondrial genome to study the evolutionary history of honeybee subspecies and their maternal lineages. Mitochondrial DNA evolves relatively quickly and is passed down from mother to child. This property makes it a fantastic choice for figuring out phylogenetic relationships and finding historical events like population bottlenecks and expansions. The tRNAleu-cox2 intergenic region is one of many markers that help us learn more about how honey bee genetics have changed over time due to geography and other factors. These markers are mostly used to identify haplotypes and differentiate lineages (Knoll et al., 2024; Meixner et al., 2013).

There are canonical protein-coding and RNA genes, but recent studies have shown that the mitochondrial genomes of different *A. mellifera subspecies* have different sequences. For example, comparative studies have indicated that there are insertions, deletions, and single nucleotide polymorphisms (SNPs) in areas like the control region and even in some protein-coding genes. Researchers can use these mutations to determine the separate evolution of two species, and they may also impact the overall functioning of mitochondria (Ilyasov et al., 2019; Knoll et al., 2024). Researchers are still looking into what these differences mean for the function of honeybees. These mutations could potentially assist scientists in developing molecular tools to safeguard honeybee populations and monitor their genetic health (Ilyasov et al., 2019; Meixner et al., 2013).

By using complete mitochondrial genome sequencing, researchers have also surpassed the limitations of partial gene sequences or restriction fragment length polymorphism (RFLP) analyses. Whole mitogenome approaches present us a clearer picture of how different species are related to each other and can show small differences between populations or subspecies that older methods might miss. This detailed sequencing plan has been very important in helping us learn more about how honeybees have changed over time and in coming up with ways to protect different genetic lineages, which is very essential because of problems like hybridization and introgression from imported bee races (D. Henriques, Chávez-Galarza, et al., 2019; Meixner et al., 2013).

The secondary structure of mitochondrial tRNAs in *A. mellifera* is another example of how genome structure and function work together. Most tRNAs form a typical cloverleaf shape that mitochondrial aminoacyl-tRNA synthetases can identify, but some tRNAs are missing a complete DHU arm, showing a structural change that has been observed. Researchers believe that these structural changes are adaptive modifications that enhance mitochondrial translation under the organelle's unique conditions (Hu et al., 2016; Karuppaiah et al., 2023). They don't necessarily make tRNA work worse.

In addition to transcription and translation, the honeybee mitochondrial genome's replication process uses complex mechanisms to protect this small but important molecule's integrity. The control region starts the replication process. Signals in the A+T-rich sequences help replication proteins bind and start DNA synthesis on both the heavy and light strands. The nucleus codes for the mtDNA polymerase, which is brought into the mitochondria. This enzyme makes new DNA strands in a way that minimizes mistakes and allows for quick turnover in response to the cell's energy needs (Hu et al., 2016; Kahnt et al., 2015).

The mechanisms that maintain the integrity of mitochondrial proteins demonstrate the interplay between the mitochondrial genome and the cell environment. Mitochondria have developed several proteolytic systems to break down proteins that are misfolded or damaged. This step is crucial for stopping the buildup of faulty subunits of the electron transport chain. This part of mitochondrial function is mostly controlled at the protein level, but the stable mitochondrial genome's ability to keep the genome intact is important for the continued production of functional proteins (Hu et al., 2016; Kahnt et al., 2015).

In brief, the *A. mellifera* mitochondrial genome's structure and function make up a finely tuned system that combines genetic information, regulatory elements, and molecular machinery to control how cells use energy. The genome has 13 protein-coding genes, 22 tRNAs, and two rRNAs, all of which are very similar to each other. This diagram shows how important these parts are for oxidative phosphorylation. The non-coding control region controls replication and transcription, which makes sure the genome is expressed and maintained efficiently (Hu et al., 2016; Kahnt et al., 2015). Changes in this genome, like point mutations, insertions and deletions, or rearrangements in tRNA clusters, give us a look into the evolutionary history of honey bees and are the basis for molecular tools used in conservation genetics. The control region and other regions with a high mutation rate show how *A. mellifera* can adapt to different environments while keeping important bioenergetic functions stable and variable (Ilyasov et al., 2019; Knoll et al., 2024).

This in-depth knowledge of how things work and how they are built is very advantageous not only for basic biological research but also for practical fields like conservation biology. For example, following mitochondrial diversity can show how human activities affect the environment, help with management practices, and protect the genetic integrity of native bee populations. New developments in whole-genome sequencing and high-resolution molecular assays are helping us learn more about the honeybee mitochondrial genome. They are also opening up new areas of research that connect evolutionary biology and practical conservation strategies (D. Henriques, Chávez-Galarza, et al., 2019; Meixner et al., 2013).

By closely examining the mitochondrial genomes of different types of *A. mellifera* and other related bee species, we can see that the basic structure and function of the mitogenome have remained the same. However, changes that have occurred over time, especially in the non-coding regulatory regions and tRNA gene arrangements, indicate that the species has adapted and changed. These changes not only help bees adapt to their surroundings, but they also give researchers important genetic markers to study how genes move between species, how they mix together, and how they diverge over time in response to environmental changes and beekeeping practices (Ilyasov et al., 2019; Knoll et al., 2024).

In conclusion, the structure and function of the *A. mellifera* mitochondrial genome show how well evolutionary flexibility and conserved bioenergetic functions can work together. The small, round mitogenome has everything it needs to make proteins that make energy and send precise signals to control its expression and replication. Because it serves as both an energy source and a record of evolutionary history, the mitochondrial genome is a major focus of research on honey bee biology and conservation. We will discover more about how the mitogenome influences the health of individual bees and the genetic makeup of entire groups through continued research on differences in gene content, how it replicates, and its control mechanisms. Such knowledge will help with conservation efforts and sustainable beekeeping practices (Meixner et al., 2013).

As new sequencing technologies and analytical methods are used in honeybee research, the ongoing improvement of mitochondrial genome data will surely provide us a deeper understanding of how *A. mellifera* uses energy, adapts to changes, and changes its population. This will help make strategies for protecting the genetic diversity and long-term health of honeybee populations around the world more effective. Such efforts will make sure that these important pollinators can keep doing their important ecological and economic jobs in a wide range of environments (D. Henriques, Chávez-Galarza, et al., 2019; Knoll et al., 2024; Meixner et al., 2013).

The mitochondrial genome of *Apis mellifera (*Western honey bee) is a circular DNA molecule ranging from 16,248 to 16,590 bp in length across subspecies. It contains 37 genes and a non-coding control region, with a high A+T content typically (84–85%). Below is a comprehensive table listing all mitochondrial markers, categorized by gene type and function.

Table 1. Mitochondrial Markers of *Apis mellifera*

Marker Name	Type	Function/Notes
Protein-coding genes13 total)		
ATP6	Protein-coding	ATP synthase F0 subunit 6
ATP8	Protein-coding	ATP synthase F0 subunit 8
COI	Protein-coding	Cytochrome c oxidase subunit I
COII	Protein-coding	Cytochrome c oxidase subunit II
COIII	Protein-coding	Cytochrome c oxidase subunit III
CytB	Protein-coding	Cytochrome b
ND1	Protein-coding	NADH dehydrogenase subunit 1
ND2	Protein-coding	NADH dehydrogenase subunit 2
ND3	Protein-coding	NADH dehydrogenase subunit 3
ND4	Protein-coding	NADH dehydrogenase subunit 4
ND4L	Protein-coding	NADH dehydrogenase subunit 4L
ND5	Protein-coding	NADH dehydrogenase subunit 5
ND6	Protein-coding	NADH dehydrogenase subunit 6
Transfer RNA genes22 total)		
tRNA-Ala	Transfer RNA	Alanine tRNA
tRNA-Arg	Transfer RNA	Arginine tRNA
tRNA-Asn	Transfer RNA	Asparagine tRNA
tRNA-Asp	Transfer RNA	Aspartic acid tRNA
tRNA-Cys	Transfer RNA	Cysteine tRNA
tRNA-Gln	Transfer RNA	Glutamine tRNA (lacks cloverleaf arm in some subspecies) (Nakagawa et al., 2018)
tRNA-Glu	Transfer RNA	Glutamic acid tRNA
tRNA-Gly	Transfer RNA	Glycine tRNA

tRNA-His	Transfer RNA	Histidine tRNA
tRNA-Ile	Transfer RNA	Isoleucine tRNA
tRNA-Leu1	Transfer RNA	Leucine tRNA (UUR codon)
tRNA-Leu2	Transfer RNA	Leucine tRNA (CUN codon)
tRNA-Lys	Transfer RNA	Lysine tRNA
tRNA-Met	Transfer RNA	Methionine tRNA
tRNA-Phe	Transfer RNA	Phenylalanine tRNA
tRNA-Pro	Transfer RNA	Proline tRNA
tRNA-Ser1	Transfer RNA	Serine tRNAAGY codon; lacks clover (leaf arm in some subspecies) (Nakagawa et al., 2018)
tRNA-Ser2	Transfer RNA	Serine tRNA (UCN codon)
tRNA-Thr	Transfer RNA	Threonine tRNA (lacks cloverleaf arm in some subspecies) (Nakagawa et al., 2018)
tRNA-Trp	Transfer RNA	Tryptophan tRNA
tRNA-Tyr	Transfer RNA	Tyrosine tRNA
tRNA-Val	Transfer RNA	Valine tRNA
Ribosomal RNA genes2 total)		
16S rRNA	Ribosomal RNA	Large ribosomal subunit
12S rRNA	Ribosomal RNA	Small ribosomal subunit
Non-coding region		
Control Region	Non-coding	A+T-rich region involved in replication/transcription; length varies by subspecies

Key Features

Strand Distribution:

Heavy strand: Encodes 9 protein-coding genes and 14 tRNAs.

Light strand: Encodes 4 protein-coding genes, 8 tRNAs, and both rRNAs (Eimanifar, T Kimball, et al., 2017a; Nakagawa et al., 2018).

Start Codons: ATT most common, ATG, ATA, or ATC (Eimanifar, T Kimball, et al., 2017a, 2017b; Nakagawa et al., 2018).

Stop Codons: TAA universal across all protein-coding genes, (Eimanifar, T Kimball, et al., 2017a, 2017b; Nakagawa et al., 2018).

Genome Conservation: Gene order and arrangement are identical across all sequenced *A. mellifera* subspecies (Eimanifar, Kimball, et al., 2018; D. Henriques et al., 2022; Tihelka et al., 2020).

This genomic framework is conserved across *A. mellifera* subspecies, with minor variations in length and base composition reflecting evolutionary divergence (Boardman, 2019; Eimanifar, T Kimball, et al., 2017a, 2017b; D. Henriques et al., 2022; Nakagawa et al., 2018; Tihelka et al., 2020).

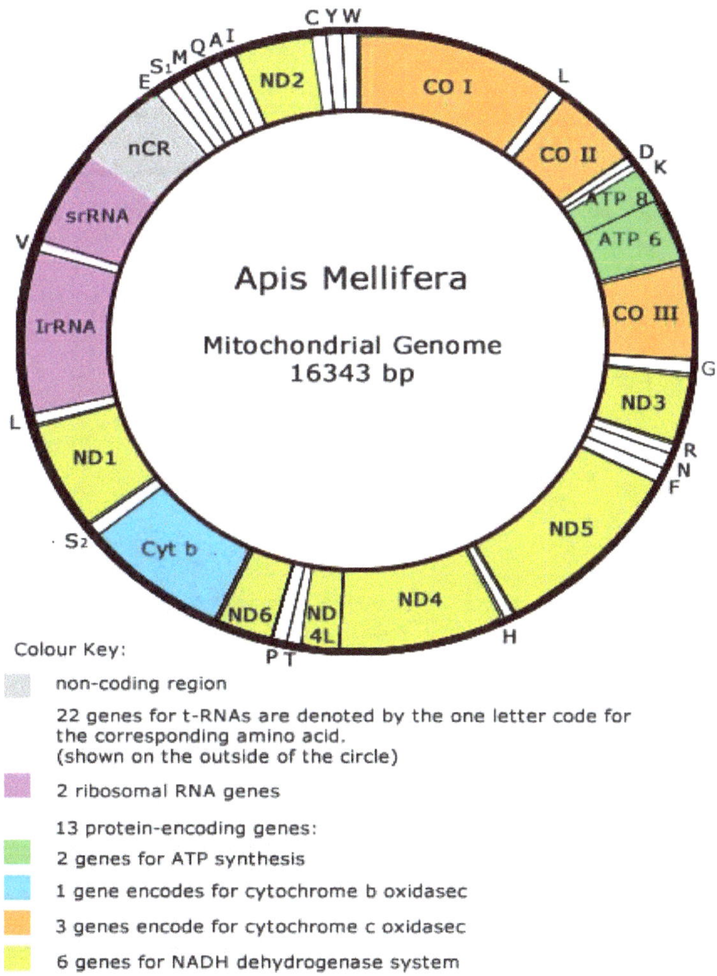

Fig 1. Mitochondrial DNA – mtDNA (https://killowen.com/genetics1H.html)

Chapter 4

4. Overview of Evolutionary Lineages: Classification and Phylogeny

Researchers have used mitochondrial DNA (mtDNA) markers to study the evolutionary history and classification of the western honeybee, *A. mellifera*. These markers provide important information about both deep and recent differences between its subspecies. mtDNA is particularly useful for tracing maternal lineages, as it passes down from one parent to another without recombination. This genetic material keeps a record of lineage-specific mutations over evolutionary time (Arias & Sheppard, 1996). Molecular studies that use mtDNA markers like the intergenic region between cytochrome oxidase subunits I and II (COI-COII) and the NADH dehydrogenase subunit 2 gene (ND2) have helped identify the main evolutionary clusters within *A. mellifera* over the past few decades. This method has found six major lineages that differ in habitat and appearance.

1. Mitochondrial DNA Markers in Honey Bee Phylogenetics

Mitochondrial DNA (mtDNA) markers have been the main focus of honey bee phylogeographic studies for a long time because they have a high mutation rate, do not recombine, and are easy to amplify and sequence (Arias & Sheppard, 1996). The COI-COII intergenic region is especially useful because its structure, which includes repeat units of different lengths and insertions and deletions indels, gives us many diagnostic polymorphisms that have been used to group honey bee populations into specific maternal lineages (Sarhaddi Avval et al., 2024). The ND2 gene, which is part of the mitochondrial NADH dehydrogenase complex, provides clearer details because it has some stable areas with differences in nucleotides that can help distinguish between similar haplotypes (Tan et al., 2011). These markers help us figure out subspecies and put together historical biogeographic patterns, which show how lineages have split and mixed over time (Techer, Clémencet, Simiand, Preeaduth, et al., 2017).

2. The Six Main Evolutionary Lineages

Recent studies that looked at individual mtDNA markers and complete mitogenome sequencing have identified six main evolutionary lineages in *A. mellifera*. Each of these lineages has its own haplotypic patterns and geographical distributions. A, C, M, O, Y, and S are the names of these lineages. Each one shows a different part of the species' global diversification.

A. Lineage (A) African

The African lineage (A) is the most diverse and has many subgroups, which are shown by names like AI through AIV. Some of the most well-known subspecies in this lineage are *A. m. scutellata,* which is also called the savannah bee, and *A. m. capensis*, which is also called the Cape bee. This lineage has adapted to a wide range of ecological conditions across sub-Saharan Africa, and there is a lot of genetic diversity between regional groups (Arias & Sheppard, 1996). It can be difficult to tell what morphological traits mean, but mtDNA analyses show that South African populations make up a monophyletic clade. However, these populations don't have subspecies-specific diagnostic markers that would make it easy to divide them into separate genetic units (Techer, Clémencet, Simiand, Preeaduth, et al., 2017). Even though there is proof of a shared maternal ancestry, the absence of clear genetic markers within the lineage indicates that there has been a history of quick changes and possible mixing of genes between nearby populations (Cleary, 2022).

B. Lineage (C) North Mediterranean

The North Mediterranean lineage (C) is mostly made up of mtDNA haplotypes called C1 and C2. These haplotypes are associated with subspecies such as *A. m. ligustica* and *A. m. carnica*, respectively. *A. m. ligustica*, which usually has the C1 haplotype, is known for being gentle and productive. *A. m. carnica*, on the other hand, has the C2 haplotype and is perfect for beekeeping in temperate areas. Traditionally, these subspecies are separated by their mitochondrial haplotypes. However, studies of COI-COII sequences have shown that the North Mediterranean group is not just a straightforward family tree, suggesting that similar populations have blended together over time (Alburaki et al., 2023; Carr, 2023). This paraphyly demonstrates how complex demographic histories and potential hybrid zones have influenced the C lineage's evolutionary path over time (Coulibaly et al., 2019).

C. Lineage (M) from Western Europe

The M lineage includes honeybee populations from western and northern Europe. *A. m. mellifera* is the main species in this lineage and has a unique set of (M) haplotypes. *A. m. mellifera* seems to be the most basic lineage in many phylogenetic reconstructions, which means it could be one of the first clades to split off from *A. mellifera* (Carr, 2023). Some phylogenetic studies indicate that the M lineage is at the base of the tree, which supports the idea that *A. mellifera's* evolutionary radiation started in Europe. However, more thorough analyses that combine nuclear and mitochondrial data have challenged this idea (Techer, Clémencet, Simiand, Preeaduth, et al., 2017). The M lineage has less haplotype diversity than the African lineage, even though it split off from the African lineage a long time ago. This conclusion could mean that there were demographic bottlenecks or selective sweeps in its evolutionary history (D. Henriques, Chávez-Galarza, et al., 2019).

D. Lineage (O) Asian

The O lineage includes the Middle Eastern subspecies of *A. mellifera*, like *A. m. syriaca*, which are well-suited to the dry and semi-dry areas of the Near East. Mitochondrial markers such as the COI-COII intergenic spacer and the ND2 gene have been instrumental in distinguishing the O lineage from its European and African counterparts (Tihelka et al., 2020). The O lineage has unique mtDNA haplotypes that show how it evolved differently from other lineages in the Mediterranean and African regions (Techer, Clémencet, Simiand, Preeaduth, et al., 2017). These haplotypes may also have diagnostic features that set them apart from other lineages in these areas. Whole mitogenome analyses that give more detail and support the idea that these Middle Eastern populations are monophyletic help further differentiate the O lineage (Techer, Clémencet, Simiand, Preeaduth, et al., 2017).

E. Lineages (Y) and (S) from Ethiopia and Syria

In addition to the well-known A, C, M, and O lineages, new molecular studies have suggested two more lineages that make the phylogenetic structure of *A. mellifera* even more precise. The Y lineage is linked to populations from Ethiopia, where there is a lot of genetic variation because of the Horn of Africa's unique environmental conditions. This evidence shows that these individuals have been isolated and adapted to their surroundings for a long time (Düttmann et al., 2022). The S lineage, which is sometimes called the Syrian lineage, includes individuals from places like Syria and Lebanon. Even though these places are close together, the S lineage is genetically different from the O lineage. This conclusion is supported by molecular signals from both mtDNA markers and whole mitogenome data (Alburaki et al., 2023; Carr, 2023). Finding these new lineages has big effects on protecting biodiversity and managing honey bee genetic resources because it indicates that there is more genetic diversity in the species than was thought before (Coulibaly et al., 2019).

III. Phylogenetic Inference and Whole Mitogenome Analyses

Although research concentrating on individual mtDNA markers like as COI-COII and ND2 has been beneficial, recent developments in whole mitogenome sequencing have offered a superior and more precise approach to elucidate the evolutionary history of *A. mellifera.* Comprehensive mitogenome analyses utilize data from all 37 mitochondrial genes, including 13 encoding proteins, 22 encoding tRNAs, and two encoding rRNAs. This offers researchers additional data to analyze when determining the relationships among various species (Carr, 2023). The enhanced resolution of whole mitogenome data indicates that *A. mellifera* originated from North Africa and the Middle East. This situation markedly contrasts with previous hypotheses suggesting that the species originated in Europe (Techer, Clémencet, Simiand, Preeaduth, et al., 2017; Tihelka et al., 2020). Robust phylogenetic analyses employing sophisticated Bayesian inference and maximum likelihood techniques support this novel scenario. These methodologies consistently identify basal clades encompassing haplotypes from North Africa and the Middle East (Carr, 2023; Tihelka et al., 2020). The hypothesis that *A. mellifera*, may have originated from a region where North African and Middle Eastern populations coexisted prompts reflection on our classification of species and the historical patterns of their dispersion (Techer, Clémencet, Simiand, Preeaduth, et al., 2017).

IV. Methodological Considerations in mtDNA Phylogenetics

Using mtDNA to figure out how *A. mellifera subspecies* are related to each other through evolution involves steps that make phylogenetic inferences more reliable and clear. Researchers first choose the right mtDNA markers based on how much they change and how quickly they evolve. The COI-COII intergenic region and ND2 gene are especially helpful because they change at rates that can show both recent and older evolutionary differences (Sarhaddi Avval et al., 2024). After selecting the markers, researchers use polymerase chain reaction (PCR) amplification to generate sufficient copies of the target regions. Then, restriction fragment length polymorphism (RFLP) analysis or direct sequencing is used to determine the genotype of individuals from different populations (Coulibaly et al., 2019). Researchers can identify nucleotide substitutions, indels, and other mutations by using bioinformatics tools to align sequences. These mutations are then used to make haplotype networks and phylogenetic trees (Cleary, 2022). Modern studies often employ whole mitogenome sequencing in addition to these methods. Our reasoning is because it gives a lot of information and helps avoid problems with homoplasy and marker saturation that can happen with single-locus analyses (Tihelka et al., 2020).

In phylogenetic investigations, when various mtDNA markers are combined, the sequences are often concatenated, followed by the use of nucleotide substitution models to ascertain the most probable tree structure. Bayesian inference and maximum likelihood techniques are frequently employed due to their capacity to manage intricate evolutionary models and offer robust support for branches through posterior probabilities and bootstrap values (Techer, Clémencet, Simiand, Preeaduth, et al., 2017). Median-joining network analysis represents an alternative method for data examination. It illustrates the link between haplotypes by indicating the number of mutational steps that distinguish them. Such networks are adept at identifying haplogroups that have recently diverged and elucidating the potential gene flow across subspecies (Coulibaly et al., 2019; Ilyasov et al., 2019).

V. Implications of Phylogenetic Findings for Honey Bee Taxonomy and Conservation

The identification of six primary evolutionary lineages in *A. mellifera* significantly impacts taxonomy and conservation efforts. The existence of distinct mtDNA haplotypes among African, North Mediterranean, West European, Oriental, Ethiopian, and Syrian (O) and (S) lineages necessitates a reevaluation of historical subspecies borders, which were only derived from morphometric evidence. The intricate representation provided by mtDNA indicates significant historical intermingling, evidenced by the paraphyly of certain lineages, such as lineage (C), and the absence of distinct diagnostic markers in others, like the African lineage (Alburaki et al., 2023; D. Henriques, Chávez-Galarza, et al., 2019). This comprehensive classification significantly influences the management of honey bee genetic resources. It enables the identification of genetically distinct populations that may possess unique adaptations, proving particularly advantageous for breeding initiatives and conservation strategies (Eimanifar, Kimball, et al., 2018; Techer, Clémencet, Simiand, Preeaduth, et al., 2017).

Finding historically important admixture events and migration routes between lineages also helps us understand how *A. mellifera*, spreads in terms of ecology and evolution. For example, the fact that whole mitogenome analyses show that the species came from Northern Africa or the Middle East means that we need to rethink the idea that modern honeybees came from Europe (Techer, Clémencet, Simiand, Preeaduth, et al., 2017; Tihelka et al., 2020). This new perspective suggests that the genetic diversity seen in European subspecies may not be due to their original European origins but rather to later colonization events and multiple waves of admixture (Carr, 2023). These kinds of insights are crucial for figuring out how things like climate change, habitat fragmentation, and human-made changes have affected the current distribution and genetic makeup of honey bee populations around the world (Alburaki et al., 2023; Arias & Sheppard, 1996; Düttmann et al., 2022).

Furthermore, the recognition of the new (Y) Ethiopian and (S) Syrian lineages shows that there is more genetic diversity in the species than was previously thought. For instance, the (Y) lineage shows how Ethiopian populations have evolved in a way that is different from other groups, while the (S) lineage shows a different group within the Middle Eastern subspecies that is usually recognized (Alburaki et al., 2023; Düttmann et al., 2022). These results are crucial for conservation because they show areas where honeybee genetic diversity is very high and may be at risk from changes in the environment. This information can help conservationists come up with plans to protect these unique maternal lineages (Coulibaly et al., 2019; Tihelka et al., 2020).

VI. Integrating mtDNA Data with Nuclear Markers and Whole-Genome Resequencing

mtDNA markers like COI-COII and ND2 give us a lot of useful information about maternal lineages, but they only show a small part of the total genomic variation in honeybees. As a result, recent studies have started to combine mtDNA data with data from nuclear markers and whole-genome resequencing to give a more complete picture of the species' evolutionary history (D. Henriques, Chávez-Galarza, et al., 2019). Researchers can compare patterns of maternal inheritance with biparental nuclear variation by combining these different genomic datasets. This approach can reveal possible differences between maternal inheritance and biparental nuclear variation that may arise due to hybridization, introgression, and sex-biased gene flow (Carr, 2023; Eimanifar, Kimball, et al., 2018). For example, looking at the entire mitogenome helps confirm the basic shape of the six evolutionary lineages, but nuclear markers can sometimes reveal more complex or subtle population details that mtDNA alone cannot detect (Techer, Clémencet, Simiand, Preeaduth, et al., 2017).

A full approach that combines mtDNA with nuclear genomic data is especially important for studying Africanized honey bee populations, where a lot of hybridization between African and European subspecies has made it difficult to tell where one lineage ends and the next begins. In these situations, mtDNA markers can show the maternal ancestry, and nuclear genes can show the overall genetic mix and traits that help colonies survive and be productive (Düttmann et al., 2022; Eimanifar, Kimball, et al., 2018). This combined method is essential for both basic evolutionary research and practical uses in beekeeping, where keeping genetic diversity and local adaptations is necessary for bees to be able to fight off pathogens and deal with environmental stressors (Cleary, 2022).

VII. Insights from Comparative Mitogenomic Studies

Comparative studies of whole mitochondrial genomes have recently become a useful way to answer long-standing questions about the evolution of honeybees. Researchers have created more accurate family trees by studying nearly complete mitochondrial genome sequences from various honeybee subspecies and locations, rather than just using single markers (Carr, 2023; Tihelka et al., 2020). For instance, research has shown that the branching order among lineages follows geographic patterns. The basal clades are mostly found in Northern Africa and the Middle East, which strongly supports the idea that the species did not come from Europe (Techer, Clémencet, Simiand, Preeaduth, et al., 2017; Tihelka et al., 2020).

These studies of mitogenomes that compare different species also show how useful high-resolution data can be for making sense of complicated evolutionary relationships that were difficult to see before, because of a lack of sequence information. Analyzing entire mitogenomes has not only helped clarify the connections between the well-known A, C, M, and O lineages, but it has also shown that the new Y and S lineages are different from the others (Alburaki et al., 2023; Düttmann et al., 2022). Using complete mitochondrial genomes makes it easier to find rare or newly created haplotypes that might help us understand important evolutionary events like rapid population growth, founder effects, and past decreases in population size (Ilyasov et al., 2019).

Whole mitogenome data also show that the substitution rate is not the same for all mitochondrial genes. For example, some regions like (COX3 or ND2) may be under different selective pressures or have different evolutionary constraints than others (Eimanifar, Kimball, et al., 2018; Ilyasov et al., 2016). This difference in evolutionary rates is crucial for getting molecular clocks and divergence times right. This process lets scientists put evolutionary events in a time frame that matches known changes in the climate and geology (Carr, 2023; Techer, Clémencet, Simiand, Preeaduth, et al., 2017).

VIII. Challenges and Future Directions in Honey Bee Phylogenetics

Even though mtDNA markers have helped us figure out the six main evolutionary lineages of *A. mellifera*, there are still many problems to solve. For one thing, using mitochondrial markers means that researchers can only look at one side of the family tree, which may not show the full picture of how honey bee evolution works, since nuclear gene flow and recombination are crucial (Techer, Clémencet, Simiand, Preeaduth, et al., 2017). Furthermore, some mtDNA regions, especially those with a lot of indel variation like the COI-COII intergenic region, can be prone to homoplasy, which can make it harder to figure out how species are related (Coulibaly et al., 2019; Ilyasov et al., 2016).

In the future, research will probably use a more integrated genomic approach, which combines whole genome resequencing data with mtDNA and nuclear markers to give a full picture of the structure and evolutionary history of honeybee populations. These kinds of studies are already starting to show how selection, drift, and gene flow affect the genetic makeup of both managed and wild populations (Eimanifar, Kimball, et al., 2018). Researchers will be able to put together bigger datasets that include not only whole mitochondrial genomes but also large panels of single nucleotide polymorphisms (SNPs) and structural variants in the nuclear genome thanks to improvements in sequencing technology and bioinformatics. This big-picture view is necessary to make sense of any differences between mitochondrial and nuclear phylogenies and to understand how evolution works to help organisms adapt to different environments (Carr, 2023; Techer, Clémencet, Simiand, Preeaduth, et al., 2017).

Africa, would undoubtedly refine existing lineage classifications and may potentially uncover previously unrecognized genetic groups. The identification of the Y and S lineages underscores the necessity of sampling diverse geographic regions and necessitates a reevaluation of traditional taxonomic concepts in light of contemporary molecular evidence (Alburaki et al., 2023; Düttmann et al., 2022). These initiatives will facilitate conservation effectively, as it is crucial to identify and safeguard distinct genetic lineages within honeybee populations to enable rapid adaptation to environmental changes (Coulibaly et al., 2019; D. Henriques, Chávez-Galarza, et al., 2019).

IX. Conclusion

In summary, research using mtDNA markers such as the COI-COII intergenic region and the ND2 gene has greatly influenced the classification and evolution of *Apis mellifera*. These molecular tools have shown that *A. mellifera*, has six main evolutionary lineages: A (African), C (North Mediterranean), M (West European), O (Oriental), Y (Ethiopian), and S (Syrian). Each of these lineages has its own unique mtDNA haplotypes that show how they have changed over time through divergence, adaptation, and mixing. In the African lineage, the subgroups AI to AIV include important subspecies like *A. m. scutellata* and *A. m. capensis*. Even though they form a monophyletic clade in South Africa, there are no clear diagnostic markers at the subspecies level. The North Mediterranean lineage also includes *A. m. ligustica* and *A. m. carnica,* whose mtDNA profiles show that they are paraphyletic and have mixed with other species in the past. Some phylogenies put the West European lineage, which includes *A. m. mellifera,* in a potentially basal position. The Oriental lineage, on the other hand, includes Middle Eastern subspecies like *A. m. syriaca*. The Y and S lineages, which are based on Ethiopian and Syrian populations, respectively, add even more complexity to the species' long and complicated evolutionary history (Arias & Sheppard, 1996; Carr, 2023).

Whole mitogenome analyses have shown that *A. mellifera*, came from Northern Africa or the Middle East, which goes against previous claims that it came from Europe. This adds another layer of understanding. This new view of how honeybees have changed over time shows how historical biogeography and molecular evolution work together in a dynamic way. It also shows how important it is to combine data from both mitochondrial and nuclear genomes to build a strong phylogenetic framework (Techer, Clémencet, Simiand, Preeaduth, et al., 2017; Tihelka et al., 2020).

Even though the current classification based on mtDNA markers has helped us learn a lot about honey bee phylogeny, we still need to keep working on it by using next-generation sequencing and collecting samples from more places. These methods will not only improve the definition of current lineages, but they may also uncover more layers of genetic structure that are important for conservation and sustainable beekeeping in a world that is changing quickly (Coulibaly et al., 2019; Düttmann et al., 2022; Eimanifar, Kimball, et al., 2018).

In the end, combining mtDNA markers with nuclear genomic data and cutting-edge computer techniques could help us figure out the full complexity of *Apis mellifera's* evolutionary history. This all-encompassing method is necessary to help with both taxonomic changes and the creation of conservation plans to protect the genetic diversity of one of the most important pollinators in the world (D. Henriques, Chávez-Galarza, et al., 2019). As we learn more about the evolutionary history of honeybees, research will continue to focus on how historical events, ecological pressures, and human actions have affected them. Such studies will help make sure that the classification of *A. mellifera*, reflects both its ancient history and its ongoing adaptive evolution (Carr, 2023; Techer, Clémencet, Simiand, Preeaduth, et al., 2017).

In summary, the section about the evolutionary lineages of *Apis mellifera*, combines many years of research using both traditional molecular markers and the latest whole mitogenome studies to build a clear family tree. By identifying six main lineages—A, C, M, O, Y, and S—and showing that the species came from North Africa and the Middle East, these studies have improved our understanding of taxonomy and given us the background we need for future research in conservation genetics, evolutionary biology, and applied apiculture. It is certain that the ongoing integration of molecular data from many genomic compartments will shed more light on the complicated evolutionary processes that have shaped the diversity and distribution of the western honeybee (Alburaki et al., 2023; Tihelka et al., 2020).

This broad view shows how powerful mtDNA markers are, especially the COI-COII and ND2 regions, in figuring out how different honeybee subspecies are related to each other. It also shows how whole mitogenome sequencing could change how we approach the origins of evolution and the branching out of lineages. As the field grows, these molecular tools will continue to be important for both theoretical models of evolution and real-world plans to protect and manage honey bee diversity around the world (Coulibaly et al., 2019; Techer, Clémencet, Simiand, Preeaduth, et al., 2017).

4.2. The Six Main Evolutionary Lineages

Recent studies that looked at individual mtDNA markers and complete mitogenome sequencing have identified six main evolutionary lineages in *A. mellifera*. Each of these lineages has its own haplotypic patterns and geographical distributions. A, C, M, O, Y, and S are the names of these lineages. Each one shows a different part of the species' global diversification.

4.2.1. Lineage A (African)

4.2.1.1. Introduction:

The topic below goes into great detail about the African evolutionary lineage of *Apis mellifera* and its known internal subdivisions. These include the different subgroups that are labeled from AI to AIV or sometimes just by letters from (A to Z in the larger molecular literatüre). This section discusses the genetic diversity, how quickly it changes, and the problems that arise when trying to define different molecular diagnostics in this group, which is very diverse and can adapt to different environments.

Beginning The western honeybee, or *Apis mellifera*, is a eusocial insect with a wide range of genetic diversity that reflects its long evolutionary history. It is crucial for both the environment and the economy. The African lineage, which is often called Lineage A, has long been known to be the most diverse of the many evolutionary lineages found in this species. This lineage includes several subspecies that have changed to fit in with many different types of environments in sub-Saharan Africa. *Apis mellifera scutellata*, also known as the "savannah bee," is one of the most well-known African subspecies. It is known for being aggressive and able to live in open, warm places. *A. mellifera capensis*, commonly referred to as the "Cape bee," possesses unique reproductive traits, including social parasitism via thelytokous parthenogenesis. Along with these subspecies, in-depth genetic studies have shown that the African lineage is divided into several subgroups, usually called AI through AIVsome authors suggest using different names, like (Z for one of the groups), which are groups of bees from different areas that have quickly evolved due to local environmental pressures (Miguel et al., 2015; Nanetti et al., 2019).

4.2.1.2. Genetic Diversity and Evolutionary Dynamics:

The subject of genetic variation and the mechanism of evolution is crucial for comprehending the African lineage. African Lineage A possesses numerous distinct mitochondrial haplotypes and a variety of nucleotides. This diversity illustrates the evolution of the population over time and the ongoing changes in the environment. Mitochondrial DNA (mtDNA) investigations are favored due to their straightforward identification in cellular samples, lack of intermixing with other DNA, and rapid evolutionary rate. They have consistently demonstrated that the African lineage is monophyletic. Despite the existence of many internal divisions, all members of this lineage originate from a common maternal ancestor (Eimanifar, Kimball, et al., 2018; D. S. M. Henriques, 2018). An intriguing observation from the genetic data is that, despite substantial evidence for a common maternal lineage, there are no definitive molecular markers to distinguish the various subgroups or potential subspecies. This pattern is believed to indicate a history of recent and rapid evolutionary radiation, wherein ecological forces have induced substantial alterations in behavior and morphology, despite the mitochondrial genome not acquiring distinct molecular markers (Nanetti et al., 2019).

Scientists typically examine regions such as the tRNAleu-cox2 intergenic element while studying mitochondria. This feature has been widely employed to classify lineages of *Apis mellifera*. The extensive variety of haplotypes observed in African populations illustrates the genetic diversity of Lineage A and underscores the challenge of differentiating discrete subgroups (Gujarati & Porter, 2010; D. S. M. Henriques, 2018). Microsatellite markers employed in supplementary analyses further substantiate the notion of significant genetic variation within these populations. Research conducted in the Republic of South Africa indicates that the average allele count per locus is approximately 10, and there is significant heterozygosity, implying considerable gene flow across groups adapted to their environments (Eimanifar et al., 2020).

4.2.1.3. Internal Subdivision

The four subgroups (AI through AIV) make up the internal subdivision. One of the most important things about African Lineage A is that it is split into several subgroups, which have been called AI, AII, AIII, and AIV in the past. These subgroups are linked to specific places and environmental conditions, and it's important to know about them in order to understand how the African lineage has evolved as a whole.

4.2.1.3.1. Subgroup AI

Many researher think that subgroup AI is one of the most common and genetically diverse groups of bees in the African lineage. Bees that belong to subgroup AI can be found in many parts of sub-Saharan Africa. We have linked them to populations that inhabit a wide range of ecological conditions, from tropical savanna areas to wetter forest edges. Genetic research has shown that AI populations have many different haplotypes in their mitochondrial genomes. This variety is a result of how these bees have adapted to different microclimates and the strong selective pressures that cause rapid evolutionary diversification in these areas (Miguel et al., 2015; Nanetti et al., 2019). Furthermore, the genetic signals are still somewhat "blurry" when it comes to diagnostic markers because the fast pace of evolution has not yet allowed for the establishment of unique markers. Instead, there is a range of allelic variation, which is caused by a lot of gene flow between populations that are close to each other (Gujarati & Porter, 2010).

4.2.1.3.2. Subgroup AII

Subgroup AII is less common in published studies than AI. It is also more likely to be found in certain geographic areas, like the center of the continent or more transitional biomes. Although AII populations possess different genes compared to AI populations, the genetic difference is not significant due to their mixing and sharing of genes with nearby groups. Interestingly, while mitochondrial studies show that lineage A is a single group, subgroup AII doesn't have clear unique features that would help identify it easily, indicating that quick changes and ongoing mixing of genes make it difficult to establish distinct markers (Nanetti et al., 2019).

4.2.1.3.3. Subgroup AIII

Subgroup AIII is usually linked to groups of bees who live in edge environments and transition zones, which are places where ecological boundaries like temperature changes or moisture gradients are obvious. When you look at the mtDNA haplotype distribution and the microsatellite allele frequencies, you can see that bees in AIII are entirely unique genetically from bees in the other subgroups. However, similar to other Lineage A groups, AIII lacks distinct molecular markers for differentiation. Admixture events complicate the genetic structure in AIII, often occurring when bees from nearby areas share mitochondrial and nuclear alleles (Miguel et al., 2015; Nanetti et al., 2019).

4.2.1.3.4. Subgroup (AIVand the Proposed Z Sub-lineage)

Subgroup AIV, also referred to as the Suggested Z Sub-lineage, suggests the existence of another group within the African lineage that is sometimes called AIV or, in some cases, the Z sub-lineage. Some studies have found this subgroup by looking at certain restriction site patterns in the tRNAleu-cox2 intergenic region. Bees in this group live in places that are very different from the places where bees live, from AI to AIII. For instance, some evidence suggests a link to more isolated or marginal habitats, where unique selective pressures may have caused small genetic changes. These changes would be great molecular diagnostic markers if they could be separated, but high levels of gene flow and recent diversification make it difficult to do so (D. S. M. Henriques, 2018; Porrini et al., 2021). Therefore, despite some indications that each region differs, definitively identifying subgroup AIV or Z remains an ongoing area of research. Many studies keep stressing how dynamic and constantly changing the African lineage is.

4.2.1.4. Molecular Diagnostic Challenges and Gene Flow

Gene Flow and Challenges with Molecular Diagnostics A prevalent issue in the examination of African Lineage A honeybees is the challenge of developing precise genetic diagnostics to differentiate bees into distinct subspecies or groupings. Comprehensive mtDNA investigations indicate that all African bees in Lineage A possess a shared common ancestor, hence affirming monophyly. Nonetheless, no individual SNP or haplotype has been identified that can consistently differentiate *A. m. capensis* from A. *m. scutellata* or distinguish the subgroups AI, AII, and AIII (Nanetti et al., 2019). It is sometimes asserted that these populations lack diagnostic markers due to their rapid adaptation to varying ecological environments. Genetic variety exists among various populations, and gene flow between these populations contributes to the homogenization of allele frequencies across geographic regions. The molecular "blur" occurring during rapid species divergence complicates the use of traditional genetic markers for distinguishing subspecies. It indicates that a few number of nuclear genes, rather than extensive alterations in the mitochondrial genome, may account for significant adaptive variations observed in behavior and morphology (Eimanifar et al., 2020; Gujarati & Porter, 2010).

For example, studies utilizing microsatellite genotyping across various regions of the Republic of South Africa revealed significant allelic diversity and provided statistics indicating that gene flow is still occurring. The high number of private alleles in some areas is more evidence of localized differentiation. However, the overall fixation indices are still low, even though there is statistically significant differentiation between regional groups. This means that while local adaptations occur, the movement of queens during swarming and human actions that move colonies help keep the entire lineage genetically connected (Eimanifar et al., 2020).

4.2.1.5. Ecological Adaptations within the African Lineage

Changes in the African Lineage's environment The African landscape is full of different types of ecosystems, from dry savannas to lush tropical forests. Bees in Lineage A have developed various physical, behavioral, and physiological traits that help them survive in these different environments. For example, *Apis m. scutellata* is known for being aggressive when defending itself, swarming a lot, and growing quickly as a colony. These are all adaptations to the open, predator-rich environments that are common in savanna habitats (Engel et al., 2021; Nanetti et al., 2019). *A. mellifera capensis*, on the other hand, is mostly found in the Cape region of South Africa. It has developed its own ways of reproducing, such as the ability to lay eggs through thelytokous parthenogenesis. This capability lets it take advantage of certain ecological niches where competition and ecological constraints are very different from those in savanna regions (Engel et al., 2021; Nanetti et al., 2019).

Moreover, these environmental differences closely correlate with the genetic differences among the various subgroups of Lineage A. For example, populations in areas with a lot of climate change tend to have more mitochondrial haplotype diversity. Their diversity is due to environmental adaptation and historical demographic events like bottlenecks and expansions during climate change. The African honey bee lineage has quickly diversified because of the interaction between ecology and genetics. This illustration shows how adaptable these bees are, which has allowed them to live in a wide range of African habitats (Miguel et al., 2015; Nanetti et al., 2019).

4.2.1.6. Implications for Taxonomy and Conservation

This information has implications for taxonomy and conservation. Because there is a lot of genetic variation and no clear molecular diagnostic markers, traditional morphological identification is still an important tool for classifying African honeybees. But morphological traits can be difficult to understand, and measurements between subspecies can sometimes be the same. The fact that evolution can happen in different ways when organisms are under similar environmental pressures makes the situation even more unclear. Taxonomists and conservation biologists are being asked more and more to combine morphometric data with high-resolution genetic data, such as mitochondrial and nuclear markers, to acquire a better picture of the population structure within African Lineage A (Eimanifar, Kimball, et al., 2018; Gujarati & Porter, 2010).

From a conservation perspective, the changing evolutionary history of African honeybees shows how important it is to protect a wide range of genetic diversity. The fact that these bees can adapt to changes in their environment, from climate change to habitat fragmentation, is directly related to the diversity that exists within and between the different subgroups. Thereby, conservation plans must consider the gene flow that links populations, even if they are still different in their areas. This method makes sure that both the local adaptations that are necessary for survival in certain habitats and the overall genetic continuity that lets species respond quickly to changes in their environment are kept (Eimanifar et al., 2020; D. S. M. Henriques, 2018).

4.2.1.7. Synthesis and Future Directions

Combining everything and looking to the future The African Lineage A of *Apis mellifera*, with its subgroups AI and AIV (also referred to as Z), is an excellent model system for researching how social insects rapidly evolve and adapt. The great diversity of mitochondrial haplotypes, the intricate patterns of gene flow, and the absence of precise molecular diagnostic markers present challenges as well as opportunities for researchers. Subspecies identification is challenging due to the large range of mitochondrial haplotypes and intricate gene flow patterns, necessitating the use of a combination of techniques that simultaneously examine physical characteristics, environmental data, and genetic research. Due in part to their genetic dynamism, bees have evolved successfully, giving us valuable insights into how species adjust to rapidly changing environments.

Future research still needs to answer some important questions. First, it is crucial to identify nuclear genetic markers or genomic regions that can provide diagnostic signals compatible with the mitochondrial data. Next-generation sequencing and genome-wide association studies (GWAS) are two promising ways to locate these markers. However, they have to confront the fact that gene flow is common and lineage sorting is not always complete in these populations (Eimanifar, Kimball, et al., 2018; Tihelka et al., 2020). Second, to understand the balance between local adaptation and gene flow on a small geographic scale, we need to take more detailed samples and do more detailed analyses in ecological transition zones. These studies will help us understand how evolution works and how to manage both wild and managed bee populations in Africa.

In addition, combining high-resolution mitochondrial genome sequencing with nuclear marker panels should help clarify the phylogenetic relationships within African Lineage A. Projects to resequence whole genomes have already started to show the small-scale population structures and historical demographic events that have shaped these bees. More research in this area may eventually lead to the creation of strong molecular diagnostic tools. Currently, these tools are lacking, despite scientists having studied the evolution of *Apis mellifera* for over 30 years (Eimanifar et al., 2020; D. S. M. Henriques, 2018).

4.2.1.8. Final Thoughts

In short, the African Lineage A of honeybees is the most genetically varied group of *Apis mellifera*. It has four internal subgroups called AI, AII, AIII, and AIVor Z). *A. m. scutellata* and *A. m. Capensis*, are two important subspecies that belong to these subgroups. Together, they show how well they can adapt to the different environments found in sub-Saharan Africa. The mitochondrial DNA evidence strongly supports the idea that this lineage is monophyletic. However, the absence of clear molecular diagnostics to distinguish its subgroups reveals a complicated evolutionary history with rapid diversification and a lot of gene flow. Overlapping haplotypes and mixed microsatellite profiles reveal this genetic "blur," further complicating taxonomy and conservation efforts. But it also shows how evolution works in real life to help organisms adapt to changing environments.

More research will be needed in the future to better understand the complex relationship between local adaptation and gene flow, to make better molecular diagnostic tools, and to protect the genetic heritage of this important ecological group. Not only do the lessons learned from African Lineage A help us understand how honeybees evolved, but they also provide us a better idea of how social insects adapt to environments that change quickly (Harman, 2021; Nanetti et al., 2019).

The studies looked at a wide range of topics, including mitochondrial genome studies, microsatellite analyses, and morphometric investigations. They all point to the same thing: even though morphological characteristics can be difficult to pin down, the molecular signatures show that adaptive traits spread quickly because of ecological diversity and ongoing reproductive connectivity. African honey bees have adapted to live in a wide range of ecological niches due to their unique mix, which has facilitated their evolution and made them an important subject for both basic evolutionary research and applied conservation strategies (Engel et al., 2021; Miguel et al., 2015; Nanetti et al., 2019).

Many studies that were mentioned show that the problems of dealing with high gene flow and overlapping genetic variation are not only problems for African Lineage A; they are problems for population genetics as a whole. Still, the quick changes in these bees are a wonderful example of how selection and migration can work together to create a mosaic of populations that are genetically linked but adapted to their local environments. As we move into a time of global environmental change that is making insect pollinators around the world more and more at risk, these kinds of insights will be very important.

Overall, the African Lineage A of *Apis mellifera*, shows how strong adaptive evolution can be when faced with different ecological pressures. It also shows how challenging it is to put natural variation into categories when a species is always changing. As scientists continue to combine genetic, morphological, and ecological data, we can expect to learn more and more about these evolutionary relationships. This knowledge will help conservation efforts that aim to protect the rich genetic diversity of these wonderful insects (Eimanifar et al., 2020; Gujarati & Porter, 2010; Nanetti et al., 2019).

To sum up, we can say that our current understanding of African Lineage A is "from A to Z" in terms of its internal diversity and ability to adapt. However, we still need to decipher the intricate details of this evolutionary puzzle. Putting together different pieces of evidence presents us with a strong base for future research that will surely shed more light on the evolutionary paths of these insects that are so important to the environment.

African A Lineage Honeybees

The African A lineage of honeybees is a large group within the *Apis mellifera* species that is mostly found in Africa (Franck et al., 2001; Kadri et al., 2016). These subspecies have developed unique traits that help them survive in a wide range of African ecosystems, from deserts and mountains to coastal areas (Frazier et al., 2024).

Table 2. Comprehensive Table of African A Lineage Honeybees

Here is a detailed table that shows all the A lineage African honeybee subspecies, along with their distribution, main traits, mtDNA lineage classification, and links to more information:

Subspecies	Common Name	Distribution	mtDNA Lineage	References
A.m. adansonii	West African honey bee	West Africa (Nigeria, Burkina Faso, Senegal to southern border of Chad, extending to Congo)	A	(Kadri et al., 2016).
A. m. capensis	Cape honey bee	Southern South Africa	A	(Rojas-Sandoval, 2022).
A.m.intermissa	Tellian honey bee	Northwestern coast of Africa (Tunisia, Libya, Morocco, north of Atlas Mountains)	A	(N. J. Haddad, 2015).
A. m. jemenitica	Arabian honey bee	Somalia, Uganda, Sudan, Yemen, Saudi Arabia	A/Y	(Alsharhi et al., 2025)
A. m. litorea	East African coastal honey bee	Low elevations of East Africa, coastal plains of Kenya, Tanzania, and Mozambique	A	(Han et al., 2012).
A. m. monticola	East African mountain honey bee	High altitude mountains (1,500-3,100m) of East Africa (Mt. Elgon, Mt. Kilimanjaro, Mt. Kenya, Mt. Meru)	A	(Abou-Shaara, Al-Ghamdi, et al., 2021).
A. m. ruttneri	Maltese honey bee	Endemic to Maltese islands in the Mediterranean Sea	A	(Abou-Shaara, Al-Ghamdi, et al., 2021).
A. m. sahariensis	Saharan honey bee/Yellow Saharan bee	Northwestern Africa, southern side of Atlas Mountains, southern Morocco (Tafilalt Oasis)	A	(Khedidji et al., 2022).

A. m. scutellata	East African lowland honey bee/African bee	Eastern and southern Africa, from Ethiopia to South Africa	A	(Chapman et al., 2015).
A. m. unicolor	Madagascar honey bee	Madagascar	A/U	(Dimbiarimanga et al., 2024).

Additional Notes on Classification

Most of these subspecies are clearly part of the A lineage based on mitochondrial DNA analysis, but some have traits that are a mix of different groups (Abou-Shaara, Al-Ghamdi, et al., 2021; Utzeri et al., 2021):

A. m. jemenitica is sometimes put in both the A and Y lineages because it is in between African and Middle Eastern populations (Abou-Shaara, Al-Ghamdi, et al., 2021; Alsharhi et al., 2025). Some scientists have put *A. m. unicolor* from Madagascar in its own U lineage, but it is still very similar to the African A lineage (Frazier et al., 2024; Han et al., 2012). Even though it is found in Africa, *A. m. Simensis (*Ethiopian honeybee) has been put in the Y lineage instead of the A lineage (Abou-Shaara, Al-Ghamdi, et al., 2021). Even though it lives in Africa, the Egyptian honeybee (*A. m. lamarckii)* is usually put in the O or L lineage instead of the A lineage (El-Seedi et al., 2022). As more genetic and morphometric studies are done, the classification of these honeybee subspecies changes. These studies provide us a better understanding of how African honeybees have evolved and adapted over time (Franck et al., 2001).

Evolutionary Dynamics and Genetic Structure of the North Mediterranean Lineage (C) of *Apis mellifera*

4.2.2. Lineage C (North Mediterranean)

4.2.2.1. Introduction to Lineage (C)

Lineage C, also known as the North Mediterranean lineage of *Apis mellifera*, is one of the most studied evolutionary groups within the western honey bee species. This is because it plays a big role in beekeeping, has a complicated phylogeographic history, and has unique mitochondrial DNA (mtDNA) signatures (Alburaki et al., 2023). The main thing that sets this lineage apart is its different mtDNA haplotypes, the most important of which are called C1 and C2. These have been well studied in both wild and managed bee populations (Alburaki et al., 2011). The C lineage is native to the temperate areas of southern and eastern Europe and parts of the Near East. Its evolutionary history has been shaped by changes in climate, geographical barriers, and human-mediated dispersal (Muñoz & De la Rúa, 2021). Lineage C is also not a single genetic unit; it has a lot of internal complexity, with paraphyly and admixture among subspecies. Such complexity has important effects on taxonomy, conservation, and apiculture (Bouga, 2011).

4.2.2.2. Mitochondrial Haplotypes and Subspecies Associations

The mitochondrial genome is a very useful tool for figuring out how Lineage C evolved because it is passed down through the maternal and shows clear patterns of nucleotide substitutions that set haplotypes apart (N. J. Haddad et al., 2009). Haplotype C1 is most often linked to *A. m. ligustica*, also known as the Italian honeybee. This bee is well-known for being calm and productive in beekeeping (Taurisano et al., 2024). On the other hand, haplotype C2 is more often linked to the Carniolan honeybee, *Apis mellifera carnica*, which is known for being able to live in colder climates and for being able to survive the winter better (Alburaki et al., 2023). Even though these connections are clear, mtDNA studies have indicated that the boundaries between these subspecies in the C lineage are not completely separate. There are also overlapping haplotypes that show complicated patterns of gene flow over time (Delaney et al., 2009).

4.2.2.3. Evolutionary Processes and Phylogeography

Using both mitochondrial markers and nuclear markers like microsatellites, researchers have found signs of admixture and introgression in Lineage C's evolutionary history. This provides a more comprehensive understanding of the gene exchange between populations (Muñoz et al., 2009). For example, studies of a specific part of the mitochondrial DNA show that C-lineage genetic types often group together in a way that suggests they are not all directly related. This shows that subspecies like *A. m. ligustica* and *A. m. Carnica,* have gone through a lot of hybridization events (Alburaki et al.,

2023). It is thought that this paraphyly is the result of being isolated in glacial refugia in the past, then recolonizing after the glaciers melted, and the ongoing effects of humans moving colonies from one area to another (Muñoz & De la Rúa, 2021).

4.2.2.4. Role of Subspecies in Apiculture

The two main subspecies in the C lineage, *A. m. ligustica* and *A. m. carnica*, are crucial for both natural evolution and commercial beekeeping (Taurisano et al., 2024). *A. m. ligustica*, which usually has the C1 haplotype, came from the Italian Peninsula. Over the past 150 years, queens have been exported all over the world, which is why it is so common in managed bee populations (Bouga, 2011). On the other hand, *A. m. carnica*, which mostly has the different subtypes of the C2 haplotype, is native to the alpine and sub-alpine zones of southeastern Europe. It is known for being very productive and calm in temperate climates (Alburaki et al., 2023).

4.2.2.5. Genetic Structure and Admixture

These mtDNA haplotypes not only show how Lineage C is related to other lineages, but they also show how the whole genome is structured, as shown by nuclear markers (Oleksa et al., 2021). Comprehensive studies that look at both mitochondrial and nuclear data have shown that the C lineage is mostly consistent, but that gene flow along hybrid zones has brought even non-native alleles into established populations (Oleksa et al., 2011). This mixing has been seen most often in places where managed bees from Lineage C have been added to populations that were originally made up of bees from other lineages. This mix has led to a genomic structure that looks like a mosaic (Muñoz & De la Rúa, 2021).

4.2.2.6. Human Influence and Hybrid Zones

Human activity has greatly affected the current genetic landscape of the C lineage. Beekeeping practices, like bringing in queens from other countries and selective breeding, have unintentionally helped C-lineage haplotypes move across traditional geographic boundaries (Delaney et al., 2009). These human-made changes have played a big role in the creation of hybrid zones, where bees of different subspecies breed with each other. This makes it harder to tell the differences between them in their native ranges (Adolph, 2016). In many parts of Europe, like the Carpathians, people have moved bees across natural barriers that used to keep different lineages apart. Such migration has led to more mixing of genes and a loss of genetic uniqueness in some native populations (Okwaro Louis Allan, 2019).

4.2.2.7. Geographic Distribution and Genetic Bottlenecks

The C lineage haplotypes are more common in the Mediterranean basin than in northern Europe (Taurisano et al., 2024). This shows a clear latitudinal gradient in the distribution of the C lineage in Europe. Most of the managed bee populations in the United States come from the North Mediterranean region, as shown by the fact that C1 and C2 haplotypes are more common in commercial stocks (Alburaki et al., 2023). This pattern, however, comes at the cost of less genetic diversity. Managed populations have focused on a small number of mitochondrial haplotypes, which has caused a genetic bottleneck that could make colonies less adaptable to new pests and changes in the environment (Alburaki et al., 2023).

4.2.2.8. Hybridization and Sub-Haplotypes

The story of Lineage C's evolution grows even more complicated by hybridization and introgression, which have been found in many studies that used mitochondrial markers (N. J. Haddad et al., 2009). There have been a few cases where hybrid zones have been found in areas where the ranges of *A. m. ligustica* and *A. m. Carnica*, overlap. Such hybridization has led to the exchange of mtDNA haplotypes and the appearance of intermediate forms (Tozkar, 2020). Not only do subspecies share haplotypes, but there are also sub-haplotypes within the C2 group, like C2d and C2j. These show that mutation and selection pressures are still happening in response to local environmental conditions (Taurisano et al., 2024).

4.2.2.9. Morphology and Molecular Correlations

The shape and size of bees in Lineage C match their genetic profiles, and morphometric studies have been used to add to the molecular data (Zammit-Mangion et al., 2017). These studies have shown that wing venation patterns and other physical traits generally support the classification of bees into *A. m. ligustica* and A. *m. carnica*. However, there are some differences that are probably due to introgressive hybridization (Ferreira, 2017). Morphological intermediates have been found in areas where these subspecies come into close contact. This discovery adds to the genetic evidence of admixture and complex evolutionary histories (Zammit-Mangion et al., 2017).

4.2.2.10. Conservation Concerns and Cross-Lineage Gene Flow

Gene flow between the C lineage and nearby lineages, especially the M lineage, which is typical of the dark European honeybee, *Apis m. mellifera*, has gotten a lot of attention because of its effects on conservation (Oleksa et al., 2021). The introduction of C-lineage mitochondrial haplotypes (Oleksa et al., 2011) has made it harder to protect the genetic integrity of native M-lineage bees in several conservation areas across Europe. These types of introgression events are particularly concerning because they can reduce the adaptability of local bee populations, making it more difficult for them to survive in their natural habitats (Muñoz & De la Rúa, 2021).

4.2.2.11. Management and Conservation Strategies

The C lineage is complicated, which makes it difficult to protect, but it also provides us new chances. On the one hand, *A. m. ligustica* and *A. m. carnica*, are used a lot in commercial beekeeping, which means that the C lineage is still the most common maternal lineage in many managed populations (Alburaki et al., 2023). On the other hand, the widespread movement of these bees by human has made the genetic pools in some areas more similar while also creating areas with a lot of mixing and hybridization (Palmer et al., 2000). This homogenization could reduce the managed bees' genetic diversity, making them more prone to illness and environmental stress (Ferreira, 2017).

4.2.2.12. Monitoring Genetic Diversity

Ongoing research shows that keeping genetic diversity in the C lineage is important for both the future of beekeeping and the preservation of evolutionary potential (Muñoz et al., 2009). To keep local subspecies safe, conservation programs use a lot of molecular and morphometric diagnostic tools to monitor how much introgression and admixture is going on (Taurisano et al., 2024). For instance, several European countries are making efforts to ensure the safety of pure *A. m. ligustica*, populations. These programs are crucial because they help us understand how populations change over time and how to manage honeybee genetic resources in a way that is advantageous for the environment (Bouga, 2011).

4.2.2.13. Evolutionary and Commercial Relevance

The current genetic structure of the C lineage (Tozkar, 2020), is still being shaped by the interaction of natural evolutionary processes and human-made changes. The C lineage's diversification began with events in the past, like post-glacial recolonization from southern refugia. Recent human activities, like migratory beekeeping and international queen trading, have sped up the mixing of haplotypes between far away places (Muñoz et al., 2009). Because of this, both ancient evolutionary history and modern management practices now affect the distribution of mtDNA haplotypes like C1 and C2 in their native range (Taurisano et al., 2024).

4.2.2.14. Phylogeographic Patterns and Research Implications

Detailed phylogeographic studies have shown that the spatial distribution of C-lineage haplotypes shows clear geographic patterns that reflect past colonization and gene flow events (Ferreira, 2017). For instance, the C1 haplotype is more common in parts of the Mediterranean basin where *A. m. ligustica*, used to live, while the C2 haplotype is more prevalent in areas where *A. m. carnica*, used to live (Alburaki et al., 2023). The existence of hybrid zones makes the geographic structuring of haplotypes even more complicated. In these zones, bees from different mitochondrial lineages breed with each other, creating a mix of genetic traits that can't be easily classified (Oleksa et al., 2021).

4.2.2.15. Evolutionary Significance and Future Directions

Researchers are still looking into why hybrid zones stay in the C lineage. These areas are like natural labs for studying introgression, local adaptation, and evolutionary divergence (Tozkar, 2020). Molecular evidence indicates that the mtDNA haplotypes don't always fit neatly into the traditional subspecies categories in these hybrid zones. Instead, they show overlapping distributions that are the result of historical gene flow events (Zammit-Mangion et al., 2017). The effects of this kind of introgression can be good or detrimental. Gene flow can bring in adaptive alleles that make a colony stronger, but it can also mess up gene complexes that are already adapted to the area, making conservation harder (Oleksa et al., 2021).

4.2.2.16. Conclusion

In conclusion, Lineage C (North Mediterranean) of *Apis mellifera*, has a long and complex evolutionary history. Its unique mtDNA haplotypes C1 and C2 are a result of natural recolonization after the last ice age, and humans have had a big impact on it through migratory beekeeping and selective breeding. This lineage's changing characteristics, which include clear groupings and areas where different types mix, highlight how difficult it is to distinguish between subspecies and how crucial it is to maintain genetic variety for the health of the environment and the economic success of beekeeping (Delaney et al., 2009). To protect the integrity and adaptive potential of Lineage C, it will be necessary to combine high-resolution genetic data with strong conservation frameworks. The combination will ensure that these important pollinators continue to thrive even as the environment changes (Muñoz & De la Rúa, 2021).

Table 3. Lineage C (North Mediterranean) Honeybee Subspecies

The C lineage of *Apis mellifera,* (North Mediterranean) is native to Central and Southeastern Europe and the northern Mediterranean. Below is a comprehensive table listing all recognized C lineage subspecies, their countries of origin, mtDNA lineage, and references.

Subspecies Name	Common Name	Country/Region	mtDNA Lineage	References
A. m. ligustica	Italian honey bee	Italy	C	(Ilyasov et al., 2020)
A. m. carnica	Carniolan honey bee	Slovenia, Austria, Croatia, Bosnia, Hungary, Romania, Serbia, Bulgaria, Poland	C	(Alburaki et al., 2023)
A. m. macedonica	Macedonian honey bee	Greece, North Macedonia, Bulgaria, Ukraine	C	(Ilyasov et al., 2020)
A. m. cecropia	Greek honey bee	Greece	C	(Alburaki et al., 2023)
A. m. cypria	Cyprus honey bee	Cyprus	C	(Ilyasov et al., 2020)
A. m. siciliana	Sicilian honey bee	Sicily (Italy)	C	(Ilyasov et al., 2020)
A. m. adami	Cretan honey bee	Crete (Greece)	C	(Ilyasov et al., 2020)
A. m. carpathica	Carpathian honey bee	Romania, Ukraine, Moldova, Bulgaria	C	(Ilyasov et al., 2020)
A. m. rodopica	Bulgarian honey bee	Bulgaria	C	(Ilyasov et al., 2020)

Table 4. Subspecies with Uncertain or Mixed Lineage (C or O)

Some subspecies are sometimes assigned to either the C or O lineage due to overlapping characteristics and genetic markers:

Subspecies Name	Common Name	Country/Region	mtDNA Lineage	References
A. m. artemisia	Russian steppe honey bee	South Russia, Ukraine	C or O	(Ilyasov et al., 2020)
A. m. sossimai	Ukrainian honey bee	Crimea, South Russia, Ukraine	C or O	(Ilyasov et al., 2020)
A. m. taurica	Crimean honey bee	Crimea, South Russia, Ukraine	C or O	(Ilyasov et al., 2020)

Notes

The main C lineage subspecies are *A. m. ligustica*, *A. m. carnica*, *A. m. macedonica*, *A. m. cecropia*, *A. m. cypria*, *A. m. siciliana*, *A. m. adami*, *A. m. carpathica*, and *A. m. rodopica* (Alburaki et al., 2023; Ilyasov et al., 2020; Utzeri et al., 2022). Subpecies with uncertain lineage (C or O) are sometimes included in the C group depending on the source (Ilyasov et al., 2020). The C lineage is characterized by gentle behavior, good overwintering, and adaptation to temperate and Mediterranean climates (Utzeri et al., 2022).

Evolutionary history and Conservation of the West European Honey Bee Lineage M: Genetic Diversity, Adaptation, and Challenges in a Changing World.

4.2.3. Lineage M from (Western Europe)

4.2.3.1. Introduction

The West European honey bee lineage (Lineage M) includes honey bee populations that are native to western and northern Europe. The dark European honeybee (Alburaki et al., 2023), is the most common species in this lineage. A group of unique mitochondrial DNA (mtDNA) haplotypes makes up this lineage. Scientists have used these haplotypes to figure out where they came from, how diverse they are, and how their populations are structured (Alsharhi et al., 2025). Researchers have conducted extensive molecular research on the mtDNA COI–COII intergenic region and other nuclear markers. The work has given us a lot of information about the M lineage's phylogenetic placement and evolutionary history. It has also shown that it is a separate phylogenetic cluster from lineages from Africa, the Middle East, and Eastern Europe (Chávez-Galarza et al., 2017).

4.2.3.2. Evolutionary History and Phylogenetic Position

Phylogenetic reconstructions based on mitochondrial sequences and nuclear markers have often placed *A. m. mellifera*, at or near the base of the *Apis mellifera* complex. This suggests that the M lineage may be one of the first diverging clades within this species (Chávez-Galarza et al., 2017). *Apis mellifera's* role in some analyses suggests it may have spread from Europe. However, later studies that used larger datasets and better evolutionary models have questioned the idea that the species originated only in Europe (Chávez-Galarza et al., 2017). Furthermore, earlier mitochondrial studies that used restriction fragment length polymorphism (RFLP) and sequence analyses backed up the idea that the M lineage is monophyletic and different from other lineages, like the African (A) and North Mediterranean (C) lineages (Cornuet, J. M. & Garnery, L., 1991). It is clear that the M lineage, which includes *Apis mellifera mellifera*, and related subspecies, is a distinct genetic group that is separate from other evolutionary branches within *A. mellifera* (Garnery et al., 1998). This conclusion is supported by phylogenetic resolution methods like maximum likelihood and median-joining network analysis.

4.2.3.3. Mitochondrial Haplotype Diversity

M lineage honeybees have unique mtDNA haplotypes that are mostly found in the tRNA^leu–COII intergenic region (Garnery et al., 1998). Some of these haplotypes, like M3, M7b, and M7c, have been found in different European populations and even a few in North America, which is not their native range (Alburaki et al., 2023). The haplotype diversity in the M lineage is intriguing because it is both complex and spatially structured. This study reveals that distance and local adaptation have shaped the M lineage's long evolutionary history (Garnery et al., 1998). The M lineage of *A. m. mellifera,* may have fewer haplotypes overall than some African lineages, but the diversity within the M lineage is still important for its ability to adapt (Hassett et al., 2018). Recent studies using mitochondrial markers have shown that even in a small area, there are different subclusters of haplotypes that may correspond to historical refugial areas and the routes that people took to recolonize after the last ice age (Jensen et al., 2005).

4.2.3.4. Population Structure and Gene Flow

In-depth research on mtDNA haplotype distribution in European populations has revealed that Lineage M has a complex history, showing evidence of both ancient separation and more recent mixing of genes. The spatial structure of M lineage honeybees in places like France, Northern Europe, and parts of Eastern Europe shows a pattern that is consistent with post-glacial recolonization from refugial zones in the Iberian and Balkan Peninsulas (Kaskinova et al., 2023). Studies that use population genetics have found intriguing phylogeographic patterns. For example, genetic differences are often linked to geographic distance and natural barriers like mountain ranges (Kaskinova et al., 2023). Furthermore, fine-scale studies in some parts of Russia have found different haplogroups within the M lineage. These include the widely spread M17 and the less common M4 groups. This suggests that there is a lot of substructuring within the M lineage that is caused by both natural dispersal and human-made movements (Kaskinova et al., 2023).

4.2.3.5. Geographic Distribution and Biogeography

The M lineage can be found all over Western and Northern Europe, from the British Islands, Scandinavia, and Central Europe to the edges of Eastern Europe (Knoll et al., 2024). *Apis m.mellifera,* has been native to these areas for a long time, and historical records and morphometric studies indicate that it spread across much of Europe after the last glacial maximum (Miguel et al., 2007). Studies show clear genetic differences between populations on either side of geographic barriers like the Pyrenees and stress how important these barriers are. This supports the idea that there have been many post-glacial recolonization routes that have shaped the current genetic landscape (Miguel et al., 2007). Molecular studies have revealed a connection between the coastal and island populations. This shows that even though there is a lot of gene flow between M lineage populations, there are still signs of isolation by distance (Oleksa et al., 2021).

4.2.3.6. Demographic Bottlenecks and Adaptive Evolution

The M group is an ancient and unique lineage, but it has less overall haplotype diversity than some other evolutionary lineages of *Apis mellifera*, especially those of African descent (Oleksa et al., 2021). This lack of diversity might be caused by past population reductions, possibly linked to climate changes during the Pleistocene, as well as recent human activities, like bringing in non-native subspecies through modern beekeeping (Parejo et al., 2016). The M lineage has less genetic diversity, which could be because of selective breedings or founder effects that have limited genetic variation over time. However, it still has a set of haplotypes that are well-suited to temperate climates (Taurisano et al., 2024). Fossil and genetic evidence indicate that these demographic events, along with ongoing gene flow between populations, contribute to a lineage that is evolutionarily strong yet genetically limited when assessed by mitochondrial diversity metrics (Taurisano et al., 2024).

4.2.3.7. Influence of Human Activity

Human activities have had a big impact on the genetic makeup of Lineage M. Historical records show that even though M lineage honeybees have a lot of genetic diversity in their native range, only a small number of these haplotypes have been brought to places like the United States (Alburaki et al., 2023). Often, selective breeding and bringing in queens from C lineage subspecies have caused a lot of introgression, which weakens the genetic integrity of pure M lineage populations in parts of Europe and other places (Alsharhi et al., 2025). This human-driven mixing of genes has not only altered where mtDNA haplotypes are found, but it has also intensified the impact of environmental and population pressures, further reducing the ability of native M lineage populations to evolve (Alsharhi et al., 2025).

4.2.3.8. Comparative Phylogenetics and Methodological Considerations

Advanced phylogenetic methods have been used in studies comparing mitochondrial genomes of *Apis mellifera*, lineages to demonstrate that Lineage M (Cridland et al., 2017), is distinct from the others. Some studies have found that the M lineage is the most basic one in the species complex, but modern methods that take into account differences in composition and model artifacts have changed this view. The evidence suggests that the perceived basic position may be affected by methodological biases (Cridland et al., 2017). Work that looks at the evolutionary history of the Western honeybee has also shown how challenging it is to figure out the exact phylogenetic relationships between the different lineages. For example, the M lineage is clearly monophyletic, but its exact position in relation to other groups, like the C and O lineages, is still up for debate as analytical tools get better (Han et al., 2012).

4.2.3.9. Standardized Methods and Integrative Approaches

M lineage honeybees have been studied in depth thanks to new molecular and morphometric techniques that help us understand their diversity and evolutionary history. Standard methods that use mitochondrial markers, such as the DraI COI–COII restriction analysis, along with nuclear markers like microsatellites and genomic SNPs, have simplified the process of identifying and protecting different subspecies (Meixner et al., 2013). Researchers can now compare data from different studies and geographic areas thanks to the creation of consensus protocols. Such information is important for checking the health of native populations in the face of widespread hybridization and introgression (Meixner et al., 2013). In practice, these standardized methods have led to a detailed list of M haplotypes that not only helps with taxonomic classification but also helps with plans to protect genetic diversity in native West European honeybee populations (Meixner et al., 2013).

4.2.3.10. Long-term trends in demographic history and evolution

There have been times when the M lineage has grown and times when it has shrunk, and these events have left a permanent mark on its genetic makeup. Studies of ancient DNA and samples from museums show that populations of the M lineage in the past had more nucleotide diversity than those of today. The evidence suggests that genetic diversity has decreased over the past century (Espregueira Themudo et al., 2020). People think that these kinds of declines are caused by smaller effective population sizes, loss of habitat, and the rise of beekeeping practices that value uniformity and high-yield traits over local genetic differences (Espregueira Themudo et al., 2020). This long-term decrease in haplotype diversity in the M lineage could have big effects on how adaptable and strong they are when faced with new problems like pathogens, climate change, and changes to their habitat (Espregueira Themudo et al., 2020).

4.2.3.11. Complexities of Introgression

One of the main topics in the genetic study of the M lineage is how introgression from non-native subspecies affects historically pure populations. There is clear molecular evidence of genetic mixing in many parts of Europe, especially in places where beekeeping practices have historically favored the importation of C lineage queens. This mixing threatens the conservation of native gene pools of *A. m. Mellifera,* (Taurisano et al., 2024). In areas where beekeeping is very active, the number of pure M lineage mtDNA haplotypes has gone down, and both molecular and morphometric analyses have found hybrid individuals (Taurisano et al., 2024). This introgression makes it harder to figure out how different species are related through evolution, and it may also weaken the traits that have evolved in the M lineage populations to help them adapt to their environment and fight off disease (Taurisano et al., 2024).

4.2.3.12. Implications for Conservation and Management

The M lineage has a unique evolutionary history and the ability to adapt, which shows how important it is to have targeted conservation efforts to keep the genetic purity and diversity of native West European honeybees. It is very important to keep pure populations of *A. m. mellifera*, alive because they have unique haplotypes and local adaptations that can't be replaced when humans put pressure on them (Hassett et al., 2018). Finding and protecting refugial populations that have maintained their M lineage genetic identity is crucial, despite the significant hybridization that has occurred (Kaskinova et al., 2023).

To achieve this, conservation programs need to use strict molecular monitoring, such as routine screening of mitochondrial haplotypes. These kinds of efforts not only help protect biodiversity, but they also help keep apiculture stable and long-lasting in areas where native bee stocks are important to farming (Knoll et al., 2024).

4.2.3.13. Role of Ecological Adaptation

The M lineage has been around for a long time because it has adapted to the temperate climates of Western and Northern Europe. *A. m. mellifera* has developed unique behavioral, physiological, and ecological traits that help it thrive in places with significant changes in the seasons and harsh winters (Meixner et al., 2013). These adaptive traits help the M lineage populations survive and do well in the long term. They include ways to control body temperature, find food quickly during short summer, and resist certain local pathogens (Meixner et al., 2013). Because the environment is changing quickly and new non-native subspecies with traits that may not be good for the M lineage are being introduced, more and more people agree that we need to protect and use the M lineage's adaptive traits as a resource for future breeding and conservation programs (Meixner et al., 2013).

4.2.3.14. Integrating Molecular Data with Traditional Approaches

Modern research on *Apis mellifera subspecies*, especially those in the M lineage, uses an integrative approach that combines molecular methods with traditional morphometric and behavioral tests. These kinds of analyses have shown that the M lineage still has a unique genetic identity, even though there is a lot of gene flow and hybridization in many places. This distinction is shown by both mitochondrial haplotype patterns and classical morphological traits (Oleksa et al., 2021). This integrated framework is important for both clearing up taxonomic confusion and coming up with reliable ways to manage and protect native populations (De la Rúa et al., 2009). The fact that data from several different molecular markers point to the same thing has strengthened the phylogenetic coherence of the M lineage. This confers conservation policies aimed at keeping native West European honeybee stocks intact more credibility (Taurisano et al., 2024).

4.2.3.15. Recent Advances in Genomic Analyses

Next-generation sequencing technologies have made it possible to look at the genetic diversity of the M lineage across the whole genome in new ways. Whole-genome sequencing has given us a better look at the genetic structure of *A. m. mellifera*. It has shown patterns of variation that match up with the historical and demographic trends that mitochondrial analyses have suggested (Parejo et al., 2016). These genomic data have helped us determine signs of selection, admixture, and localized adaptation that aren't always clear from mitochondrial markers alone (Tihelka et al., 2020). Genomic studies have indicated that certain parts of the genome in the M lineage are under positive selection. These parts may be linked to traits like disease resistance, foraging efficiency, and developmental processes (Tihelka et al., 2020). Combining mitochondrial and nuclear genomic data is helping us acquire a better picture of how evolution works in the M lineage. It is also expected to help conservation strategies that use both adaptive variation and overall genetic diversity (Tihelka et al., 2020).

4.2.3.16. Challenges and Future Directions

Despite our extensive understanding of the M lineage, we still face several unresolved issues. One of the main problems is that non-native subspecies are still mixing with native ones, which makes it difficult to tell which populations are pure M lineage and puts native gene pools at risk of dying out in the long term (Alburaki et al., 2023). Another problem is that we need a more detailed and standardized haplotype naming system that can make sense of differences between studies and assure that conservation efforts are based on reliable genetic indicators (Alsharhi et al., 2025). More research is required to fully comprehend the demographic history of the M lineage, especially since genetic diversity has been decreasing over the past century (Espregueira Themudo et al., 2020). Researchers should be able to answer questions about the evolution, adaptation, and future of the M lineage in a world that is changing quickly as genomic tools develop better and more samples are taken (Knoll et al., 2024).

4.2.3.17. Conclusion

The study of Lineage M honeybees, especially *A. m. mellifera*, tells an evolutionary story that is both old and new. Even though the M lineage is at the bottom or close to the bottom of some phylogenetic trees, it has been shaped by many events of growth, bottlenecks, and mixing that have led to a population of honeybees that are perfectly suited to the temperate climates of western and northern Europe (Chávez-Galarza et al., 2017; Miguel et al., 2007). The M lineage has less genetic diversity than some other African lineages, but it still has enough genetic variation to keep local adaptations going and to keep native bee populations alive in the long term (Hassett et al., 2018). In the end, combining old morphometric data with new molecular and genomic methods presents us a strong way to understand and protect the genetic heritage of the M lineage (Meixner et al., 2013). To protect the evolutionary history of native West European honeybees in the face of ongoing environmental and human-made challenges, we will need focused conservation strategies based on detailed genetic analyses (Cridland et al., 2017; Taurisano et al., 2024).

In short, Lineage M is not only a basic evolutionary unit in *Apis mellifera,* but it is also a living storehouse of genetic changes that have been made over time because of Europe's unique climate and ecology. Its study continues to provide us useful information about how evolution, domestication, and conservation work in one of the most important pollinators in the world (Han et al., 2012; Tihelka et al., 2020). It is essential to keep the genetic integrity of this lineage for both biodiversity and the long-term health of beekeeping. Such an endeavor requires ongoing research, standardized methods, and proactive conservation measures that take into account both the past and the future (Alsharhi et al., 2025; Espregueira Themudo et al., 2020).

By continuing to work together across fields to connect traditional beekeeping with cutting-edge molecular genetics, we will learn more about *Apis mellifera's* past and come up with better ways to protect it in the future (Knoll et al., 2024; Oleksa et al., 2021).

Table 5. Lineage M (Western European) Honeybee Subspecies

Lineage M encompasses honeybee subspecies native to western and northern Europe. These bees are adapted to temperate climates and are known for their dark coloration and overwintering abilities.

Subspecies Name	Common Name	Country/ Region	mtDNA Lineage	References
A. m. Mellifera	European dark honey bee	France, UK, Ireland, Switzerland, Belgium, Netherlands, Germany, Denmark, Norway, Sweden, Poland, western Russia	M	(Wragg et al., 2022).
A. m. İberiensis	Iberian honey bee	Spain, Portugal, Gibraltar	M	(Wragg et al., 2022).
A. m. Sinisxinyuan	Xinyuan honey bee	Xinjiangnorthwest (China)	M	(Ilyasov et al., 2020).

Notes:

A. m. mellifera and *A. m. iberiensis* are the two primary and widely recognized M lineage subspecies in Europe (Wragg et al., 2022).

A. m. sinisxinyuan is a recently described subspecies from western China, genetically grouped with the M lineage (Ilyasov et al., 2020).

These subspecies are characterized by their dark coloration, strong adaptation to cold and wet climates, and generally defensive behavior (Meixner et al., 2013).

No other subspecies are currently recognized as part of the M lineage in major taxonomic revisions (Ilyasov et al., 2020; Wragg et al., 2022).

4.2.4. Lineage (O) Honeybees

4.2.4.1. Introduction to Lineage (O) Honeybees

Lineage O honeybees, also known as the Oriental lineage of *Apis mellifera,* are a separate group of honeybee subspecies that live mostly in the Middle East and nearby areas. These groups have unique genetic and physical traits that have helped them survive and thrive in dry and semi-dry areas, such as deserts and steppes. A number of molecular, morphometric, and phylogeographic studies have indicated that their evolutionary history is very different from that of the other major honeybee lineages, which are represented by the African (A), Western and Northern European (M), and Eastern European (C) groups (Abou-Shaara, Al-Ghamdi, et al., 2021).

4.2.4.2. Phylogenetic and Biogeographic Origins

The O lineage is closely linked to the Near Eastern homeland from both a phylogenetic and biogeographic perspective. This is where honeybees have faced unique climate pressures that have led to adaptive evolution. This lineage seems to have come from an area with long periods of dryness and harsh environmental conditions that required specific physiological and behavioral adaptations. Molecular studies have indicated that many subspecies in this group, including *Apis mellifera syriaca,* and others, possess mitochondrial DNA (mtDNA) markers and haplotypes unique to their Eastern origin (Alattal et al., 2014).

4.2.4.3. Mitochondrial DNA and Genetic Differentiation

Mitochondrial markers have been very advantageous in separating the O lineage from its European and African relatives. Studies using mtDNA segments, such as the area between cytochrome oxidase I (COI) and cytochrome oxidase II (COII) and parts of the ND2 gene, have revealed clear differences in the DNA sequences that set the Oriental bees apart from other groups of *Apis mellifera,* (Pierre Franck et al., 2000). For example, researchers have used the presence or absence of certain mutations, like nucleotide transitions in important parts of the mtDNA COI–COII region, as a molecular signature for these bees (Kandemir et al., 2006).

4.2.4.4. Whole Mitogenome Analysis and Phylogenetic Distinctiveness

Whole mitogenome analyses help us learn more about the genetic identity of Lineage O honeybees. Recent studies that have looked at the complete mitochondrial genomes of different subspecies have always found a monophyletic clade that represents this lineage. This evidence supports the idea that it is unique and has evolved in a consistent way (Carr, 2023). These studies not only back up the idea that the Oriental lineage is separate from the African A, C, and M lineages, but they also show that the O lineage has less mitochondrial diversity than some of its neighbors. Such results could mean that the O lineage has either recently expanded or that strong purifying selection is keeping important adaptive traits in these environments (Tihelka et al., 2020).

4.2.4.5. Phylogenetic Algorithms and Analytical Approaches

Molecular studies that use modern phylogenetic algorithms to combine mtDNA sequence data with these results have made them even stronger. Bayesian inference and maximum likelihood analyses of both partial gene sequencing and whole mitochondrial genomes have created strong phylogenies that always put the Oriental honey bees in a separate group (Tozkar, 2020). These kinds of in-depth phylogenomic studies are crucial because they can identify small genetic differences that might be hidden by gene flow or recent mixing with nearby subspecies (Carr, 2023).

4.2.4.6. Morphological Characteristics and Integrative Taxonomy

Morphologically, some traditional methods for identifying honey bee subspecies based on wing venation patterns and body measurements have helped us classify them in the past. However, more recent methods that combine morphometric data with molecular markers have helped us better understand what makes the O lineage unique. The small differences in morphometric traits between these bees are often the result of adaptations to their local environment. However, without molecular data to back them up, these traits could lead to misclassification because they look like bees from other lineages (Palmer et al., 2000).

4.2.4.7. Molecular Signatures and Diagnostic Tools

The mitochondrial DNA data, particularly from the COI–COII intergenic region, reveals that Lineage O subspecies possess unique sequence motifs that distinguish them from other lineages. This makes them a reliable way to diagnose. Researchers have found specific nucleotide changes in this area that set these bees apart from others. For example, studies have shown that these bees lack certain restriction enzyme sites that are common in European and African honeybees (Muñoz et al., 2009). These markers can be used to identify subspecies and also provide us with information about the historical biogeography and dispersal routes in the Middle East and nearby areas (Doctor, 2009).

4.2.4.8. Monophyly and Mitogenomic Support

Studies that use complete mitochondrial genome sequences provide more evidence for the monophyly and unique evolutionary path of the Oriental lineage. Whole mitogenome analyses show that the O lineage is closely related to groups like the C lineage, but it is still a separate evolutionary unit with a clear boundary. This conclusion is supported by high bootstrap values and posterior probabilities in Bayesian analyses (Carr, 2023). This kind of detailed genetic information also helps clear up taxonomic problems that have come up because of similar physical traits and names that have been used in different ways over time.

4.2.4.9. Adaptive Evolution and Environmental Pressures

The unique selective pressures of the dry and semi-dry environments that are common in the Middle East and Near East have led to the adaptive evolution of Lineage O honeybees. These bees have changed over time to be able to handle very hot or cold weather, not much water, and flowers that bloom only occasionally. Mitochondrial markers show that certain haplotype groups in the O lineage are connected to traits that help the colony survive and thrive in harsh environments (Alattal & Algamdi, 2022). In these environments, the mitochondrial genome may help with more than just making energy; it may also help with metabolic changes that make the organism more resistant to heat and lack of food (Alsharhi et al., 2025).

4.2.4.10. Regional Genetic Differentiation

The O lineage has genetic differences between regions that show how it has adapted to its environment. For instance, populations of *A. m. jemenitica*, which are often put in the Oriental group, have been found to have unique mitochondrial haplotypes that show they have been isolated and evolved on their own for a long time in the Arabian Peninsula (Alsharhi et al., 2025). These haplotypes show that there is a fine-scale genetic structure within the larger O lineage that is shaped by geographic barriers, changes in climate, and historical gene flow events. Sometimes, subspecies within the O lineage may even have different ecological adaptations that can be seen at the molecular level through different mtDNA haplotypes (Abdulla et al., 2023).

4.2.4.11. Genomic Tools and Nuclear-Mitochondrial Integration

Recent progress in next-generation sequencing and whole-genome resequencing has also helped us learn a lot about the genetic structure of Lineage O. Researchers can now figure out how much admixture there is, determine subtle genetic boundaries, and even trace historical dispersal routes by combining mitochondrial DNA data with nuclear single nucleotide polymorphism (SNP) analyses. These kinds of studies have indicated that there is some gene flow between the O lineage and nearby lineages because of migratory beekeeping and natural dispersal. However, the genetic integrity of the Oriental group has mostly stayed the same (Carr, 2023; Pierre Franck et al., 2000).

4.2.4.12. Non-Coding mtDNA Regions and Their Importance

The study of mitochondrial non-coding regions is another part of the research on Lineage O honeybees. People used to think that these areas were evolutionarily neutral, but new research has started to show that they might have important functions, especially when it comes to adaptation. Research on the non-coding mtDNA sequences has shown that the relative conservation and specific nucleotide composition of these areas may have something to do with how mitochondrial genes are turned on and off. This knowledge could be important for figuring out how Lineage O bees respond to extreme environmental conditions (Abou-Shaara, Syed Abbas, et al., 2021).

4.2.4.13. Conservation and Hybridization Threats

Field studies and conservation programs have also shown how important it is to keep Lineage O honeybees' genes pure. In places like Saudi Arabia, North Africa, and parts of the Levant, native populations of the Oriental lineage are at risk because they are mixing with imported subspecies. Such interactions can weaken the unique genetic traits that have developed over thousands of years of adapting to local conditions. To protect the evolutionary history of these bees, it is important to use conservation methods that carefully identify mitochondrial haplotypes and nuclear markers (Abou-Shaara, Al-Ghamdi, et al., 2021; Kandemir et al., 2006).

4.2.4.14. Integrative Population Genetic Approaches

The use of both mitochondrial and nuclear DNA analyses together has made a significant difference in population genetic studies. Mitochondrial markers are great for tracing maternal lineages and learning about historical demographic events. Nuclear markers like microsatellites and SNPs, on the other hand, present us a different view of gene flow and population structure today. This combined method has been very advantageous in finding small genetic differences within the O lineage, figuring out how it mixes with other lineages, and figuring out how environmental pressures lead to local adaptation (Dogantzis & Zayed, 2019; D. Henriques, Chávez-Galarza, et al., 2019).

4.2.4.15. Low mtDNA Diversity and Evolutionary Implications

One of the most important things that recent studies have found is that the Oriental lineage has less mitochondrial diversity than other lineages, like the African or European groups. This could be a sign of a bottleneck event during the north-to-south expansion, or it could mean that a strong selective pressure has kept a certain set of adaptive haplotypes around for a long time. The pattern of mtDNA diversity in the O lineage is very useful for figuring out how these bees evolved and how they adapt to different environments, regardless of what the cause is (Tihelka et al., 2020).

4.2.4.16. Environmental Factors That Cause Genetic Differentiation

Research on Lineage O honeybees often looks at how genetic markers and environmental adaptation work together. Genetic differences have been linked to ecological factors like extreme temperatures, seasonal changes, and the distribution of resources. In the hot, dry places where these bees live, parts of the mitochondrial genome that help them use energy more efficiently and keep their body temperature stable are probably under positive selection. This selective regime not only affects the coding regions of mtDNA that are involved in oxidative phosphorylation, but it can also affect non-coding regions that control gene expression (Tihelka et al., 2020).

4.2.4.17. Hybrid Zones and the Persistence of Molecular Signatures

Researchers have also done in-depth genetic studies on groups of bees from the Oriental lineage that live in areas where they come into contact with other subspecies. In these areas, hybridization happens often, which can cause complicated patterns of genetic mixing. Even in places where there is known introgression from European or African lineages, distinct mtDNA haplotypes that are typical of the O lineage often stay the same. Native populations can monitor their genetic health using this molecular signature (Ilyasov et al., 2020; Tozkar, 2020).

4.2.4.18. Diversity within a lineage and adaptation to microclimates

Research in places like Turkey, Syria, and the Arabian Peninsula has shown that there are multiple mitochondrial haplotypes within the O lineage. This means that these populations have a lot of diversity within their lineage, even though their average nucleotide diversity is low. This difference is probably because different groups have adapted to different microclimatic conditions in their areas. To implement effective conservation management, we need to conduct more research on this topic (Kaskinova et al., 2023; Özdil et al., 2009).

4.2.4.19. Historical Dispersal and Phylogeographic Patterns

Mitochondrial analyses reveal phylogeographic patterns indicative of the historical dispersal routes of the Oriental lineage. The data imply that populations migrated from the Near Eastern core region via various pathways, with contemporary populations representing the last vestiges of previously more extensive gene pools. These patterns are crucial for understanding the temporal alterations in subspecies boundaries within the O lineage and for predicting the responses of these bees to climate change and anthropogenic influences (Tihelka et al., 2020; Tozkar, 2020).

4.2.4.20. Classifying subspecies and using multiple markers

One of the most significant problems with working with honey bee subspecies, including those in the Oriental lineage, is that they can be misclassified because their morphometric traits and gene flow can be similar. Researchers have stressed how important it is to use a multi-marker approach that includes both mitochondrial and nuclear markers to correctly define subspecies boundaries. This method has worked well for showing hidden diversity within the O lineage and for confirming that these bees have a different evolutionary history (Ostroverkhova & Konusova, 2022).

4.2.4.21. Ecological and Behavioral Adaptations (Changes in behavior and the environment)

Along with genetic evidence, ecological observations and behavioral studies also support the idea that different types of Oriental honeybees are well-suited to their environments. Bees from the O lineage are usually able to handle high temperatures, low humidity, and sporadic floral resources. These traits are in line with the selective pressures that desert and steppe ecosystems put on them. These kinds of changes probably include a number of physiological and metabolic changes, many of which may be seen in the mitochondrial genome, since this organelle is important for thermoregulation and energy metabolism (Oleksa et al., 2021; Qiu et al., 2023).

4.2.4.22. Conservation Priorities and Molecular Monitoring

From a conservation perspective, it is essential to keep the O lineage's genetic uniqueness. Local adaptation means that bees from this lineage are especially well suited to the conditions in their native ranges. If you replace them with non-native subspecies, you could lose valuable adaptive traits. Therefore, conservation programs should concentrate on identifying and safeguarding pure-lineage populations through the use of robust molecular markers, particularly those derived from mtDNA, to detect and monitor hybridization (Abou-Shaara, Al-Ghamdi, et al., 2021).

4.2.4.23. Genomic Perspectives and Future Directions

Lastly, combining mitochondrial and nuclear genomic data is helping us learn how honeybee subspecies evolve and differ. High-throughput sequencing and genome-wide association studies will help us better understand the adaptive significance of lineage-specific markers, such as those in the O lineage. These methods help us understand not only how the genetic structure of these populations has changed over time but also how they will change in the future as climate change and human-caused disturbances continue to affect them (Oleksa et al., 2021; Qiu et al., 2023).

4.2.4.24. Conclusion

In short, Lineage O honeybees, which are also known as the Oriental lineage of *A. mellifera*, are a genetically and ecologically unique group that lives only in the Middle East and Near East. Studies of mitochondrial DNA show that they are evolutionarily different by consistently finding unique haplotypes and diagnostic mutations in the COI–COII intergenic regions, ND genes, and even in complete mitogenome analyses. These molecular markers, along with information about the bees' physical appearance and habitat, show how well they have adapted to the harsh, dry places they live. The O lineage is still a defined evolutionary unit, even though it has problems with hybridization and overlapping geographic distributions with other subspecies. It is vital to the future and history of apiculture in its native areas. To protect the genetic integrity and adaptive potential of these unique honeybee populations, it is important to do more research and conservation work, especially those that combine genomic, morphometric, and ecological data (Ilyasov et al., 2019; Qiu et al., 2023)23.1, 24.1).

Table 6. Lineage O (Oriental) Honeybee Subspecies

Lineage O (Oriental) of *Apis mellifera,* includes subspecies native to the Middle East, northeastern Africa, and parts of Central Asia. These bees are adapted to diverse climates, from arid deserts to mountainous regions.

Subspecies Name	Common Name	Country/Region	mtDNA Lineage	References
A. m. lamarckii	Egyptian honey bee	Egypt, Sudan	O	(Ilyasov et al., 2020).
A. m. pomonella	Tien Shan honey bee	Tian Shan Mountains (Kazakhstan, Kyrgyzstan)	O	(Ilyasov et al., 2020).
A. m. remipes	Armenian honey bee	Armenia, South Russia, Iran, Georgia	O	(Ilyasov et al., 2020).

Table 7. Subspecies with Uncertain or Mixed LineageC or O):

Subspecies Name	Common Name	Country/Region	mtDNA Lineage	References
A. m. artemisia	Russian steppe honey bee	South Russia, Ukraine	C or O	(Ilyasov et al., 2020).
A. m. sossimai	Ukrainian honey bee	Crimea, South Russia, Ukraine	C or O	(Ilyasov et al., 2020).
A. m. taurica	Crimean honey bee	Crimea, South Russia, Ukraine	C or O	(Ilyasov et al., 2020).

Notes:

The three core O lineage subspecies are *A. m. lamarckii*, *A. m. pomonella*, and *A. m. remipes*.

Subspecies with uncertain status (C or O) are sometimes included in the O lineage in certain taxonomic treatments.

O lineage bees are generally adapted to arid, semi-arid, or mountainous environments and show a range of behaviors and morphologies.

Genetic and Ecological Distinctiveness of the Y and S Lineages in *Apis mellifera*

Introduction

4.2.5. Lineages Y and S (Ethiopia and Syria)

Researchers have known for a long time that the Western honeybee, *Apis mellifera*, has four main evolutionary lineages: A, C, M, and O. However, new molecular and morphometric studies have found two more lineages that make the species' phylogenetic structure even clearer. Scientists are especially interested in the Y and S lineages, which are made up of Ethiopian and Syrian honeybee populations, respectively. This is because they have different genetic signatures and adaptive traits that make them stand out. This report goes into great detail about the genetic, morphometric, and ecological traits of the (Y lineage mostly found in Ethiopia, and the S lineage found in Syria and nearby areas). It also talks about what these findings mean for protecting biodiversity and managing honeybee genetic resources in a way that lasts.

4.2.5.1. An overview of the diversity of honeybee lineages

***Apis mellifera*,** can be found all over Africa, Europe, and Western Asia. Classical studies have indicated that there are four main evolutionary lineages: African (A), North Mediterranean (C), West European (M), and Oriental (O). However, more detailed studies using mitochondrial DN (AmtDNA) markers, microsatellites, and nuclear single nucleotide polymorphism (SNP) data have revealed clear signs of additional separate genetic groups. In the last few years, two of these groups, the Y and S lineages, have come to light. The Y lineage, associated with Ethiopian populations, exhibits significant genetic diversity. This difference is mostly because of isolation and adaptation to the unique environmental conditions in the Horn of Africa. The S lineage, also called the Syrian lineage, is found in honeybees in places like Syria and Lebanon. The Oriental (O) lineage, despite its close proximity, differs from it. These extra lineages show how complicated *Apis mellifera's* evolutionary history is, and point out genetic diversity hotspots that were not known before but are important for long-term ecological resilience (Alburaki et al., 2011).

4.2.5.2. The Ethiopian Y Lineage

4.2.5.3.1. Molecular Markers and Genetic Differentiation

A combination of molecular, morphological, and ecological studies has indicated that the Y lineage honeybees, which are mostly found in Ethiopia, are different from other African subspecies. Extensive research using mitochondrial markers, especially the areas between COI and COII, along with nuclear markers like microsatellites and gene fragments from chromosome 7, has revealed that Ethiopian honey bee populations have a unique set of haplotypes that do not match those in the classical African (A) lineage. For instance, mitochondrial studies have found haplotype groups that are stable and only found in Ethiopian honeybees. The result suggests that these bees have been genetically isolated for a long time and have adapted to the different agro-ecological zones in the Ethiopian highlands, midlands, and lowlands (Hailu et al., 2020).

4.2.5.3.2. Morphometric Distinctiveness

Morphometric studies add to the evidence for this difference. Detailed geometric morphometric analyses of the wings show that Ethiopian honeybees have unique venation patterns and forewing configurations that set them apart from other African species, like *A. m. scutellata* and *A. m. jemenitica*. These studies often look at differences in forewing centroid size and landmark configurations, which are statistically significant and related to changes in the environment, especially elevation. For example, Ethiopian highland honeybees tend to have bigger wings, which some scientists think is a way for them to adapt to cooler temperatures and less dense air at higher altitudes (Hailu et al., 2021).

4.2.5.3.3. Adaptive Genetic Variation and Environmental Correlation

Nuclear markers have also been very useful at the genetic level. The r7-frag locus, which is linked to the octopamine receptor beta-2R gene, is one that has been used to show nucleotide diversity and allelic variation within a population that is linked to local adaptation. There is a strong link between nucleotide variation in the r7-frag region and elevation across different Ethiopian ecozones. Natural selection in these diverse environments has caused genetic differentiation in the Y lineage, as this study supports. The presence of specific genetic changes and variations in the r7 region highlights how Ethiopian honeybees have adjusted differently compared to other honeybee populations in Africa or the Near East (Hailu et al., 2021).

4.2.5.3.4. Ecological Adaptation and Behavioral Diversity

Ethiopian honeybees live in various climates, from the cool, mountainous areas of the highlands to the warmer, semi-arid lowlands. This ecological variability has led to changes in behavior and physiology that may help organisms adapt, as well as changes in shape. For example, altitude and local flora have been linked to how they swarm, raise their young, and forage. In places where temperature, precipitation, and resource availability change a lot, selective pressures are probably the ones that cause these kinds of changes. The Y lineage has become a separate evolutionary unit within *Apis mellifera* because of these ecological pressures and the genetic differences that exist between them (Legesse & Getu, 2022).

4.2.5.3.5. Defining Mitochondrial Lineage

It has been crucial to define the Y lineage using mitochondrial DNA studies. Research on the tRNA^leu-cox2 intergenic region, often using DraI CAPS analysis, has shown that Ethiopian populations have unique haplotype patterns that are not seen in other regions. The high haplotype diversity and the discovery of several new variants in these groups suggest that they have a complicated evolutionary history with times of population growth and isolation. Often, Ethiopia is the exclusive home of certain haplotypes, sometimes referred to as Y1, Y2, or Y3. This evidence supports the idea of a separate Y lineage (Alsharhi et al., 2025).

4.2.5.3.6. Geographic Isolation and Allopatric Divergence

Geographic isolation has also made the lineage unique, along with genetic markers. The rough terrain of Ethiopia, with its high mountains and deep valleys, has acted as a physical barrier that keeps genes from moving between groups. This kind of separation makes allopatric divergence possible and helps keep the Y lineage intact. The genetic differences seen in Ethiopian bees are a result of both historical biogeography and current ecological processes. This situation makes the Y lineage a perfect example of how geographic isolation and local adaptation can affect the genetic diversity of species that are found all over the world (Cridland et al., 2017).

4.2.5.3.7. Conservation Implications and Biogeographic Significance

From a conservation perspective, it is critical to recognize the Y lineage. Ethiopia is home to a wide variety of honeybees, and the Y lineage is a unique genetic resource that could be crucial as the climate changes and new pests and diseases appear. Keeping this genetic diversity alive can help beekeeping last for a long time, both locally and around the world. Researchers have prepared for local breeding programs and conservation efforts focused on the special traits of these bees by identifying and thoroughly studying the Y lineage using a combination of genetic and physical analysis (Hailu et al., 2020).

The Y lineage honeybees could also help us learn more about the evolutionary history of *Apis mellifera.* The Y lineage may be one of the first to split from the main species group, according to some studies. This may be due to its older ancestral genetic makeup than other African lineages. Comparing Ethiopian bees to bees from other parts of the Middle East and Africa supports the idea that modern honeybees came from northeastern Africa. This idea significantly impacts our understanding of the biogeographic history of the species (Cridland et al., 2017).

Table 8. Ethiopian Y Lineage and Syrian S Lineage Honeybees

The Y lineage (Ethiopian) and S lineage (Syrian) represent distinct genetic branches within *Apis mellifera.* The Y lineage is endemic to the Horn of Africa and Arabian Peninsula, while the S lineage is confined to the Levant region. Below is a comprehensive table of all subspecies in these lineages:

Subspecies Name	Common Name	Country/Region	Main Characteristics	mtDNA Lineage	References
Apis mellifera simensis	Ethiopian honey bee	Ethiopian highlands (Simien Mountains)	Large size, dark coloration, long body hair, high-altitude adaptation. Longer wings than other African subspecies.	Y	(Franck et al., 2001).
Apis mellifera jemenitica	Arabian honey bee	Arabian Peninsula (Yemen, Saudi Arabia), Horn of Africa (Somalia, Sudan, Uganda)	Small size, yellowish abdomen, extreme heat/drought tolerance. Efficient forager in arid conditions.	Y	(Al-Ghamdi et al., 2013).
Apis mellifera syriaca	Syrian honey bee	Syria, Lebanon, Jordan, Palestine, Southern Israel	Distinct nuclear genome (S lineage). Avoids predation by remaining in hive. Resistant to *Cryptophagus hexagonalis* beetle.	(S,Z mtDN)	(Alburaki et al., 2013).

Key Notes:

Lineage Distinctions:

The Y lineage (simensis and jemenitica) is genetically distinct from neighboring A (African) and O (Oriental) lineages (Franck et al., 2001; Tihelka et al., 2020).

A. m. syriaca forms the S lineage, with unique nuclear DNA and mtDNA Z haplotype (Alburaki et al., 2013).

Geographical Isolation:

A. m. simensis is restricted to Ethiopian mountains >1,500 m elevation.

A. m. syriaca shows minimal genetic introgression at Turkish/Iraqi borders but remains distinct in southern populations (Alburaki et al., 2013).

Evolutionary Significance:

The Horn of Africa (Y lineage) and Levant (S lineage) are critical zones for honeybee diversification, showing ancient divergence (Franck et al., 2001; Tihelka et al., 2020).

4.2.5.4. The Syrian S Lineage

4.2.5.4.1. Genetic Differences and Molecular Markers

In addition to the Ethiopian Y lineage, researchers have identified a distinct genetic group of honeybees in Syria, commonly referred to as the S lineage. Historically, Syrian honeybees (*Apis mellifera syriaca*) were classified within the Oriental (O) lineage due to their morphological similarities and geographical proximity. Recent molecular investigations, however, have challenged this conventional perspective, demonstrating that Syrian bees constitute a distinct evolutionary group with genetic characteristics that differentiate them from the O lineage. The application of DraI restriction enzyme digestion to analyze the mitochondrial COI–COII intergenic region has revealed that certain haplotypes previously attributed to the Oriental lineage are, in fact, part of a distinct subgroup. This subgroup, previously referred to as the Z subgroup and now predominantly identified as the S lineage, exhibits distinct mtDNA sequence variations and restriction patterns that consistently differ from those of the Oriental lineage (Alburaki et al., 2011).

Microsatellite loci and nuclear markers provide more evidence that the S lineage and the Oriental lineage are different. For example, analyses using structure, principal coordinates, and neighbor-joining trees all show that the honeybee populations in Syria and Lebanon are genetically connected but distinct from other groups. In some studies, when the number of assumed genetic clusters (K) is set to five, Syrian bees stand out as a separate group that isn't closely related to either the African (A) or Oriental (O) lineages. This strengthens their unique genetic identity (Cridland et al., 2017).

It's intriguing to note that the S lineage is genetically very different from other lineages, especially in populations in the interior regions of Syria and Lebanon. On the other hand, populations in northern Syria, which are closer to the Turkish border, often show signs of introgression from the Oriental lineage. Queen imports and migratory beekeeping practices can accidentally introduce alleles from subspecies such as *A. m. anatoliaca* or *A. m. caucasica* into local Syrian populations, leading to gene flow. Even though these mixing events occurred, the S lineage can still be recognized by its special mitochondrial haplotypes and steady microsatellite allele frequencies, which distinguish it from the Oriental cluster (Cridland et al., 2017).

4.2.5.4.2. Morphometric Distinctiveness

Morphometric studies have also been crucial in proving that the S lineage is different from other lineages. There are clear differences between Syrian honeybee populations and those in nearby areas when you look closely at things like wing venation patterns, body size, and other phenotypic traits. When combined with molecular data, morphometric data make it easy to tell the S lineage apart from the Oriental (O) and African (A) lineages. This clears up any previous taxonomic confusion. These studies indicate that Syrian bees are a separate evolutionary group because they have a unique set of molecular and morphological traits, even though they look a lot like Oriental bees (Cridland et al., 2017).

4.2.5.4.3. Adaptive Genetic Variation and Environmental Correlation

The S lineage's ecological context is also crucial for describing it. Syrian honeybees commonly inhabit dry and semi-dry environments in the Near East. Their behavior, like being more defensive, running away a lot when it grows really hot, and changing how they raise their young when there isn't enough nectar, shows how they have evolved to adapt to the tough climate where they live. The selective pressures that cause these changes are probably very different from those that affect the Oriental lineage, which lives in areas with more stable or temperate climates. The unique ecological niche of Syrian honeybees has been a big part of how the genetic and phenotypic differences in the S lineage have developed (Zakour & Bienefeld, 2014).

Next-generation sequencing studies have also found functional genomic differences between *A. m. syriaca* and other subspecies of *Apis mellifera*. Genes involved in immune responses, resistance to varroa mites, and overall stress tolerance exhibit sequence polymorphisms indicating that Syrian bees have developed unique strategies to combat common pathogens and manage stress. These unique genomic signatures not only show how different the S lineage is, but they also show how valuable it could be as a genetic resource that could add useful traits to global beekeeping if it is kept safe and well-managed (N. Haddad et al., 2016).

4.2.5.4.4. Ecological Adaptation and Behavioral Diversity

Recognizing the S lineage has important effects on protecting biodiversity. The introduction of non-native queen stocks and the resulting gene flow from the Oriental lineage are putting the genetic integrity of native Syrian honeybees at risk. To protect the evolutionary history of honeybees in the Near East, it is important to keep the S lineage alive. Conservation plans should prioritize safeguarding pure, locally adapted genetic stocks, establishing genetic reserves or isolated mating apiaries, and closely monitoring introgression levels to ensure the preservation of the S lineage's unique evolutionary adaptations (Alburaki et al., 2011).

4.2.5.5. Comparative Discussion: Y versus S Lineages

Even though the Y and S lineages have only been recognized recently, their evolutionary and ecological contexts are entirely unique because of the different environmental pressures in Ethiopia and the Near East. The Ethiopian Y lineage is a group of honeybees that have changed over time in the Horn of Africa, which has high elevations, a wide range of microclimates, and long periods of being cut off from the rest of the world. The S lineage, on the other hand, is found in Syria and nearby areas, and it is in a contact zone where the Oriental lineage is common. Its genetic uniqueness is mostly preserved, even though genes are still flowing between the two lineages. These differences show how important it is for *Apis mellifera,* to adapt to different regions to increase its genetic diversity (Cridland et al., 2017).

The Y lineage has special features at the molecular level, including unique mitochondrial haplotypes and nuclear markers, which indicate it has developed differently over a long time due to environmental differences. Research on mtDNA in Ethiopian populations has discovered several new haplotypes unique to that region. On the other hand, nuclear markers like the r7-frag show that there are allelic variants that are strongly linked to elevation and ecological niche. The Y lineage has isolated itself from other lineages and adapted to its environment (Hailu et al., 2020).

On the other hand, the S lineage has a unique set of mtDNA haplotypes that were first thought to belong to the Oriental lineage. More detailed analyses later showed that they were not. Both microsatellite data and morphometric analyses show that the S lineage is separate from the Oriental group. This is because the genetic structure is always consistent, even when there are cases of mixing in border areas. Consequently, even though both the Y and S lineages are the result of localized adaptation, the Y lineage seems to have evolved in relative isolation because of environmental factors in Ethiopia. On the other hand, the S lineage has come from an area where different lineages interact with each other, which has led to a genetic signature that shows both divergence and admixture (Cridland et al., 2017; Meixner et al., 2013).

To manage honeybee genetic resources, it is essential to know the differences between them. It is important to protect the Y lineage not only to keep a unique part of African honey bee diversity alive, but also because it could help breeding programs that aim to make bees more resistant to environmental stressors. Ethiopian honeybees of the Y lineage may have traits that help them adapt to climate change, like being able to handle high temperatures, changing their foraging patterns, and using resources more efficiently. The S lineage, on the other hand, is critical for beekeeping in the Near East because its unique adaptations to dry conditions and scorching summers make it a valuable genetic resource that could help keep bees alive in places with similar environmental problems (Legesse & Getu, 2022).

The discovery of these new lineages makes it harder to classify *Apis mellifera,* in the traditional way, but it also opens up exciting new ways to learn about the species' evolutionary history. The Y and S lineages show that the four-lineage model was not enough to capture the diversity of honeybees. Instead, they show that extensive genetic differences and local adaptations have created separate evolutionary units within what was once thought to be a fairly homogeneous species. This new knowledge means that conservation strategies need to be more detailed and take into account all types of genetic diversity, not just broad taxonomic groups (Cridland et al., 2017).

4.2.5.6. Geographic Isolation and Allopatric Divergence

The discovery of the Y and S lineages has big effects on both protecting biodiversity and managing honeybee genetic resources around the world. First, these results show how important it is to protect populations that have adapted to their local environment and developed unique genetic traits over time. Thereby, conservation efforts need to include plans that are specific to the unique ecological situations in the Ethiopian highlands and the dry areas of the Near East. Setting up protected breeding programs, separate mating apiaries, and strict monitoring of gene flow from introduced commercial stocks could help keep the Y and S lineages' genetic identities separate (J. D. Evans et al., 2013; N. Haddad et al., 2016).

Furthermore, the fact that these lineages are genetically unique could be useful in some way. For instance, the Y lineage's ability to adapt to highland environments and possibly tolerate extreme temperatures could help people breed bee strains that can handle climate change. The S lineage may also be a beneficial source of alleles for improving honeybee health and productivity in stressful conditions because it has been shown to do well in very dry conditions and is genetically resistant to some pests and diseases. These targeted breeding programs would work better if these lineages were carefully characterized at the genomic level. Such an effort would assure that the valuable adaptive traits are not lost when they mix randomly with non-native stocks (Cridland et al., 2017; Tihelka et al., 2020).

From an ecological perspective, keeping the genetic diversity of the Y and S lineages helps guarantee that honeybee populations can still adapt to changes in the environment in the future. Honeybees play a crucial role in maintaining ecosystem functionality and ensuring food availability by pollinating both wild plants and agricultural crops. Losing unique genetic lineages would not only reduce biodiversity, but it could also make the bees less adaptable, which would put both natural ecosystems and agricultural systems that rely on pollination services at risk (Cridland et al., 2017; Engel et al., 2021).

Furthermore, it's important to remember that recognizing these lineages shows how important it is to do full genetic surveys and to combine molecular, morphometric, and ecological data in future studies. Continuing to map the full genetic landscape of *Apis mellifera* will help us better understand how species are related to each other, identify more hidden diversity, and ultimately help us make better decisions about what to focus on when it comes to conservation. Because the environment is changing so quickly and species that are important to agriculture, like the honeybee, are under more and more stress (Alsharhi et al., 2025; Hailu et al., 2020), these kinds of research methods are necessary.

4.2.5.7. Conclusion

To sum up, the recent discovery and detailed description of the Ethiopian Y and Syrian S lineages are big steps forward in our understanding of the diversity of *Apis mellifera*. The Y lineage has a set of unique mitochondrial and nuclear genetic markers, as well as unique morphometric traits and clear ecological correlates that show how it has adapted to the Horn of Africa's diverse and often difficult environments over a long period of time. On the other hand, Syria and nearby areas host the S lineage, which differs genetically and morphologically from the Oriental lineage. This study shows how complicated the effects of mixing and isolation are on the genetic structure of honeybee populations in the Near East.

These new findings have big effects on how we protect biodiversity and manage the genetic resources of honeybees. It is important to recognize and protect the genetic uniqueness of the Y and S lineages not only to protect the evolutionary history of *Apis mellifera* but also to make sure that breeding programs and conservation strategies can use a wide range of adaptive traits to deal with the problems caused by environmental change, disease, and genetic introgression caused by humans.

Future research should build on these results by taking more samples from different areas where these lineages live, adding more genomic markers, and looking into the physical and behavioral traits that help them adapt to their new environments. These kinds of efforts will not only help us learn more about the evolutionary history of honeybees, but they will also help protect honeybee populations around the world so they can stay strong and productive.

It is clear that the Y and S lineages should be included in the taxonomic framework of *Apis mellifera* based on the evidence from molecular analyses, morphometrics, and ecological studies. Recognizing them adds to our understanding of honeybee diversity and gives us important information for conservation efforts that can help make sure that beekeeping is sustainable in various environments (Cridland et al., 2017; Hailu et al., 2021).

The Y and S lineages serve as crucial reminders of the hidden biological diversity present in well-studied species. They also show how important it is to continue doing molecular research and monitoring biodiversity as the world changes quickly. It will be very important to carefully protect these unique lineages so that honeybees can keep providing their ecological services and so that future breeding programs can use their genetic material in a world that is changing quickly (Cridland et al., 2017; Tihelka et al., 2020).

The changing view of the genetic structure of *Apis mellifera*, which now includes the classical A, C, M, and O lineages as well as the newly recognized Y and S lineages, gives us a more detailed picture of how honeybees have evolved. This knowledge is important for making plans to protect honeybee populations that take into account and reflect the full range of their biodiversity. The results will help these important insects continue to thrive in their natural habitats and in managed landscapes around the world (Legesse & Getu, 2022; Meixner et al., 2013).

In conclusion, the separation of the Ethiopian Y and Syrian S lineages is a big step forward in the fields of apicultural research and genetic resource management. Researchers have shed light on the many different evolutionary paths that have led to the formation of these distinct lineages by combining detailed molecular data, morphometric evidence, and ecological context. For scientific, cultural, and ecological reasons, it is vital to protect them (J. D. Evans et al., 2013; N. Haddad et al., 2016).

These kinds of in-depth studies show how useful advanced molecular techniques and careful field sampling can be for finding out more about the evolution of one of nature's most important pollinators. The Y and S lineages will continue to be important in talks about genetic conservation, sustainable apiculture, and the long-term health of honey bee populations around the world (Cridland et al., 2017; Hailu et al., 2020).

Table 9. A comparative table summarizing the key differences and similarities between the Ethiopian Y lineage and the Syrian S lineage

Feature	Ethiopian Y Lineage	Syrian S Lineage
Geographic Region	Ethiopian Highlands (Horn of Africa)	Syria and parts of the Near East
Environmental Context	High elevations, diverse microclimates, long-term geographic isolation	Arid and semi-arid regions; contact zone with the Oriental lineage
Genetic Lineage	Recently identified as a unique lineage, not part of A, C, M, or O	Formerly classified within Oriental (O); now recognized as distinct (S lineage)
Mitochondrial DNAmtDNA)	Unique haplotypes exclusive to Ethiopian region (Hailu et al., 2020)	Distinct COI–COII haplotypes; misattributed to O lineage in past (Alburaki et al., 2011)
Nuclear Markers	Novel allelic variantse.g., (r7-frag) associated with elevation and local niches (Hailu et al., 2020)	STRUCTURE, PCoA, and NJ trees show strong genetic cohesion separate from O lineage (Cridland et al., 2017)
Morphometric Traits	Morphological differentiation associated with ecological adaptation(not deeply described)	Clear morphometric differences (wing venation, body size) from O and A lineages (Cridland et al., 2017)
Behavioral Adaptation	Adapted to elevation, temperature, and resource variability	Defensive, absconding under heat, modified brood patterns (Zakour & Bienefeld, 2014)

Adaptive Genes	Suggested adaptations for temperature and foraging patterns (Legesse & Getu, 2022)	Genomic adaptations to heat, stress, and disease resistance (N. Haddad et al., 2016)
Genetic Introgression	Less affected due to isolation	Subject to introgression from Oriental bees *(A. m. anatoliaca, A. m. caucasica)* especially near borders) (Cridland et al., 2017)
Taxonomic Implication	Challenges traditional four-lineage model (Cridland et al., 2017)	Similarly challenges traditional lineage classification (Cridland et al., 2017)
Conservation Importance	High value for climate-resilient breeding and unique biodiversity (Legesse & Getu, 2022)	Critical to Near East apiculture; at risk from non-native gene flow (Alburaki et al., 2011)
Role in Sustainable Apiculture	Source of heat-resilient traits for global breeding programs (Legesse & Getu, 2022)	Valuable for dry-region beekeeping and genetic conservation (N. Haddad et al., 2016)
Evolutionary History	Long-term isolation and environmental divergence (Hailu et al., 2020)	Shaped by both divergence and admixture at lineage boundaries (Cridland et al., 2017)

This table highlights the shared significance of the Y and S lineages as newly identified evolutionary units within *Apis mellifera*, while also distinguishing their unique, ecological, genetic, and conservation profiles.

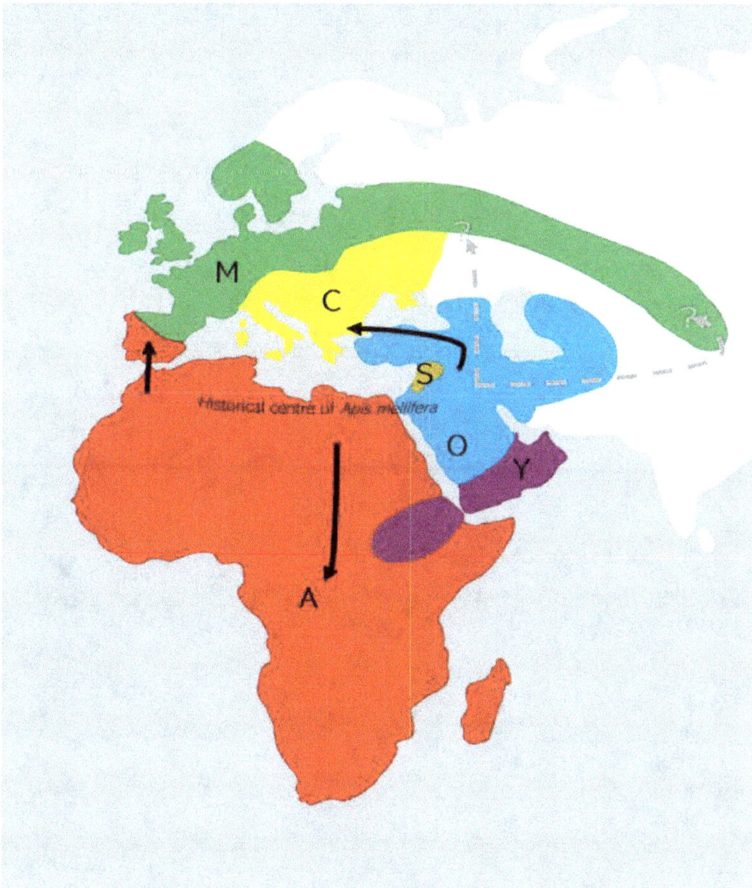

Fig. 2. Hypothesis of *Apis mellifera* origin and dispersal routes.

Solid black arrows are based on well-supported relationships inferred from Bayesian analysis of mitochondrial genomes while dashed grey arrows represent hypothetical dispersal routes. Lineage ranges are based largely on Ruttner, and Dogantzis and Zayed (Tihelka et al., 2020).

Chapter 5

5. Phylogeography and Global Dispersal

This chapter goes into great detail about the historical and modern dispersal routes, evolutionary radiation, and genetic differences of the Western honeybee, *Apis mellifera*. It does this by looking closely at African and European lineages and events that caused the bee to spread through human activity. The study brings together a lot of mitochondrial and nuclear genetic data from different parts of the world. It looks at unique insular divergences, hybrid zones, and how human activities affect genetic diversity. The next sections present an expert-level story that brings together recent results from several molecular studies.

5.1. African Lineage Radiation

A. mellifera's evolutionary history in Africa is marked by a complicated radiation process that has led to the development of many different subspecies and private haplotypes. This is especially clear in the Southwest Indian Ocean (SWIO) islands. Madagascar is home to a unique honeybee subspecies called *Apis mellifera unicolor*. This species is part of the African subgroup, which is sometimes called the AI lineage. A lot of research on mitochondrial DNA has shown individuals on Madagascar have unique haplotypes that show how the island's ecosystems have changed over time because of being cut off from the rest of the world and having a small number of founders (Techer, Clémencet, Simiand, Turpin, et al., 2017). When compared to populations in continental Africa, these genetic markers are different. The evidence indicates that unique micro-environmental conditions and historical colonization events have shaped the evolutionary path of Malagasy bees.

Islands in the SWIO region, like the Mascarenes and Comoros, which are archipelagos near Madagascar, also show signs of insular radiation. Mitochondrial analysis on these islands has identified unique haplotype groups. The evidence shows that there was an ancient colonization event, and then the islands evolved separately over time (Techer, Clémencet, Simiand, Turpin, et al., 2017). Nuclear microsatellite studies have also shown that the genetic makeup of these island populations is both varied and structured. This pattern supports the idea that isolation helps create unique genetic variants (Techer, Clémencet, Simiand, Turpin, et al., 2017).

The African lineage is also known for having stable hybrid zones between some subspecies, in addition to being diverse within its species. The hybrid zone between the Cape honey bee *(A. m. capensis)* and the African honey bee *(A. m. scutellata)* is a well-known example. These two subspecies live together in South Africa. Even though there are parasitic behaviors that are often called the "capensis calamity," where workers of *A. m. capensis* can take on a reproductive role and parasitize nearby colonies, these hybrid zones stay very stable over time. Genetic markers, such as mitochondrial haplotypes and microsatellite allele frequencies, have been used to map out the size and shape of these zones. The study has shown that there is a dynamic balance that lets both subspecies live together even though they behave differently (Techer, Clémencet, Simiand, Turpin, et al., 2017). A mix of behavioral, ecological, and genetic factors probably keeps this stability going. These factors stop either subspecies from completely replacing the other. Evidence suggests that these hybrid zones could serve as storage sites for adaptive genetic variation during environmental changes (Chávez-Galarza et al., 2015).

More recent studies indicate that the African hybrid zones not only have a lot of mixing, but they also keep separate genetic lineages in certain areas. For example, research in areas near the South African mainland has found mosaic genetic patterns where admixture happens but private lineages stay the same. This trend is because of historical separations that were broken up by gene flow every so often (Chávez-Galarza et al., 2015). The genetic differences between these groups are crucial because they may make colonies more resistant to parasites and environmental stress, which is something that has gotten a lot of attention in the context of global honeybee declines.

Overall, the African lineage radiation, especially in the SWIO islands and hybrid zones in South Africa, is a positive example of how geographic isolation and occasional gene flow have caused honey bee populations to evolve in unique ways. These groups not only hold a lot of genetic diversity that hasn't been used yet, but they also show how evolution on islands can keep adaptive traits around for a long time (Techer, Clémencet, Simiand, Turpin, et al., 2017).

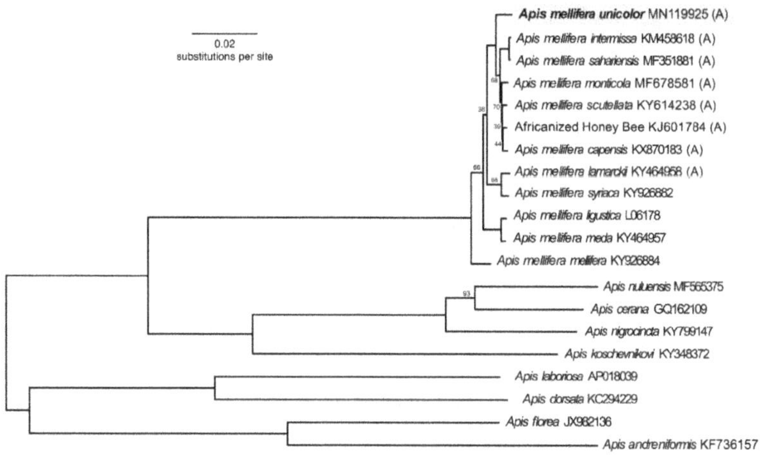

Fig. 3. Phylogenetic tree showing the relationship between *Apis mellifera unicolor* GenBank: (MN119925) and 19 other *Apis* honey bee mitochondrial genomes (Gen Bank accession numbers are listed after species names).

A) indicates honey bees from the African A-lineage. The tree is midpoint rooted. Node labels indicate bootstrap values, and unlabeled lineages are 100% (Boardman, 2019).

5.2. European Lineages

The European part of *A. mellifera* is made up of several evolutionary lineages that have had different histories. The C and M lineages are the most well-known. The C lineage, which includes subspecies like the Italian bee (*Apis mellifera ligustica)* and the Carniolan bee (*Apis mellifera carnica),* has become the most common type of bee in many parts of the world, especially in the United States. Molecular phylogenetic studies that look at the whole mitochondrial genome have shown over and over again that most of the honeybee colonies in the U.S. are part of the Mediterranean C lineage (Carr, 2023). Specifically, surveys have shown that about 94% of bees in the U.S. belong to the C lineage. Haplotype breakdowns show that about 38.8% belong to the C1 group and 37.6% to the C2j group (M. Taylor & Sheppard, 2014).
The C lineage is so common in the U.S. because of historical events when beekeeping started in North America. At that time, beekeepers chose European subspecies known for being calm, productive, and resistant to some diseases over others. *A. m. ligustica* and *A. m. carnica* are both very popular in commercial beekeeping. Such selection has led to a small genetic base in U.S.

populations, which is shown by the fact that mitochondrial haplotype diversity is low (Alburaki et al., 2023). Founder effects and selective breeding focused on certain desirable traits instead of a wider range of genetic diversity are likely to limit genetic variability in these populations (Alburaki et al., 2023).

In contrast, the M lineage, which used to include western and northern European subspecies like *A. m. mellifera*, has seen a big drop in numbers in the last few decades. There are many pieces of evidence that point to the fact that the M lineage is less common because it is more vulnerable to pathogens like the Varroa destructor mite, which has had terrible effects on honeybee health around the world. It is well known that varroa infestation is a major cause of colony losses. Researchers have also linked it to the gradual replacement of more susceptible lineages by more resistant or better managed ones (M. Taylor & Sheppard, 2014). The M lineage used to be common in Europe, but disease, habitat loss, and the competitive edge of imported C lineage bees have all caused native M lineage populations to drop sharply (Techer, Clémencet, Simiand, Turpin, et al., 2017).

The decline of the M lineage in Europe has important effects on conservation. Native M lineage subspecies are important genetic pools that may contain alleles that help them survive better in their environments. However, the M lineage has lost some of its genetic material because there isn't much gene flow, and C lineage bees are taking over. This picture is especially clear in parts of Europe, where the introduction of foreign queens and inter-lineage hybridization have made the genetic integrity of native subspecies even worse (Chávez-Galarza et al., 2015).

To sum up, the phylogeography of the European honeybee shows a split between the C lineage, which is growing, and the M lineage, which is dying out. The C lineage is dominant in places like the United States because Italian and Carniolan bees are brought in and bred selectively. The M lineage is shrinking in Europe because of environmental factors like Varroa mite infestations and the ongoing mixing of genes from non-native stocks (M. Taylor & Sheppard, 2014). This ongoing interaction between lineage replacement and hybridization is still changing the genetic makeup of European honeybee populations. It creates both problems and chances for conservation and sustainable beekeeping.

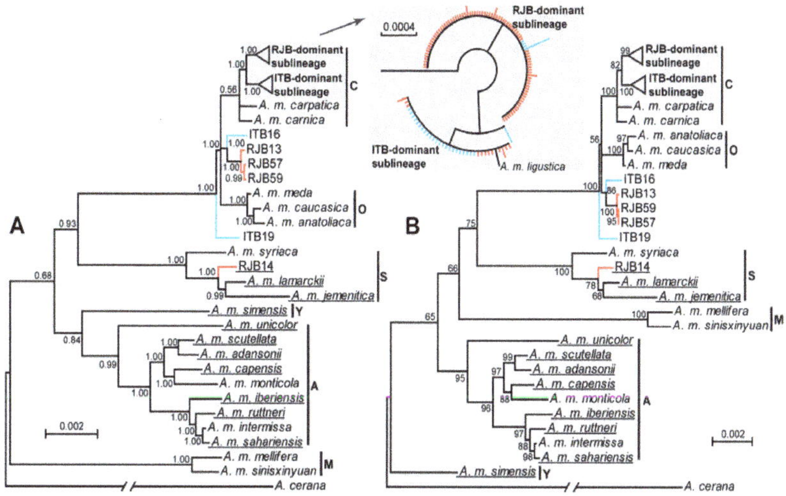

Fig 4. The phylogenetic trees of *A. mellifera subspecies*.

Posterior probabilities for the Bayesian inference tree A) and bootstrap values for the maximum likelihood tree B) are shown at nodes. The branch of the outgroup *A. cerana* is truncated. Subspecies with long insert in the intergenic spacer between cox2 and trnL2 are underlined. The previously designated six lineages (A, C, M, O, S, and Y) for *A. mellifera* are indicated. The clades for the RJB- and ITB-dominant sublineages are shaded. The clustering pattern of the two sublineages is similar in the two trees and is shown in the insetred color for (RJBs and blue for ITBs) for the Bayesian inference tree (C. Ma et al., 2022).

5.3. Human-Mediated Dispersal

Human actions have significantly influenced the global spread of honeybee genetic lineages today. The intentional introduction of certain subspecies to new areas has caused giant changes in the genetic diversity of many native species. This shift has had effects on the stability of ecosystems and the way beekeepers do their jobs.

In the past, beekeepers in the United States primarily introduced European subspecies, particularly those from the C lineage, to establish commercial apiaries. Recent genetic studies of honeybee populations in the U.S. show that there are only 27 different mitochondrial haplotypes in the whole country. The overall haplotype diversity H is about 0.597 (Alburaki et al., 2023). Interestingly, 14 of these 27 haplotypes were previously unknown. This finding suggests that small founder populations and selective breeding have led to the creation of new variants within a small genetic pool (Alburaki et al., 2023). The C lineage makes up 94% of all colonies in the U.S., and the overall genetic diversity is low. This graph shows how much historical importation and modern breeding practices have affected the genetic profile of U.S. honeybees (Alburaki et al., 2023).

In Italy, things are both the same and different because there are regional laws in place to protect the native subspecies *A. m. ligustica*. In the Emilia-Romagna region, detailed studies of mitochondrial DNA haplotypes have shown that around 86.5% of honeybees have the C1 haplotype, which is common for the *A. m. ligustica subspecies* (Taurisano et al., 2024). In addition to this primary haplotype, the C2 haplotype, associated with *A. m. carnica*, constitutes approximately 11.0% of the population. African lineage haplotypes, on the other hand, are still present at a low frequency of about 1.3% (Taurisano et al., 2024). The fact that these African and C2 haplotypes are present, even if only in small numbers, shows how complicated gene flow is in Italy. Despite conservation efforts, imported stock and natural introgression have altered the current haplotype distribution (Taurisano et al., 2024).

This law in Italy is meant to stop the possible genetic dilution that could happen if non-native and hybrid queens were used a lot. These steps are meant to protect the genetic integrity of *A. m. ligustica* and keep a well-defined gene pool that can handle things like disease outbreaks and changes in the environment. The plan calls for strict rules on bringing in queens and breeding practices that help protect and restore regional genetic resources. The legal system in Italy is an example of how targeted policy changes can help protect native bee lineages from the effects of global trade and inter-lineage hybridization (Taurisano et al., 2024).

From a global perspective, it is evident that human-mediated dispersal has had both positive and negative effects on honeybee populations. On the one hand, bringing strong, productive subspecies like the C lineage to places where they don't belong, like the United States, has been key to building big apiculture industries. On the other hand, the genetic bottlenecks and narrow genetic bases that have resulted have made people worry about how well these populations will be able to adapt to new threats like the Varroa mite and related viruses (Carr, 2023). Furthermore, the fact that there is low mtDNA diversity and fewer haplotypes in the United States is very different from the fact that there is more haplotype diversity in the Old World, especially in areas where there has been less breeding and importation (Alburaki et al., 2023).

The combination of historical importation, selective breeding, and modern laws shows how important it is to include genetic monitoring in beekeeping management. Researchers can spot changes in which bee lineages are more common and early signs of genetic decline by regularly checking variations in mitochondrial DNA and nuclear genetic markers. To ensure that honeybee populations can thrive in their native ranges and in areas where they have been introduced (Taurisano et al., 2024), taking these kinds of proactive steps is important. Additionally, new genomic technologies make it easier to learn more about how different honeybee lineages can adapt. For instance, recent whole-mitogenome studies have shed light on the genetic basis of disease resistance, stress tolerance, and productivity traits. These are important for breeding programs that aim to make colonies more resilient (Carr, 2023). These improvements are especially important because honeybees are facing more and more problems around the world, such as climate change, habitat loss, and the quick spread of harmful pests and diseases. In this situation, protecting genetic diversity through planned human actions, like those taken in Italy, is a crucial step toward making sure that this important pollinator can survive for a long time (M. Taylor & Sheppard, 2014).

In brief, human-mediated dispersal has changed the genetic makeup of *Apis mellifera* on a global scale. In places like the United States, C lineage bees are very common, but there isn't much genetic diversity overall. This disparity is because of both historical import events and selective breeding (Alburaki et al., 2023). In Italy, local efforts are helping to maintain a good number of the native *A. m. ligustica* bees while still allowing some bees from other lineages to exist (Taurisano et al., 2024). These changes necessitate monitoring genetics and devising adaptable management strategies to address issues such as low genetic diversity and external stressors such as Varroa infestations (M. Taylor & Sheppard, 2014).

5.4. Final thoughts

This chapter has discussed the big phylogeographic patterns and global dispersal mechanisms that have led to the current distribution of *A. mellifera*. The spread of the African lineage, especially in the SWIO islands, has created unique genetic differences on those islands and has helped maintain stable areas where different lineages meet, keeping them separate from one another (Chávez-Galarza et al., 2015; Techer, Clémencet, Simiand, Turpin, et al., 2017) . The C lineage has been important in Europe for a long time, and its persistence has helped it become the most common lineage in places like the United States, where it is not native. The M lineage, on the other hand, has been declining quickly because of the detrimental effects of Varroa mite infestations (Taurisano et al., 2024; M. Taylor & Sheppard, 2014; Techer, Clémencet, Simiand, Turpin, et al., 2017). Human actions, like the low variety of haplotypes in U.S. populations and strict laws in some parts of Italy, demonstrate that people can both limit genetic diversity and help protect species (Alburaki et al., 2023; Taurisano et al., 2024).

When you look at all of these results together, they show how important it is to know the phylogeography and global dispersal of honeybees in order to create effective breeding and conservation programs. The interaction between natural evolutionary processes and human activities shows how complicated honeybee population dynamics are and how important it is to use genetic research and targeted management policies together. In the future, whole-genome sequencing and fine-scale landscape genetic analyses will help us learn more about these processes and the genetic factors that assist honey bee populations adapt and stay strong (Carr, 2023; Techer, Clémencet, Simiand, Turpin, et al., 2017).

In the end, the chapter calls for a multidisciplinary approach that uses molecular genetics, traditional breeding, and proactive policy measures. This integrated approach is important not only for protecting the rich genetic heritage of both African and European honeybee lineages but also for making sure that future beekeeping practices can keep this important pollinator species healthy and productive (Taurisano et al., 2024; M. Taylor & Sheppard, 2014).

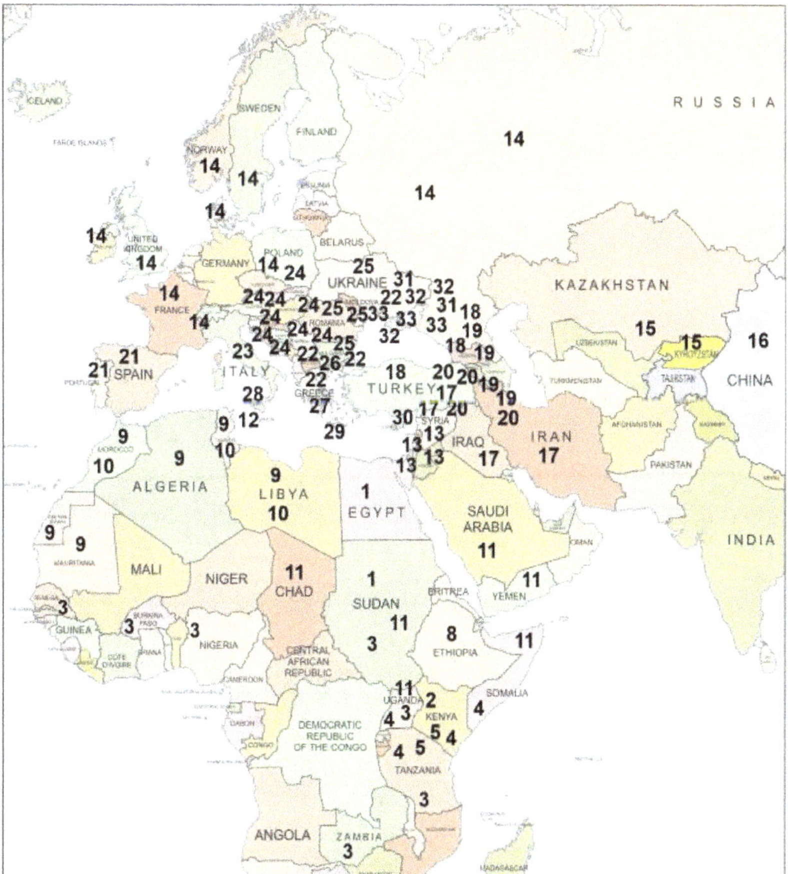

Fig 5. The spatial distribution of 33 subspecies of honey bee *A. mellifera* on geographic map. The numbers correspond to numbers of subspecies in the table (Ilyasov et al., 2020).

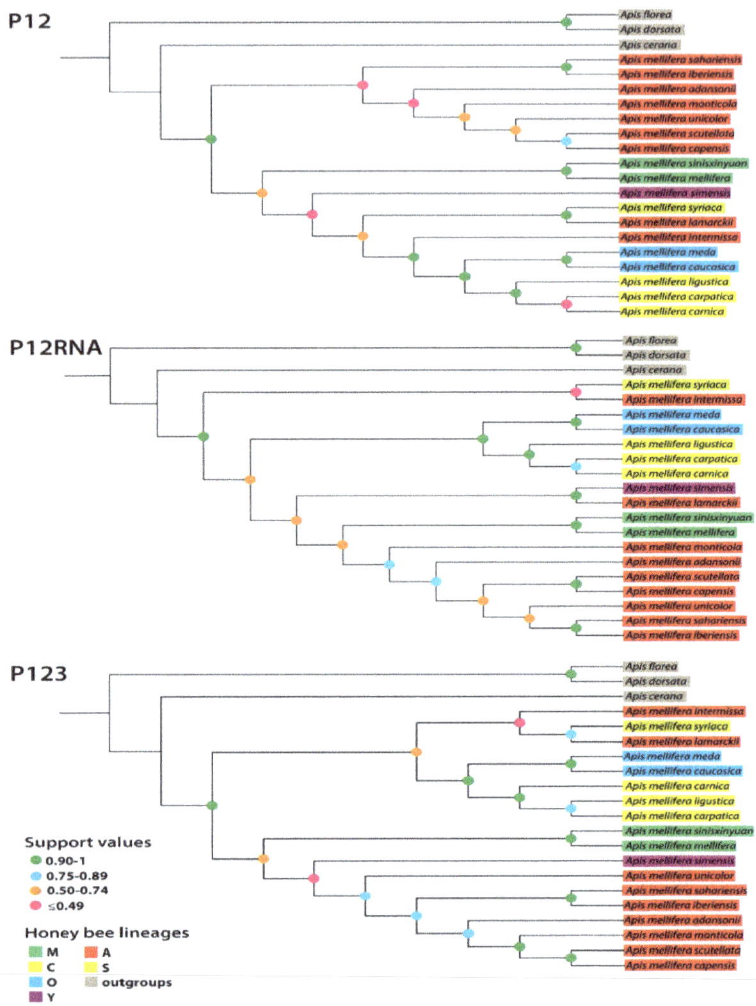

Fig 6. Three competing topologies recovered based on different gene sampling approaches with the site-heterogeneous CAT-GTR + G model ignoring branch length for easier legibility.

Subspecies are grouped into proposed molecular lineages primarily after (Ruttner and Meixner et al), Support values represent Bayesian posterior probabilities. Abbreviations: P12, first and second codon positions of protein coding mitochondrial genes; P12RNA, first and second codon positions of protein coding mitochondrial genes and two ribosomal rRNA) genes; P123, protein coding mitochondrial genes without the third codon position excluded (Tihelka et al., 2020).

Chapter 6

6. mtDNA Methodologies Research

6.1. Overview of Methodologies in mtDNA

Research gives a full picture of the experimental, computational, and analytical methods that have been used to study the diversity, evolution, and conservation effects of mitochondrial DNA (mtDNA) in *Apis mellifera* populations (Arias & Sheppard, 2006). This chapter starts with a look at how important mtDNA studies are for figuring out how honeybees pass on traits from mother to offspring, how they spread across the world, and how different subspecies are from each other (Abou-Shaara, Al-Ghamdi, et al., 2021). Researchers have chosen mtDNA as a strong genetic marker because it mutates quickly, is passed down from maternal to offspring, and is easier to amplify than nuclear DNA. These factors make it widely used in evolutionary studies and conservation biology (Arias & Sheppard, 2006).

Choosing samples carefully and keeping high-quality genetic material is an important part of any mtDNA research project. Samples are usually taken from worker bees or other life stages where the colony's identity is known. To keep the DNA intact, the samples are often stored in absolute ethanol or frozen at –80°C (J. D. Evans et al., 2013). Researchers may collect individual bees for direct mtDNA extraction or even use environmental DNA (eDNA) from honey, which has traces of bee DNA that show colony genetic signatures (Bovo et al., 2020). This process depends on the goals of the study. It is critical to choose the right tissue. For example, thoracic tissues or legs can be used for quick DNA extraction methods that keep contamination and degradation to a minimum, giving you high-quality templates that can be used in later steps (J. D. Evans et al., 2013).

Researchers have improved DNA extraction methods for honey bee tissues, like using CTAB-based methods, Chelex extraction, or enzymatic digestion, to better suit the small size of bee tissues (J. D. Evans et al., 2013). Researchers have changed protocols in some studies so that they can successfully extract DNA from older or damaged specimens. The procedure makes sure that there is high-quality genetic material available for both PCR amplification and sequencing (Kandemir et al., 2006). We regularly check the quality of DNA extracts by measuring their purity and yield with spectrophotometry and agarose gel electrophoresis (J. D. Evans et al., 2013).

Once high-quality DNA is isolated, the next important step for targeting specific mtDNA regions is PCR amplification. Researchers create specific primers to copy certain parts of the DNA, such as the cytochrome oxidase subunit I (COI), the space between cytochrome oxidase I and II (COI–COII), cytochrome bcytb), and NADH dehydrogenase components (Arias & Sheppard, 2006). It is especially easy to tell the difference between evolutionary lineages in the area between tRNA^Leu and COII because it has many different structures and is useful for diagnosis (J. D. Evans et al., 2013). When making these primers, a lot of care is taken to make sure that the region chosen has enough variation for phylogenetic analysis (Soares et al., 2019) and that it doesn't amplify anything else.

We carefully change the PCR settings to use the right temperatures, number of cycles, and types of polymerase that are effective for the low amounts and high A+T content found in many mitochondrial regions (Wu et al., 2017). Some studies have used multiplex PCR methods to amplify several mitochondrial fragments at once. Such an approach speeds up the process and makes it possible to analyze complicated genetic signatures from small samples (J. D. Evans et al., 2013). It is also common to examine the amplification products by using gel electrophoresis to see the expected band sizes. This approach makes sure that the PCR conditions produced a single, clear product without any nonspecific bands (Kandemir et al., 2006).

After amplification, the next step is to sequence these mtDNA fragments to find out about nucleotide variation and haplotype diversity. Sanger sequencing has been widely used for single-locus analyses because it is accurate and dependable (Arias & Sheppard, 2006). As technology has improved, next-generation sequencing (NGS) platforms like Illumina have become more popular for making complete mitochondrial genome sequences. This method lets researchers look at the whole mitochondrial genome as well as single gene regions in a high-throughput way (Eimanifar, Rebecca, et al., 2017). The use of NGS has changed the field by allowing for the collection of large amounts of sequence data, which has given us a deeper understanding of the structure and evolutionary history of mitochondrial populations (Eimanifar, Kimball, et al., 2018).

When using NGS, strict quality control steps are taken, such as using Trimmomatic to eliminate rid of adapter sequences and low-quality bases from raw reads. This approach makes sure that only data of superior quality are used for assembly and later analysis (Eimanifar, Rebecca, et al., 2017). We use sequence assembly algorithms like SPAdes to put together the whole mitochondrial genome from the short sequencing reads. Then, we confirm our work by comparing it to reference sequences that can be found in public databases like GenBank (Eimanifar, Kimball, et al., 2018). Researchers also use tools like Bowtie to match the reads to a standard mtDNA sequence to ensure even coverage and identify any mistakes in assembly (Abou-Shaara, 2019).

Another common method is PCR-RFLP Polymerase Chain Reaction-Restriction Fragment Length Polymorphism) analysis. This method uses different cutting points in the copied mtDNA regions to distinguish between haplotypes and subspecies (Wu et al., 2017). These studies use enzymes such as DraI, EcoRI, or HinfI to cut the PCR products at certain locations in the mitochondrial genome. The DNA pieces made are then organized on agarose gels, forming a unique pattern that helps identify samples from specific mtDNA lineages (Kandemir et al., 2006). This method is quick and cheap, which is great for doing large-scale surveys of honeybee populations in many different places (Wu et al., 2017).

High-resolution melting (HRM) analysis used with real-time PCR is a powerful method for detecting small differences in mtDNA sequences, alongside traditional sequencing and PCR-RFLP (Soares et al., 2019). HRM analysis separates sequence variants by looking at how amplified DNA fragments melt. This procedure is a very sensitive way to tell apart mitochondrial haplotypes in honeybees that are very similar to each other (Soares et al., 2019). Conservation programs find these fast screening methods especially useful in identifying introgression and monitoring the genetic integrity of native honey bee subspecies (Pinto et al., 2014).

Bioinformatics is very important for aligning sequences, finding polymorphisms, and making phylogenetic trees that show how populations have evolved over time (Arias & Sheppard, 2006). To accurately align mtDNA sequences, tools like MUSCLE, MAFFT, or ClustalW are used. Any areas that are not clearly aligned, especially in AT-rich noncoding regions, are either fixed or left out of further analysis (Sarhaddi Avval et al., 2024). These alignments are the starting point for phylogenetic inference, which uses software tools like MEGA and RAxML to make maximum likelihood trees that show how *Apis mellifera subspecies* have evolved over time (Eimanifar, Kimball, et al., 2018).

Researchers often use both Bayesian inference and maximum likelihood methods to compare evolutionary trees and ensure there is strong statistical backing for the branching patterns observed in the mitochondrial phylogeny (Eimanifar, Kimball, et al., 2018). Bayesian methods used in programs like MrBayes let you use complicated evolutionary models to obtain posterior probability estimates for clade support. Maximum likelihood methods give you bootstrap replicates to check how reliable the inferred tree structure is (J. D. Evans et al., 2013). These two methods work well together to draw the phylogenetic conclusions stronger and to give a more detailed picture of how mtDNA changes in honeybees (Eimanifar, Kimball, et al., 2018).

The examination of single-nucleotide polymorphisms (SNPs) has markedly improved the investigation of mitochondrial variations in honeybees *(Apis mellifera)* (Kim, Ah Rha, et al., 2019). Analyzing the full mitochondrial genomes of several individuals facilitates the identification of SNP markers unique to distinct strains or evolutionary lineages(Kim, Ah Rha, et al., 2019). Employing high-throughput genotyping platforms such as the Agena BioScience iPLEX Mass ARRAY system, which enables simultaneous examination of mitochondrial and nuclear markers to assess genetic integrity and introgression (D. Henriques, Parejo, et al., 2019), facilitates the identification of these specific SNPs. These SNP markers are crucial for conservation management, as they serve as precise molecular instruments for detecting non-native alleles in protected populations of Apis mellifera mellifera (Pinto et al., 2014).

Using both nuclear and mitochondrial markers in one genotyping test helps make population genetic studies clearer by showing a complete view of both maternal and biparental inheritance patterns (Behura, 2007). These kinds of approaches are especially important for keeping an eye on conservation programs where hybridization between native and introduced honeybee subspecies can put genetic purity at risk (Abou-Shaara, Al-Ghamdi, et al., 2021). Using both types of markers in conservation studies allows researchers to more accurately measure cytonuclear congruence and introgression proportions (D. Henriques, Parejo, et al., 2019).

New methods for environmental DNA (eDNA) have made mtDNA research even more intriguing by allowing researchers to collect honey bee genetic material from honey without hurting the bees (Bovo et al., 2020). eDNA taken from honey has small amounts of bee mtDNA and a lot of other genomic information from microorganisms and plants that are related to bees. This method gives a complete picture of the honeybee hologenome (Bovo et al., 2020). This method not only helps to verify the authenticity of honey by confirming its entomological origin, but it also helps to find mtDNA haplotypes that are specific to certain subspecies across large areas (Utzeri et al., 2021). Researchers have been able to get around the problems that come with targeted PCR amplification and get a complete picture of the genetic diversity in honey samples (Bovo et al., 2020), by combining shotgun metagenomic sequencing with well-developed bioinformatic pipelines.

There are usually several steps in bioinformatics pipelines used in mtDNA research. These steps include checking the quality of the raw sequence data, trimming the sequences, putting them together into contigs, adding annotations, and calling variants (Abou-Shaara, 2019). To process high-throughput sequencing data, people often use standard software tools like Trimmomatic, SPAdes, and Bowtie along with custom scripts. Such processing makes sure that the mitochondrial sequences that come out are of high quality and can be used for phylogenetic analysis (Eimanifar, Rebecca, et al., 2017). Once we combine the mitochondrial genomes, we annotate them to identify protein-coding genes, rRNA, tRNA genes, and noncoding regions. Tools like tRNAscan-SE help accurately identify the shapes of transfer RNA, which is a key part of understanding how mtDNA works (Eimanifar, Rebecca, et al., 2017).

After being annotated, the aligned mtDNA sequences are put through statistical tests to see how genetically diverse they are. These tests include haplotype diversity, nucleotide diversity, and Tajima's D, which provide us with information about demographic history and selection pressures (Eimanifar, Kimball, et al., 2018). These metrics aid in understanding the growth of populations, their encounters with obstacles, and the impact of purifying or balancing selection on mitochondrial genes. Their use puts the evolution of *A. Mellifera* (Sarhaddi Avval et al., 2024) in context. Haplotype networks, which show how mtDNA haplotypes are connected, are often used to visualize genetic relationships. One such method is the median-joining algorithm (Sarhaddi Avval et al., 2024).

Phylogenetic reconstruction using mtDNA data is a key part of honey bee evolution studies. Many studies combine data from 13 protein-coding genes, rRNA genes, and complete mitogenomes to create reliable phylogenetic trees (Eimanifar, Kimball, et al., 2018). The trees that came out of this show the evolutionary history of *A. mellifera* and show that African, European, and Middle Eastern lineages are very different from each other (Abou-Shaara, Al-Ghamdi, et al., 2021). These datasets are frequently employed in conjunction with maximum likelihood and Bayesian inference techniques. The procedure ensures that the phylogenies constructed are statistically robust and accurately reflect genuine evolutionary relationships (J. D. Evans et al., 2013).

Along with traditional phylogenetics, comparative genomic analyses have helped find areas of the mtDNA that are evolving paternal than others. These areas may be linked to traits that help organisms adapt, like disease resistance and adapting to new environments (Kim, Ah Rha, et al., 2019). Finding specific SNP markers that are linked to phenotypic traits, like resistance to Varroa destructor, is a useful tool for breeding programs that want to make honeybee colonies stronger (Kim, Ah Rha, et al., 2019). Researchers use advanced DNA sequencing and careful statistics to find these markers by looking at mitochondrial DNA sequences from both resistant and vulnerable strains (Kim, Ah Rha, et al., 2019).

Using both conventional molecular techniques and cutting-edge genomic and bioinformatic methods has made it much easier for us to study mtDNA diversity in *A. mellifera* (Sarhaddi Avval et al., 2024). New methods like SNP genotyping, PCR-RFLP, and HRM analysis make it possible to quickly and cheaply screen large sample sizes. This lets us keep an eye on genetic integrity over large geographic areas (D. Henriques, Parejo, et al., 2019). Strong bioinformatics pipelines that combine data from nuclear and mitochondrial markers add to these methods. They give a full picture of the population structure and admixture events in honeybees (Pinto et al., 2014).

In mtDNA research, the standardization of methods is a common topic. Over the past few decades, detailed protocols for collecting samples, extracting DNA, setting up PCR conditions, and doing bioinformatics analyses have been created and improved to make sure that studies can be repeated and compared (J. D. Evans et al., 2013). Researchers all over the world can work together and compare their results thanks to these standard methods. Such collaboration speeds up our understanding of how honeybees have changed over time and makes it easier to put conservation plans into action (J. D. Evans et al., 2013). Comprehensive methodological guidelines not only improve the research, but they also help a lot with the creation of diagnostic tools for checking the genetic purity of native *Apis mellifera* populations (Arias et al., 2006).

Furthermore, new studies have started to look into new ways to use mtDNA research in unusual situations, like checking the authenticity of honey. Researchers can find out where a product came from by combining mitochondrial markers with eDNA taken from honey samples. They can also find out if certain subspecies are present (Soares et al., 2019). This new method is important not only for verifying food but also for keeping track of where native and exotic honeybee lineages are found, which helps with conservation management (Utzeri et al., 2021).

In simple terms, studying mtDNA in *A. mellifera* uses various methods, starting with careful sampling and DNA extraction and moving on to more advanced techniques like PCR-based tests, next-generation sequencing, bioinformatics, SNP genotyping, and creating evolutionary trees (Arias & Sheppard, 2006). To make sure that the data we get is strong, repeatable, and useful for learning about the genetic diversity and evolutionary history of honeybees (Abou-Shaara, Al-Ghamdi, et al., 2021), each step in this pipeline is crucial. As these methods get better and new technologies are added, they will help us learn more about honey bee mtDNA and support ongoing conservation and breeding programs to protect these important pollinators for the environment and the economy (Kim, Ah Rha, et al., 2019).

Looking forward, new techniques like combining nuclear and mitochondrial genotyping tests, improved eDNA extraction methods, and more advanced computer processes are expected to lead the next wave of mtDNA research in honeybees (Behura, 2007). These advancements will enable us to monitor population changes more closely, identify introgression events more effectively, and make more informed conservation management decisions (D. Henriques, Parejo, et al., 2019). In the end, combining these different methods gives us a complete picture that not only helps us learn more about how *Apis mellifera* evolved but also helps us use genetic information in beekeeping and protecting biodiversity (Pinto et al., 2014).

In conclusion, the methods used in mtDNA research, which are briefly described in this chapter, show how molecular biology and modern bioinformatics work together to help us learn more about honey bee genetics. The careful planning of experiments, from gathering samples and extracting DNA to PCR amplification, sequencing, and computational analysis, is what lets us look at mitochondrial diversity in ways that have never been possible before (Arias & Sheppard, 2006). Standard methods that have been improved over decades have made it possible to do comparative studies that show how *Apis mellifera* populations have evolved, how they adapt, and what they need to be protected (J. D. Evans et al., 2013). As the field moves forward, it is important for researchers to stick to strict methodological standards while also using new technologies to better understand the complicated factors that cause mtDNA variation in honeybees (Abou-Shaara, 2019).

This chapter has shown how important it is to combine wet-lab protocols with computational strategies. This makes sure that every part of the experimental pipeline helps us get a strong and reliable understanding of mtDNA diversity (Sarhaddi Avval et al., 2024). Researchers can deal with the problems that arise because honey bee mtDNA is so unique by using the methods described here, which range from the first steps of preserving genetic material from field-collected samples to the last steps of phylogenetic inference using both nucleotide sequence data and SNP markers (Eimanifar, Kimball, et al., 2018). Using these methods together not only makes it easier to repeat studies, but it also makes it easier to compare results from different geographic areas and ecological settings (Arias & Sheppard, 2006).

In general, improving our methods for studying mtDNA will continue to be critical for figuring out the evolutionary history of *A. mellifera* and for helping conservation efforts that aim to protect native subspecies from the dangers of introgression and genetic erosion (Soares et al., 2019). The research community can make sure that honeybee mtDNA studies stay at the cutting edge of evolutionary genetics and conservation biology (Kim, Ah Rha, et al., 2019), by keeping a strong base in well-known laboratory and bioinformatics methods while being open to new ideas.

Finally, combining old molecular methods with new high-throughput technologies is a big step forward in our ability to study the complicated mitochondrial genome of honeybees (J. D. Evans et al., 2013). The methods discussed in this chapter show how using multiple disciplines can help us learn important things about the genetic structure and evolutionary history of *A. mellifera.* We can then utilize this information to devise strategies for safeguarding these crucial pollinators for upcoming generations (Pinto et al., 2014). As these methods improve and become standardized, they will help us learn more about mtDNA variation and how it can protect honeybees. This will ensure that the results of research lead to real benefits in beekeeping and biodiversity management (Abou-Shaara, Al-Ghamdi, et al., 2021).

6.2. Optimizing Sample Selection and DNA Preservation for High-Quality Mitochondrial Analysis in Honey Bees

6.2.1. Integrative Approaches Using Tissue and Environmental DNA Sources

In the examination of mitochondrial DNA (mtDNA) in honeybees, it is crucial to meticulously select appropriate samples and ensure the preservation of high-quality genetic material to prevent any damage or contamination. The significance of these factors in a comprehensive investigation cannot be overstated, as the quality of mtDNA collected influences subsequent processes such as PCR amplification, sequencing, phylogenetic analysis, and biodiversity monitoring (Card et al., 2021).

6.2.1.I. Introduction

In the past ten years, molecular studies of honeybee populations have made a lot of progress. This is because there is a need for strong genetic markers and high-quality DNA sequences that accurately show differences between colonies. The type of sample used has a direct effect on how reliable genetic analyses are. Because of this, researchers usually choose worker bees or other life stages that reliably show the genetic signature of a single colony. In addition to taking samples of tissue directly, non-invasive methods like extracting environmental DNA (eDNA) from honey have become popular ways to acquire colony-specific genetic information (Ribani et al., 2021). Preservation methods, designed to maintain DNA integrity during collection, transport, and storage, closely align with these strategies. The best ways to do this are to store it in absolute ethanol or at very low temperatures (–80°C) (Card et al., 2021).

6.2.1.2. Sample Selection: Rationale and Methods

Choosing the right samples that accurately represent the genetic diversity within and between colonies is an important part of any mtDNA research project. Researchers often collect worker bees because their genetically similar backgrounds in each hive make them a reliable way to tell which colony they belong to (Chibani Bahi Amar et al., 2024). Researchers can make accurate guesses about colony lineage or sub-lineage partitions when they need to establish colony genetic signatures by collecting worker bees from a single colony. In some studies, however, it is important to keep the colony's overall genetic context without losing individual bees. In these cases, non-invasive sources like honey are often used because honey contains traces of bee DNA that the bees excrete into the matrix during processing (Indahsari & Purwanto, 2024).

Researchers also use individual-based sampling, in which they collect individual bees and then follow direct mtDNA extraction protocols on them. This kind of sampling is especially helpful when the study is looking at population genetic structure, lineage differentiation, or finding hidden species (Smith et al., 2024). Using both direct tissue sampling and environmental DNA (eDNA) methods not only improves the accuracy of genetic analyses, but it also lets you double-check results from different types of samples (Soares et al., 2023).

6.2.1.3. Preservation of Genetic Material

It is just as important to choose the right samples as it is to keep genetic material safe, since DNA is naturally prone to damage from enzymatic activity, oxidative stress, and hydrolytic reactions. Researchers usually use two main methods to keep things safe: storing them in absolute ethanol and freezing them at –80°C. Ethanol preservation works by drying out the tissues, which stops enzymes from working and stops microbes from breaking down the tissues. This approach makes it a cheap and easy way to preserve things in the field (Soares et al., 2023). Furthermore, putting something in absolute ethanol right away lowers the risk of cross-contamination because the ethanol works as both a disinfectant and a preservative (Kjærstad et al., 2022).

To preserve high-molecular-weight DNA suitable for sophisticated genomic applications such as next-generation sequencing, freezing at –80°C is typically regarded as the optimal method. Ultra-low temperature storage halts molecular movement and enzymatic activity. This technique is particularly advantageous for prolonged storage when samples require extensive transportation from field collection to laboratory analysis (Dahn et al., 2022). Logistical issues and downstream uses often determine which of these two methods to use. Some studies have used a combination of ethanol fixation right after collection and then storage at -80°C to get the best results for both short-term and long-term stability (Soares et al., 2023).

6.2.1.4. Tissue Selection and Its Impact on DNA Quality

The quality and quantity of mtDNA you get depends on the tissue you choose. For instance, bee thoraxes and legs are good targets because they have low levels of PCR inhibitors and high numbers of mitochondria, which makes them good candidates for quick DNA extraction methods (Smith et al., 2024). The thoracic tissue has a lot of muscle fibers and is metabolically active, which means it has a lot of mitochondria. This property makes it a reliable source of high-quality mtDNA. Worker bees' legs can be sampled because they are external appendages that can be cleaned before extraction (Kjærstad et al., 2022). Worker bees can also be used for sampling because they are external appendages that can be easily cleaned before extraction (Kjærstad et al., 2022). Furthermore, protocols for quickly extracting DNA from these types of tissue have been improved to reduce damage that can happen when they are exposed to harmful temperatures or contaminants for too long during processing. These methods use detergent-based lysis buffers and sometimes bead-beating to verify that the tissue is completely mixed up, which helps cells break down and DNA be released. Standardized protocols like this make sure that experiments are consistent and that the DNA extracted is of high quality, which is important for tests extracted is of excellent quality, making it suitable for important tests like PCR amplification, restriction fragment length polymorphism (RFLP) analysis, and high-throughput sequencing (Corrales et al., 2023).

6.2.1.4. Environmental DNA (eDNA) from Honey

Besides taking samples of tissue directly, using environmental DNA (eDNA) from honey has become a useful way to obtain bee mtDNA without hurting the bees. Honey is a mixture of bee cells, secretions, and even tiny bits of hive debris, which makes it possible to detect mtDNA signatures at the colony level. The study of honey eDNA is a non-invasive option that is especially useful in long-term studies or when keeping the bee population safe is a top priority (Ribani et al., 2021).

To extract DNA from honey, you must dilute it to make it less thick and then centrifuge it to concentrate the eDNA. Commercial DNA extraction kits have been modified so that they can be used on honey samples to effectively separate the DNA even when PCR inhibitors like polysaccharides and polyphenols are present. eDNA-based methods only work if there is no contamination when samples are taken and processed in the lab, as even tiny amounts of contaminant DNA can change the results (Indahsari & Purwanto, 2024). The extracted eDNA often reflects a blend of the colony's genetic information. It can supplement the data obtained from direct worker bee tissue samples (Soares et al., 2023).

6.2.1.5. Contamination Avoidance and Quality Control

mtDNA studies must collect high-quality samples and avoid contamination. When using direct tissue sampling or eDNA methods, contamination can come from the environment, reagents, or by transferring samples from one to another. Using sterilized tools and separate workspaces for pre- and post-extraction procedures is a common practice. Researchers often follow rules like using disposable gloves, sterilizing dissection tools with a flame, and cleaning laboratory surfaces with ethanol, bleach, or UV light every so often (Dahn et al., 2022).

Furthermore, it is common practice to use negative controls during DNA extraction and PCR amplification to verify for contamination. To make sure the DNA is good enough for later tests, quality control steps like checking the purity ratios (A260/A280 and A260/A230) and measuring the DNA concentrations with fluorometry are used (Indahsari & Purwanto, 2024). These steps are essential because each cell has many copies of mtDNA, and even small amounts of contaminant DNA can cause the wrong amplification, the wrong haplotype identification, and the wrong conclusions about genetic structure.

6.2.1.6. Practical Considerations in Field and Laboratory Settings

Field sampling has its own issues, the most important of which is how to quickly and effectively preserve the DNA after collection. Researchers may choose to use ethanol preservation because it is simple to move and doesn't need any special equipment when they are working in remote or resource-limited areas. However, because ethanol is so unstable and body fluids can dilute it, it's important to make sure the tissue-to-ethanol volume ratio is rightusually at least (3:1) for effective preservation (Kjærstad et al., 2022).

On the other hand, if there are labs nearby, flash freezing with liquid nitrogen or portable freezing devices can be used to freeze the samples right away at –80°C. This quick-freezing process stops ice crystals from forming, which could otherwise damage DNA and cellular structures. Using ethanol or freezing quickly processes samples, which shortens the time frame during which enzymatic degradation can happen. This approach makes sure that high-quality, intact mtDNA is available for molecular analyses later on (Corrales et al., 2023).

6.2.1.7. Integration of Multiple Sample Types for Comprehensive Analysis

Combining data from both direct tissue samples and eDNA from honey presents a full picture of the colony's genetic structure in many mtDNA research projects. Directly taking genetic material from worker bee tissues usually provides you high-resolution genetic material that is great for studies of population genetics, phylogeography, and evolution. On the other hand, eDNA from honey provides us more information about the hive as a whole, how bees forage, and the presence of pathogens or parasites that may be found with bee mtDNA (Soares et al., 2023).

Because these two approaches work well together, it's important that the methods used for both types of samples are the same and that strict preservation protocols are followed. For example, ethanol preservation might be the best way to quickly collect samples in both cases, but the analytical pipeline for tissue-derived DNA and honey eDNA may be somewhat different when it comes to extraction buffers, steps to remove inhibitors, and conditions for amplification. However, the primary objective remains obtaining high-quality, representative mtDNA for robust genetic analyses (Corrales et al., 2023).

6.2.1.8. Advanced Applications and Future Directions

Next-generation sequencing technologies have made it possible to do a lot more research on bee mtDNA. High-throughput sequencing methods not only let us quickly sequence whole mitochondrial genomes, but they also make it easier to discover rare variants in mixed samples, like those from honey eDNA. These new technologies have made it even more important to keep DNA safe from the time it is collected until it is finally sequenced. Even small amounts of damage or contamination can cause amplification bias, lower sequence coverage, and problems with correctly assigning haplotypes (Smith et al., 2024).

Furthermore, modern protocols are starting to include bioinformatic pipelines that can look at data from both traditional tissue samples and eDNA sources. These pipelines use techniques to filter out mistakes, identify mixed sequences, and apply strict rules for classifying species to distinguish between genuine mtDNA sequences and potential contaminants. As research moves forward, there is more and more emphasis on creating integrated methods that combine detailed morphological identifications with high-throughput molecular data to help with full biodiversity assessments and colony health monitoring (Smith et al., 2024).

In the future, researchers may focus on enhancing non-destructive sampling methods that utilize both bee tissue and honey-derived eDNA to monitor bee populations in real time. These kinds of studies not only help us learn more about genetic diversity, but they also help protect bee populations that are already in danger by having less of an effect on them. New portable sequencing devices and DNA extraction kits promise to simplify the process even further. They will allow for immediate genetic assessments in the field and lower the need for long-term storage, which could lead to degradation risks (Ruppert et al., 2022).

6.2.1.9. Case Studies Illustrating Best Practices

Several studies have shown that these best practices work well in real life. For instance, one study utilized worker bee samples from a single colony to accurately identify the mtDNA lineages for that colony. The researchers used a combination of preserving the samples in the field with ethanol and then moving them to storage at -80°C when they got to the lab. Because of this strict method, we got high-quality mtDNA sequences that we used to tell the difference between the main mtDNA lineages in the population (Chibani Bahi Amar et al., 2024).

In another study, scientists looked at how to get eDNA out of honey. They were able to obtain bee mtDNA that matched well with direct tissue extractions by carefully diluting the honey, centrifuging the sample to make it more concentrated, and using DNA extraction kits that were made for complex matrices. The study provided insights into non-invasive monitoring methods and demonstrated how the combination of honey eDNA and worker tissues can provide a comprehensive understanding of an individual's genetic makeup (Indahsari & Purwanto, 2024).

Furthermore, protocols that simplify quick extraction from bee legs work well to reduce contamination. We used a lysis buffer to process the legs of individual bees that had been washed to remove surface dirt. The procedure made it easy to obtain DNA back quickly. This method ensured a high mtDNA yield and made PCR amplification and sequencing easier. This instance shows how important it is to choose the right types of tissue for bee mtDNA research (Smith et al., 2024).

6.2.1.10. Challenges and Mitigation Strategies

Even though the current methods for sampling and preserving are excellent, there are still some problems. One of these problems is that DNA yield and quality can vary naturally because of differences in tissue type, age of the specimen, or adverse field conditions. For instance, if you don't preserve bee tissues collected in extreme weather right away, they may break down faster. Some ways to lessen the damage are to use preservation media right away and move samples to ultra-low temperature storage as quickly as possible, preferably in portable coolers or with cryoprotective agents when possible (Dahn et al., 2022).

Another problem is the risk of cross-contamination, which could happen during the collection process or later in the lab. Because environmental DNA is so common, even small amounts of contaminant genetic material can mess up sensitive analyses later on, which can lead to wrong conclusions. It is critical to follow contamination-avoidance rules exactly, like using sterile tools, negative controls during extractions, and setting aside clean areas for pre-PCR work. Furthermore, using the same protocols in different studies makes it easier to compare and trust the results of mtDNA research (Kjærstad et al., 2022).

Additionally, both freezing at –80°C and preservation in absolute ethanol are widely used methods, but each has its problems. If not done carefully, ethanol fixation can cause tissue to shrink or change shape. On the other hand, freezing needs reliable cold storage infrastructure that may not always be available in remote field settings. Researchers are still looking into other preservation buffers, like RNAlater and DNAgard Tissue, which might work better for stabilizing DNA and RNA. However, cost is still a concern (Soares et al., 2023).

6.2.1.11. Implications for Genetic and Ecological Studies

The methods used to choose and store samples go beyond just lab protocols; they have a big impact on how we understand bee genetics and ecology. Researchers can determine the evolutionary relationships between different bee populations, monitor the spread of invasive mtDNA lineages, and track the genetic health of colonies over time using high-quality mtDNA sequences. For example, studies that combine mtDNA data with ecological observations have provided us useful information about how environmental stressors affect bee populations and possible ways to focus conservation efforts (Corrales et al., 2023).

Additionally, the ability to extract mtDNA from honey provides a unique method to monitor bee health and biodiversity without causing harm to them. Researchers can make detailed models to predict whether a colony will succeed or fail by comparing genetic data from honey eDNA with environmental factors like the quality of the food or the number of pathogens present. These kinds of all-encompassing approaches are essential for dealing with the world's worries about the decline of bee populations and coming up with ways to protect these important pollinators (Ribani et al., 2021).

6.2.1.12. Conclusion

In short, picking the right samples and keeping high-quality genetic material safe are two of the most important parts of any successful mtDNA research project in honeybees. Researchers have to balance several important factors, such as choosing tissues that are typical of the colony, like worker bee thoraces or legs, and choosing preservation methods, like freezing at -80°C or using absolute ethanol, that stop harmful degradation processes. Furthermore, using environmental DNA from honey is a powerful, non-invasive way to supplement traditional tissue sampling and provide a complete picture of the genetic signatures of a colony. Following strict rules for avoiding contamination and using standardized extraction methods makes sure that the mtDNA that comes out is suitable for advanced uses like high-throughput sequencing and detailed phylogenetic analysis. These best practices not only improve the quality of the data and the accuracy of the analysis, but they also help us learn a lot more about bee genetics, evolution, and how they adapt to their environment (Card et al., 2021; Ribani et al., 2021; Smith et al., 2024).

The use of different types of samples and preservation methods, along with cutting-edge molecular techniques, makes mtDNA research an important tool for learning about and protecting honeybee populations in a time of environmental change and more human-caused stress. As methods develop better and more consistent, future studies should supply us even more information about the genetic structure of bee colonies. Such knowledge will help us come up with better ways to protect these important pollinators and manage them in a way that is advantageous for the environment (Corrales et al., 2023; Soares et al., 2023).

Researchers can obtain the most and best mtDNA by continuing to improve the rules for collecting samples, choosing tissues, and preserving DNA. This way, each study adds strong and reproducible data to the overall knowledge of bee genomics and biodiversity. The fact that both field and laboratory practices are constantly changing shows how dynamic mtDNA research is and how important it is to use strict methods to make progress in this field (Indahsari & Purwanto, 2024; Kjærstad et al., 2022).

In the end, combining new ways of sampling with careful ways of preserving samples will help us make progress in understanding honeybee genetics, how pathogens work, and how bee populations and their environments affect each other in complex ways. This all-encompassing method is necessary for both academic research and real-world uses in apiculture, conservation biology, and ecosystem management. It shows how important high-quality mtDNA data is for protecting biodiversity and keeping pollinator populations healthy for future generations (Corrales et al., 2023; Smith et al., 2024).

Chapter 7

7. Optimization of DNA Extraction Protocols for Honey Bee (*Apis mellifera)* Tissues

Researchers are working diligently to obtain high-quality genetic material from honeybee tissues that are tiny and often old or damaged. One important area of research is improving DNA extraction protocols for these tissues. Honeybee specimens have small tissue masses that weigh only a few milligrams. This means that extraction methods need to be very sensitive so that they can address problems like low DNA yield, proteins or polysaccharides getting in the way, and PCR inhibitors. Recent research has made great strides in getting reproducible and high-quality DNA that can be used for downstream applications like PCR amplification and sequencing (Balkanska et al., 2018). The improvement was done by changing CTAB-based methods, Chelex extraction, and enzymatic digestion techniques.

7.I. CTAB-Based Extraction Methodologies

For a long time, people have known that the CTAB (cetyltrimethylammonium bromide) reagent is a useful detergent for breaking down cell membranes and sticking to polysaccharides. This makes it easier to separate nucleic acids from other substances that might become in the way. CTAB-based methods were first adapted from plant DNA extractions for honeybee tissues because bee tissues and hive materials often contain many sugars, proteins, and other substances that can stop the extraction process. To make up for the lower starting material and lower the risk of DNA damage that can't be fixed, researchers have changed important factors like the temperature of the incubation, the length of the reaction, and the makeup of the lysis buffer (Balkanska et al., 2018; Kek et al., 2018).

Some protocols reduce the lysis temperature from the standard 65 °C, commonly used in many plant extractions, to lower temperatures such as 37–45 °C. This keeps the DNA from breaking up even more in already fragile samples while still allowing for effective cell lysis (S. A. Jain et al., 2013). Furthermore, CTAB-based methods for honeybee tissues often use extra chemicals like (polyvinylpyrrolidone PVP) and (dithiothreitol DTT) to bind polyphenolics and break disulfide bonds, respectively. These additives are crucial for getting rid of things that would otherwise make PCR amplification and sequencing less accurate (Kek et al., 2018; Lalhmangaihi et al., 2014).

Mechanical disruption is a crucial step in CTAB-based methods when handling bee tissues. Due to the limited number of samples, some laboratories have been employing bead pounding or grinding in liquid nitrogen to thoroughly disrupt cells while preserving the integrity of the nucleic acid fraction (Balkanska et al., 2018; Holloway et al., 2013). This mechanical method not only boosts the amount of material extracted by ensuring more cell contents come into contact with the lysis buffer, but it also reduces the risk of leftover tissue that could hold onto DNA and make extraction harder. Adding these mechanical pre-treatment steps has been especially useful for processing samples that are hard to work with because of what they're made of, like pollen stuck in honey or tissues with many polysaccharides (Kek et al., 2018; Soares et al., 2015).

7.2. Chelex Extraction Protocols

Researchers have developed the Chelex extraction method to handle small insect tissue samples, such as those from honey bees. Protocols that use Chelex resin typically involve boiling the sample in a Chelex solution with added proteinase K, as the resin grabs onto metal ions that damage DNA. People like this method because it doesn't require organic solvents like phenol and chloroform, which makes it quick and cheap. Also, Chelex extraction is very useful when you need to process many samples at once, which is an important part of high-throughput genotyping studies (Bubnič et al., 2020; J. D. Evans et al., 2013).

Protocols utilizing Chelex for small specimens, such as honeybee legs or wing clippings, typically necessitate minimal tissue manipulation, which is advantageous. Numerous investigations have demonstrated that robust DNA can be extracted from a single leg of an *Apis mellifera*, sufficient for many PCR reactions (Bubnič et al., 2020; J. D. Evans et al., 2013). The Chelex method is easy to use, so it doesn't need as many purification steps. This approach lowers the risk of losing DNA during sample processing. But these protocols do have some problems. For example, DNA obtained through Chelex extraction may still have inhibitors in it that need to be removed before sequencing (Bubnič et al., 2020; J. D. Evans et al., 2013).

Table 10. Working Solutions (Gautam, 2022)

Solution	Preparation	Storage
0.5 M EDTA	Add 186.1 g of EDTA to 800 mL of distilled water, stir vigorously on a magnetic stirrer and adjust pH to 8 by using NaOH	Room temperature
5% Chelex suspension	Add 4 g of Chelex resin in 40 mL TE buffer and add 40 mL of sterilized water	4 °C
10X TE bufferpH 8)	Add 200 µL EDTA into 1 mL of tris buffer and add 8.8 mL of autoclaved Milli-Q water	RT
7.5 M ammonium acetate	Use as purchased; no preparation needed	4 °C
3 M sodium acetate	Use as purchased; no preparation needed	4 °C
70% ethanol	Add 737 mL of 95% ethanol, and then add deionized water to bring the total volume to 1 L	4 °C
10 N NaOH	Add 40 g of NaOH pellets to 80 mL of water slowly, with a continuous stir. Maintain final volume to 100 mL by adding more water	RT

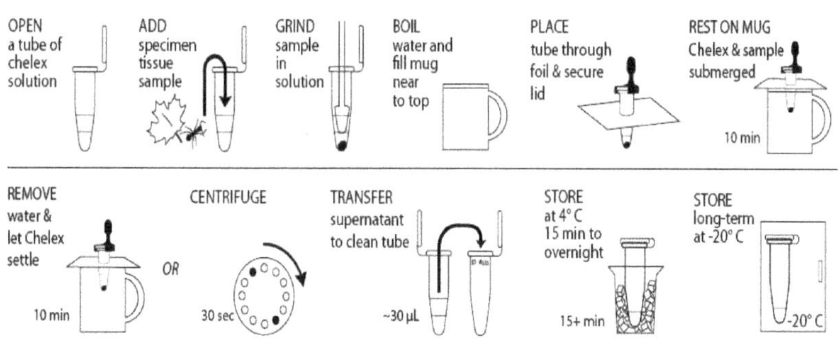

Fig. 7. DNA Learning Center Barcoding 101 (Gautam, 2022)

7.3. Enzymatic Digestion Methods

Enzymatic digestion methods work well with CTAB's chemical lysis or Chelex extractions' resin binding. These methods usually use proteinase K, sometimes with other enzymes like RNase A, to break down proteins and free DNA from cells and extracellular matrices. Enzymatic methods are sometimes better for honeybee tissues, which are small and often composed of various materials like the gut or the spermatheca, because they break down cellular structures without causing significant damage to the DNA (Holloway et al., 2013; S. A. Jain et al., 2013).

Researchers have optimized digestion protocols by adjusting enzyme concentration, incubation time, and temperature to maximize the yield from the limited amount of available bee tissue. Longer incubations at moderate temperatures (55–60 °C) facilitate the complete digestion process and inhibit nuclease activity, which could otherwise degrade the extracted DNA (Bubnič et al., 2020; Holloway et al., 2013). Researchers have altered protocols to include a two-step digestion process. The procedure initiates with the incubation of proteinase K and concludes with RNase treatment to eliminate any interfering RNA. The latter step ensures that spectrophotometric measurements reflect the purity of DNA rather than elevated levels due to RNA (J. D. Evans et al., 2013; Lado, 2023).

Enzymatic digestion is essential for addressing degraded or older specimens that exhibit partial cellular structure damage. The gentle nature of these enzymes allows for the recovery of intact genomic fragments essential for effective PCR amplification and sequencing (Hazır & Bock, 2019; Moškrič et al., 2023). Enzymatic methods may require more time than chemical methods; however, they offer the significant benefit of preserving DNA integrity. This advantage is significant in the context of museum or archival specimens, where DNA fragmentation remains a concern (Al-Dhafer et al., 2025; Bubnič et al., 2020).

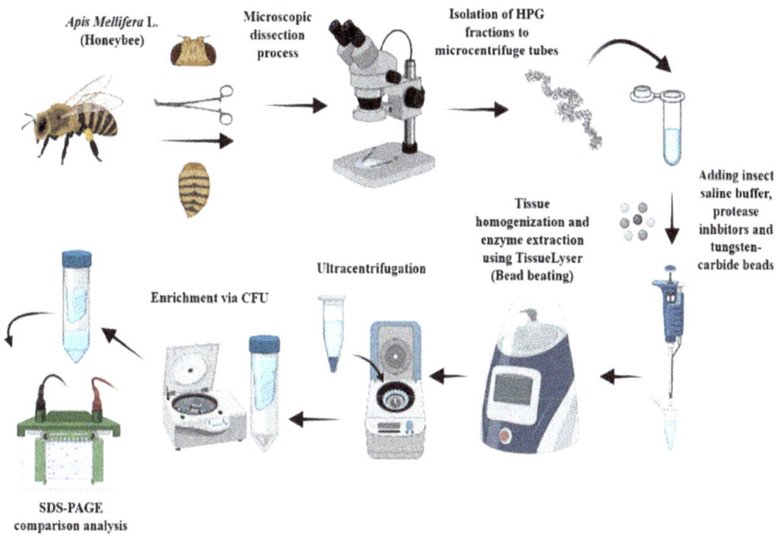

Fig. 8. ***Apis mellifera L.*** **dissection and protein extraction from the isolated HPGs. Created with BioRender.com (Akyildiz et al., 2022)**

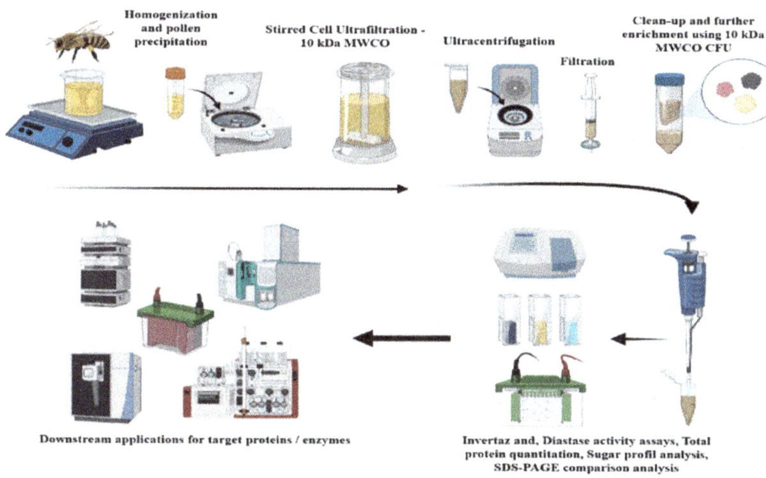

Fig. 9. Determined protocol for the total protein extraction from pine honey. Created with BioRender.com (Akyildiz et al., 2022)

7.4. Modifications Tailored to Very Small Tissue Sizes

Honeybee tissues can be difficult to work with because there isn't much material available, sometimes only a few micrograms or less. This necessitates the careful adjustment of standard extraction volumes and reagent concentrations in protocols designed for larger tissue samples. Several research groups have changed extraction protocols by using less reagent and more effective ways to break up cells, like bead beating in microcentrifuge tubes (Bubnič et al., 2020; MOREIRA et al., 2013).

In real life, scaling down a CTAB-based extraction means lowering the volume of lysis buffer to match the amount of tissue being used while still keeping the right amounts of CTAB, PVP, and other important reagents. This change keeps the sample from getting too diluted, which could lower the final DNA concentration below the level needed for reliable downstream analysis (Balkanska et al., 2018; Kek et al., 2018). Researchers have also improved the Chelex extraction process by using small amounts of Chelex resin suspension and lowering the incubation volume to keep the DNA released highly concentrated (Bubnič et al., 2020; J. D. Evans et al., 2013). This miniaturization is crucial because losing even small amounts of DNA can greatly lower the success rate of PCR amplification. This is especially true for microsatellite genotyping or SNP analysis, where the amount of template is a limiting factor (Holloway et al., 2013; MOREIRA et al., 2013).

Protocols that use enzymatic digestion can also benefit from these changes. For example, lowering the reaction volumes while raising the enzyme concentration can cause quick and complete lysis with very little dilution of the DNA that was extracted (Holloway et al., 2013; S. A. Jain et al., 2013). These kinds of changes are especially important when working with tissues like the bee spermatheca or tiny pieces of leg tissue, where even small increases in yield can greatly affect the success rate of amplification (Bubnič et al., 2020; MOREIRA et al., 2013).

7.5. Adaptations for Older or Degraded Specimens

Along with improving protocols for small tissue sizes, a lot of work has gone into changing extraction methods so that they can be used on older or otherwise damaged specimens. These changes are crucial because many studies of honeybee genetics use samples that have been stored in museums or other places that aren't ideal for them, and these samples have naturally broken down and changed chemically while they were being stored (Cilia et al., 2022; Hazır & Bock, 2019).

Changing the times and temperatures of incubation is one of the main changes made to make it easier to get DNA from old samples. Lower lysis temperatures and longer incubation times in CTAB-based protocols help keep the DNA from breaking down even more while still breaking down the cells completely (S. A. Jain et al., 2013; Soares et al., 2015). Adding reducing agents like DTT to the lysis buffer not only helps break down protein cross-links, but it also keeps DNA from oxidizing, which is a common problem with specimens that have been stored for a long time (Al-Dhafer et al., 2025; Lalhmangaihi et al., 2014).

When employing Chelex extraction methods on degraded samples, it is necessary to modify the boiling duration and incorporate pre-treatment steps to eliminate inhibitors that have accumulated during storage. Certain protocols incorporate an initial washing step using ethanol or a buffered solution prior to Chelex treatment to eliminate residual contaminants that may co-extract with the DNA. This enhances the yield and amplifiability of the extracted material (Bubnič et al., 2020; Hazır & Bock, 2019). Enzymatic digestion protocols are advantageous for older samples, as the mild action of proteinase K prevents further degradation. High-molecular-weight fragments can be recovered from samples stored for extended periods by meticulously optimizing enzyme concentrations and incubation durations (Al-Dhafer et al., 2025; Moškrič et al., 2023).

In the analysis of degraded specimens, it is crucial to account for the potential co-purification of inhibitors with DNA. The presence of these inhibitors can markedly diminish the efficacy of PCR. The combination of CTAB-based precipitation and organic extractions using chloroform and isoamyl alcohol effectively removes various contaminants. Incorporating additional enzymatic treatments during digestion can enhance the purity of the DNA. Multi-step methods ensure that DNA extracted from damaged samples meets the quality standards necessary for sensitive downstream applications, such as high-throughput sequencing and genotyping (Bubnič et al., 2020; Hazır & Bock, 2019).

7.6. Quality Control of DNA Extracts

Spectrophotometric measurements are combined with agarose gel electrophoresis. Spectrophotometry checks how much light is absorbed at 260 nm to determine the amount of nucleic acids and calculates the 260/280 and 260/230 ratios to look for protein and organic contamination, respectively. Most people think that DNA extracts with a 260/280 ratio close to 1.8–2.0 are very pure, which is useful for PCR amplification and sequencing (Holloway et al., 2013; Kek et al., 2018).

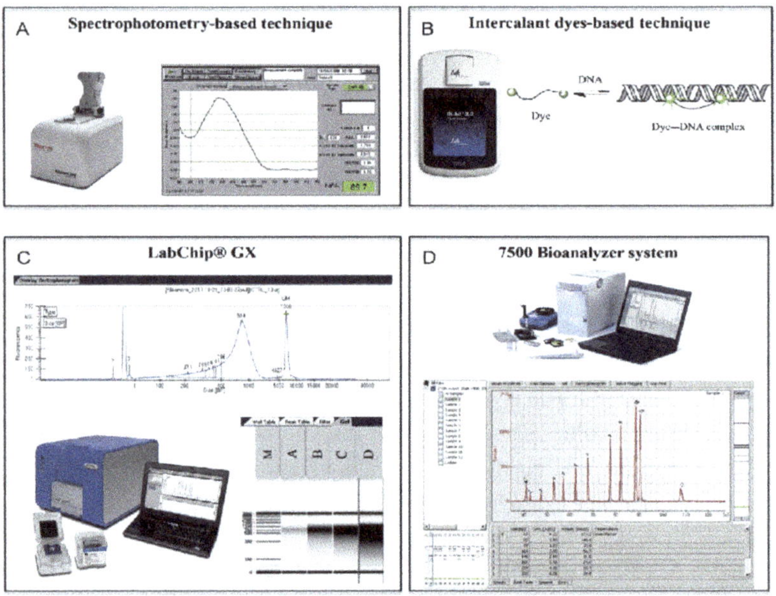

Fig. 10. Examples of methods used for evaluation of DNA quantity/quality.

A) Spectrophotometry-based techniques as NanoDrop 2000 TM (Thermo Scientific): DNA concentration is calculated from the value of absorbance at 260 nm.B) Fluorometers as Qubit ® 3.0 (LifeTechnologies) in association with Qubit ® quantitation assays: DNA concentration is calculated from the fluorescence resulting from the binding of a reporter dye. C) and D) Electrophoresis-based techniques: C) LabChip ® (GXPerkinElmer) performs DNA electrophoresis separation and provides DNA size and concentration; D) Agilent 2100 Bioanalyzer System (Agilent Technologies) provides an on chip-electrophoresis tool for sizing, quantitation and quality control of nucleic acids. A) Copyrigth 2014 Thermo Fisher Scientific Inc. All Rights Reserved. B) Copyrigth Life Technologies All Rights Reserved. C) Copyrigth PerkinElmer All Rights Reserved. D) Copyrigth Agilent Technologies All Rights Reserved (Didelot et al., 2013).

Agarose gel electrophoresis remains a critical component of quality assessment, in conjunction with spectrophotometry. Researchers can assess the integrity of DNA by analyzing DNA extracts using gel electrophoresis. They can assess high-molecular-weight bands and confirm that degradation has not resulted in excessive fragmentation. This step is crucial when handling small or damaged samples, as it provides a direct visual representation of any smearing, indicating that the sample is fragmented and unsuitable for sensitive applications (J. D. Evans et al., 2013; Torricelli et al., 2016). Furthermore, consistent verification through gel electrophoresis can confirm that the extraction process worked before moving on to PCR-based assays. This saves time and lowers the risk of downstream assay failure because the template quality is poor (J. D. Evans et al., 2013; Soares et al., 2015).

Often, these quality control steps follow the extraction protocol, facilitating prompt identification of any issues. Many protocols combine both methods. First, the DNA is measured spectrophotometrically, and then a sample is electrophoresed on an agarose gel to make sure that the spectral data matches with restricted degradation and high molecular weight (J. D. Evans et al., 2013; Moškrič et al., 2023). This two-pronged method of quality control is standard in labs that work with honeybee tissues. It is very important to confirmt that the DNA is suitable for use in downstream applications like PCR, sequencing, and different types of genotyping (Holloway et al., 2013; Lado, 2023).

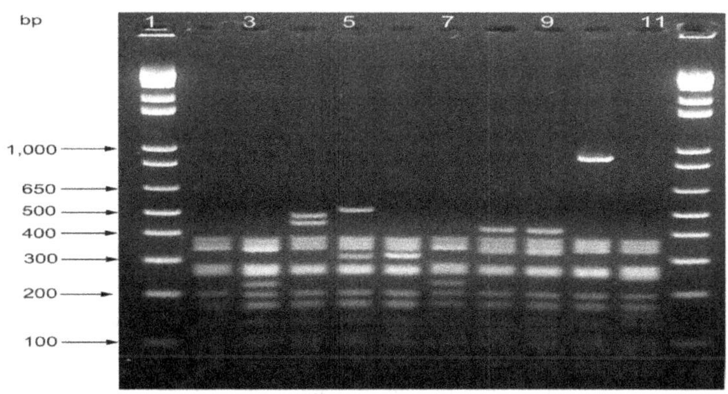

Fig. 11. Representative RFLP analysis. Plasmid DNA isolated from clones produced in the functional gene archives was digested with HinPII.

Two hundred fifty nanograms of each digest was separated by using 2% LMP GTG agarose run at 5 V cm^{-1} for 1.75 h in 1% Tris-acetate-EDTA. Lanes 1 and 12 contain 500 ng of 1-kb Plus DNA molecular weight marker. Lanes 2 and 3, 9-661; lanes 4 and 5, 9-682; lanes 6 and 7, W9-661; lanes 8 and 9, W9-682; lanes 10 and 11, W10-682. The gel was stained with ethidium bromide. Images were captured and recorded by using Kodak 1D software and UV transillumination (Erwin et al., 2005).

7.7. Comparative Analysis and Integrated Approaches

Multiple studies have conducted direct comparisons of various extraction methods to determine the most effective approach for honeybee tissues under diverse conditions. CTAB-based methods can yield highly pure DNA when properly implemented; however, they often require more effort compared to Chelex extraction protocols. The latter are more time and cost-efficient but may result in lower purity if residual inhibitors are not thoroughly eliminated. Researchers determined that CTAB methods require meticulous optimization of various pre-treatment steps to effectively manage the intricate matrix of honeybee tissues. The overall yield and integrity surpassed those of simpler extraction methods **(Balkanska et al., 2018; Kek et al., 2018).**

On the other hand, Chelex-based extractions have been very popular in high-throughput studies where hundreds of small samples need to be processed quickly. But the trade-off is often that there are some leftover contaminants that can stop PCR from working if they aren't properly neutralized by further dilution or quick processing (Bubnič et al., 2020; J. D. Evans et al., 2013). Enzymatic digestion methods are a great option for handling a small number of samples that are damaged, especially when the correct conditions for proteinase K and RNase are applied. In these cases, the enzymes gently free DNA from tiny tissues without damaging the structure needed for later genotyping (Holloway et al., 2013; S. A. Jain et al., 2013).

Researchers have developed integrated approaches that combine elements of various methods. Certain protocols initiate with an enzymatic digestion step to ensure complete cell lysis, followed by a CTAB-based precipitation step to eliminate polysaccharides and other contaminants that hinder the process. Finally, one may add a Chelex-based cleanup after extraction to further enhance the quality of the DNA sample. These kinds of hybrid methods can maximize both yield and purity while also saving valuable sample material. This feature is critical when working with bee tissues that are minimal and may not be able to be replaced in some ecological or forensic studies (Bubnič et al., 2020; Moškrič et al., 2023).

Researchers employing these enhanced procedures have successfully PCR amplified and sequenced samples ranging from freshly acquired to many years old, even when stored under suboptimal circumstances. It is particularly remarkable that high-quality DNA can be extracted from ancient or compromised materials. This approach facilitates numerous new research opportunities, including historical genetic investigations and retrospective analyses of bee populations. To get these results, it was essential to implement modifications such as prolonging the lysis durations and incorporating additional purification procedures. (Al-Dhafer et al., 2025; Hazır & Bock, 2019) (Al-Dhafer et al., 2025; Hazır & Bock, 2019).

7.8. Practical Considerations and Future Directions

Despite significant advancements, practitioners must carefully consider several practical factors when selecting an extraction protocol for honeybee tissues. The choice is usually a balance between the available resources, the number of samples that need to be processed, and the uses that will be made of the results. Chelex extraction is a quick and cheap way to do large-scale population genetics studies. However, improved CTAB methods that include careful preparation and physical breaking of the samples may be more suitable for research that requires very high-quality DNA for next-generation sequencing (J. D. Evans et al., 2013; Holloway et al., 2013).

When handling antiquated or compromised specimens, it is essential to employ protocols tailored to their condition. Processing these sample types necessitates increased time and reagents; however, the resultant DNA is more stable and amenable to amplification. This enhances the reliability of the analyses and their resultant conclusions (Bubnič et al., 2020; Moškrič et al., 2023). In the future, we may utilize more sophisticated purification techniques, such as magnetic bead-based technologies or microfluidic systems, to enhance extraction efficiency and minimize sample loss.

Automation of the extraction protocols is an area of active development in the context of high-throughput applications. Robotic extraction systems could make the steps for CTAB-based or enzymatic methods the same for everyone, leading to more reliable results across many samples and less variation between different people doing the work. Additionally, current research shows that using real-time quality checks during the extraction process such as monitoring with spectrophotometers could lead to better and more adaptable protocols (Neumann et al., 2013; Torricelli et al., 2016).

In general, the progress made in the past ten years has changed the way molecular studies are done on honeybees. Researchers now have strong genetic data that helps with studies in evolution, ecology, breeding, and conservation. They succeeded by creating protocols that work for both the small tissues of bees and the difficult conditions of older or degraded samples. The many changes made to CTAB-based methods, Chelex extraction protocols, and enzymatic digestion techniques show how dedicated the scientific community is to finding new ways to get around technical problems (Bubnič et al., 2020; J. D. Evans et al., 2013).

7.9. In conclusion

In conclusion, improving DNA extraction methods for honey bee tissues is a complex problem that has been solved in new ways with CTAB-based methods, Chelex extraction, and enzymatic digestion. These protocols have been carefully designed to work with the tiny sample sizes that are common in bee tissues, as well as the problems that arise with older or damaged specimens. Researchers have significantly increased the amount and quality of DNA they can extract by using methods that break apart the cells, adjusting the temperatures and times for breaking down the samples, and adding substances that counteract harmful effects. These kinds of improvements not only make PCR amplification and sequencing more reliable, but they also open up new possibilities for high-throughput molecular studies and genetic analyses in apiculture and related fields (Bubnič et al., 2020; Holloway et al., 2013).

Quality control steps are still an important part of the extraction process. For example, spectrophotometric quantification and agarose gel electrophoresis are two important ways to confirm the purity and integrity of DNA. These steps assure that the final extracts meet the strict standards needed for sensitive downstream applications, even though the input material is difficult to work with. The improvements in extraction efficiency, quality control, and protocol miniaturization are a big step forward in honeybee molecular research. They give us reliable methods for both fresh and archived specimens (J. D. Evans et al., 2013; Lado, 2023).

Researchers will probably continue to investigate how to combine these improved protocols with automated systems and next-generation sequencing platforms. This will make it even easier to do large-scale genetic studies on honeybees. Researchers will be able to learn even more about the genetic basis of honey bee biology, ecology, and evolution (Moškrič et al., 2023; Torricelli et al., 2016), if they keep improving these protocols and adapting them to new molecular technologies.

In short, the combined work on improving DNA extraction from honeybee tissues using CTAB-based methods, Chelex extraction, and enzymatic digestion has led to protocols that are strong, scalable, and can be used on tiny samples as well as older, damaged specimens. These kinds of protocols not only assure that there is high-quality genetic material available for PCR amplification and sequencing, but they also support the rapid progress of molecular apiculture research. This combination of new methods and strict quality control is making it possible to do more accurate genetic research on honeybee populations, their breeding programs, and how they deal with environmental problems (Bubnič et al., 2020; J. D. Evans et al., 2013; Holloway et al., 2013).

Chapter 8

8. Integrated DNA Extraction from Small Bee Tissues

A Hybrid Protocol Combining Enzymatic Digestion, CTAB Precipitation, and Chelex Cleanup

Here is a detailed, step-by-step guide on how to create an integrated DNA extraction protocol for small bee tissues that uses enzymatic digestion, CTAB-based precipitation, and Chelex-based cleanup. This hybrid method uses a number of well-known protocols and techniques from the literature to extract the most DNA from a sample while using as little of it as possible.

8.1. Overview and Rationale

In the analysis of small bee tissues, which are irreplaceable in ecological or forensic contexts, achieving both a high yield and a high level of purity is crucial for subsequent molecular applications such as PCR amplification and high-throughput sequencing. The integrated protocol consists of three primary components: an enzymatic digestion phase to facilitate the breakdown of cells and structural proteins; a CTAB-based precipitation step to eliminate polysaccharides, polyphenols, and other potential PCR enzyme inhibitors; and a Chelex-based cleanup to capture metal ions and additional substances that may disrupt PCR enzyme function. Each step aims to collaborate with others to address the issues associated with single-method protocols (Ali et al., 2017).

8.2. Module 1: Mechanical Disruption and Enzymatic Digestion

To get to the contents inside the cells, the tissue must be homogenized before the extraction process can begin. If you have minimal bee samples, you can mechanically break them up by gently crushing the tissue with a sterile scalpel or bead beating with minimal amounts of liquid and gentle movement. This kind of initial physical damage makes it easier for lysis buffers to get in, which is especially important because insect tissues are protected by a strong chitinous exoskeleton (Greenslit, 2018). After the tissue is broken up, an enzymatic digestion step is used to make sure that all of the cells are broken down. In this step, we prepare a lysis buffer containing Proteinase K at a concentration of approximately 20 mg/ml. If needed, it may also have other enzymes like lysozyme in it to break down both cell wall parts and proteins inside cells (Jangra & Ghosh, 2022).

The lysis buffer should also include a detergent like SDS, which helps break down membrane fats, and EDTA, which attaches to certain metal ions that could help enzymes that cut DNA. Thereafter, the sample is kept at 55–65 °C for 1 to 3 hours, with gentle shaking to make sure that all of the tissue fragments are evenly exposed (Adema, 2021). At the end of this phase, many protocols call for extra incubation steps or even a short heating cycle at around 98–99 °C to denature the enzymes and make sure that all the cells are broken down (Ali et al., 2017).

The combination of mechanical disruption and enzymatic digestion ensures complete breakdown of even the toughest tissues, such as those with chitinous exoskeletons. This step is essential because if lysis isn't complete, it could make DNA recovery less efficient and amplification later on less effective (Greenslit, 2018).

8.3. Module 2: CTAB-Based Precipitation

The next step in the protocol uses the cetyltrimethylammonium bromide (CTAB) method, which is excellent at separating nucleic acids from other substances like polysaccharides, proteins, and polyphenols. The first step is to add the same amount of a preheated CTAB extraction buffer to the lysate. This CTAB buffer typically contains 2% CTAB, 1.4 M NaCl, 100 mM Tris-HCl (pH 8.0), 20 mM EDTA, and a fresh addition of β-mercaptoethanol or another reducing agent to help prevent damage from oxidation and to bind polyphenolic compounds (Adema, 2021).

Next, thoroughly mix the sample and let it sit at 60–65 °C for 30 minutes to an hour. During this time, the CTAB interacts with nucleic acids and makes complexes that effectively separate DNA from other substances. At this point, organic solvent extractions are often done with chloroform:isoamyl alcohol in a (24:1) ratio to get rid of proteins and lipids and make the aqueous phase even cleaner (Ali et al., 2017). You get a clear aqueous supernatant with the nucleic acid complexes after spinning it at a high speed (10,000–12,000 g) for 10–15 minutes. This step of purification is crucial because CTAB precipitation (Malik et al., 2021) selectively removes many inhibitory compounds, like polysaccharides that are common in plant-based contaminants that bees eat.

Now that most of the nucleic acids are in the CTAB–nucleic acid complexes, DNA is usually re-precipitated from the water phase by adding isopropanol or ethanol and then spinning it to make a pellet. Washing the pellet with 70% ethanol ensures the complete removal of any dissolved salts and leftover contaminants. The precipitation process leaves behind a DNA pellet that is mostly free of inhibitors but may still have small amounts of salts, leftover CTAB, or other compounds that can slow down the process. This implies the necessity for a final cleanup step (de la Cruz-Ramos et al., 2019).

8.4. Module 3: Chelex-Based Cleanup for Optimal Purity

The last step in the cleanup process is to use Chelex resin, which is a chelating resin that binds divalent metal ions like Mg2+ and Ca2+. If these ions are left in the sample, they could make nuclease activity stronger or stop downstream polymerase reactions like PCR (A. Evans, 2022). The DNA pellet that was dissolved after the CTAB precipitation step is cleaned further by adding a mix of 5–10% Chelex 100 resin in water or a low-salt buffer that doesn't contain any nucleases. This step typically requires heating the DNA-Chelex mixture to 95–100 °C for 10–15 minutes while shaking it occasionally to help the unwanted ions attach to the resin (A. Evans, 2022).

After boiling, the sample is spun at a high speed for example, (10,000 g for 5 minutes) to pellet the Chelex resin and any contaminants that were attached to it. The purified DNA stays in the supernatant. This last step of purification makes the DNA much better by getting rid of any chelators and inhibitors that could mess up sensitive downstream applications (A. Evans, 2022). Chelex is especially useful when working with small amounts of samples, like bee tissues (Felczykowska et al., 2015), because it doesn't need organic solvents and reduces the amount of sample lost during multiple transfers.

8.5. Integration Benefits and Considerations

The method outlined integrates the optimal aspects of enzymatic digestion, CTAB precipitation, and Chelex cleanup, compensating for their respective deficiencies. Enzymatic digestion ensures the complete breakdown of cellular material, hence releasing all nucleic acids. The subsequent step, CTAB, selectively isolates DNA from several contaminants, particularly detrimental for samples containing high levels of polysaccharides or polyphenols. Ultimately, Chelex-based purification enhances the sample's integrity by sequestering residual metal ions and PCR inhibitors from preceding stages. This technique enhances yield and purity while employing minimum sample material. This process is crucial when handling diminutive and nonrenewable bee tissue specimens (Greenslit, 2018).

8.6. Below is a step-by-step summary of the integrated protocol, providing specific details for each phase

Step 1: Sample Collection and Pre-Treatment

- Please obtain a small piece of bee tissue, such as a 2 mm leg clip, and minimize handling as much as possible. Then, store it right away at −20 °C or in a preservation solution.
- To prevent ethanol from interfering with storage, dry the tissue a slightly if needed. It is best to dry as little as possible to keep the most nucleic acid (Greenslit, 2018).

Step 2: Mechanical Disruption and Enzymatic Lysis

- Use a sterile scalpel, a pestle, or a bead beater to break up the tissue in a small amount of lysis buffer that has (100 mM Tris-HClpH 8.0), 20 mM EDTA, 1% SDS, and Proteinase K at a concentration of about 20 mg/ml.
- Keep the homogenized sample at 55–65 °C for 1–3 hours to make sure that all of the cellular and nuclear membranes are broken down and the enzymatic digestion is finished. Gentle vortexing every so often might help the enzymes get deeper into the solution (Jangra & Ghosh, 2022).

Step 3: CTAB Extraction

- Heat the CTAB extraction buffer, which is made up of 2% CTAB, 1.4 M NaCl, 100 mM Tris-HCl pH 8.0, 20 mM EDTA, and β-mercaptoethanol that has just been added, to 60–65 °C.
- Combine the digested lysate with the same amount of CTAB buffer and let it sit at 60–65 °C for 30–60 minutes. This will help CTAB form complexes with nucleic acids and help to precipitate polysaccharides and polyphenols that get in the way.
- Add the same amount of chloroform:isoamyl alcohol (24:1) to the mixture. Invert the tubes a few times to mix them well, and then centrifuge them at 10,000–12,000 g for 10–15 minutes to separate the phases.
- Cautiously move the DNA-containing upper aqueous phase to a new tube (Ali et al., 2017).

Step 4: DNA Precipitation and Washing

- To make the DNA fall out of solution, mix 0.6–1 volume of ice-cold isopropanol or ethanol with 0.1 volume of 3 M sodium acetate or ammonium acetate.
- Gently mix the sample by turning it upside down and letting it sit at −20 °C for at least 30 minutes or overnight for the best precipitation. Then, spin it at 10,000–12,000g for 10–15 minutes to pellet the DNA.
- Use 70% ethanol to wash the DNA pellet to get rid of any leftover salts and CTAB. Then, let the pellet air dry for a short time to get rid of the ethanol without over-drying it, which can make it difficult to resuspend (de la Cruz-Ramos et al., 2019).

Step 5: Resuspension and Chelex Cleanup

- Put the DNA pellet back in a small amount (20–50 µl) of water or buffer that doesn't contain nucleases.
- Make a 5–10% Chelex 100 resin suspension in water that doesn't have any nucleases in it. Then, add the same amount of this suspension to the DNA that has been resuspended.
- Heat the mixture to 95–100 °C for 10–15 minutes, stirring it every so often to make sure that all of the divalent metal ions and any leftover inhibitors are completely chelated.

- To pellet the Chelex resin, spin the sample at 10,000 g for 5 minutes. Then, carefully move the clear supernatant with the purified DNA to a new tube. This last DNA solution should no longer have any PCR inhibitors in it and should be good for use in later steps (A. Evans, 2022).

Step 6: Quality and Quantity Assessment

- Use a spectrophotometer or fluorometer to verify the DNA's concentration and purity. The ideal 260/280 ratio should be around 1.8, and the 260/230 ratio should be high to show that there is little contamination.
- You can also run an aliquot on an agarose gel to check for integrity and make sure there isn't any major degradation or inhibition (A. Evans, 2022).

8.7. Optimization Considerations

Depending on the specific properties of the bee tissue and the needs of the downstream application, it may be necessary to optimize several parameters. For instance, you could adjust the temperature and length of the enzymatic digestion step to suit various tissue preservation methods (Greenslit, 2018). In the same way, the amount of CTAB and salt in the extraction buffer can be adjusted to find the right balance between getting rid of contaminants and keeping nucleic acid yield (Adema, 2021). Changing the ratio of isopropanol to the aqueous phase during precipitation may also be very important, especially when working with tiny sample volumes where maximum recovery is significant (Ali et al., 2017).

Another area that needs to be improved is the Chelex cleanup step. Chelex works well to get rid of divalent metal ions, but the incubation time and resin concentration should be carefully set so that DNA isn't lost due to non-specific binding. To find the best conditions for the highest PCR sensitivity and reproducibility, it is a beneficial idea to do test extractions with different Chelex concentrations and incubation times (A. Evans, 2022).

8.8. Pros and Cons

One of the primary advantages of combining these methods is the ability to conserve sample material while still obtaining a high yield and integrity of DNA. The breakdown of even the most resilient bee tissues is guaranteed by the enzymatic digestion phase. This stage is essential when there are only tiny quantities of tissue available. The CTAB-based precipitation method selectively bonds and separates nucleic acids, thereby eliminating a diverse array of contaminants that may disrupt downstream processes such as PCR and sequencing. The final stage in the purification process is the Chelex-based cleanup. It removes any inhibitors that may still be present and reduce the effectiveness of enzymes in subsequent reactions by chelating metal ions (Greenslit, 2018).

The trade-offs with this combined protocol mostly have to do with the fact that it has more steps for handling and incubation than simpler, single-method approaches. Every extra step raises the risk of losing a sample and may make the whole process take longer. But when working with rare and valuable samples, like those from endangered bee species, these trade-offs are okay because the benefits of getting high-quality DNA that is free of inhibitors are greater than the small increase in processing difficulty (Adema, 2021; Jangra & Ghosh, 2022).

8.9. Troubleshooting and Advice

- If the DNA yield seems low, make sure that the tissue has been thoroughly mixed and that the enzymatic digestion has been given enough time at the right temperature. Incomplete lysis often causes low yield (Greenslit, 2018).
  If the 260/280 ratio is much lower than 1.8, the result could mean that there is protein contamination. You might want to increase the concentration or incubation time of Proteinase K and add another chloroform:isoamyl alcohol extraction step (Ali et al., 2017).
- If the 260/230 ratio indicates that there are still sugars or plant compounds in the sample, you may want to improve the CTAB buffer by increasing the salt level or extending the incubation time (Adema, 2021).

- You need to be cautious with the Chelex step. If you use too much Chelex resin or let the DNA sit at a high temperature for too long, it could accidentally bind to it. Running a pilot experiment with different amounts of resin can help improve this phase (A. Evans, 2022).
- If PCR amplification keeps failing even though the purity looks fine, you may need to dilute the DNA several times to eliminate any leftover inhibitors (Greenslit, 2018).

8.10. Scaling the Protocol

The integrated protocol can be changed to a 96-well format for labs that need to process more than one bee tissue sample at the same time, like when they are doing population genetics studies. To make this work, each step needs to be smaller so that the volumes are lower, and multichannel pipettes need to be used to ensure that reagents are added in the same way every time. To ensure that each sample receives the same amount of enzymatic digestion buffer and CTAB solution, researchers need to spread the tissue evenly in the wells. The Chelex step can also be changed to fit a plate format, where the resin is added, the plate is heated with a heating block, and a plate centrifuge is used to spin it. Changing these steps to a high-throughput format can save a lot of time and make sure that the valuable bee tissue is processed the same way in all samples (Greenslit, 2018; Jangra & Ghosh, 2022).

8.11. Uses and Applications

This integrated DNA extraction method is particularly effective for forensic bee conservation studies, population genetics, and next-generation sequencing (NGS) in genomic research, where the quantity and quality of DNA are crucial. Combining enzymatic digestion, CTAB precipitation, and Chelex cleanup results in a very pure product. The purity ensures that leftover contaminants, a major problem in PCR and sequencing reactions, won't affect downstream applications. In forensic situations where bee tissues may be the only thing available, this protocol makes the most of each sample and makes molecular identifications more reliable (Ali et al., 2017; Greenslit, 2018).

Using this method to prepare NGS libraries usually leads to a higher quality sequencing library with fewer unwanted pieces and blockers, which enhances the overall quality of the genomic data produced. This is especially important when changes are made to keep small mammal or insect samples safe when there aren't enough of them to start with (Jangra & Ghosh, 2022).

8.12. Final Thoughts

In short, the best way to get high-quality DNA from small bee tissues is to use an integrated method that includes enzymatic digestion, CTAB-based precipitation, and Chelex-based cleanup. The protocol starts with a gentle but effective method of breaking up tissues and using enzymes to break down cell walls and membranes completely. Next, it goes to a CTAB precipitation step that is designed to get rid of tough contaminants like polysaccharides and polyphenols that could cause problems with later steps. Finally, we use a Chelex-based cleanup to eliminate any remaining metal ions and enhance the purity of the DNA. To keep the tiny and valuable bee samples (Greenslit, 2018; Jangra & Ghosh, 2022), safe, each step is carefully optimized for temperature, incubation time, and reagent concentration.

Some methods, like commercial kits or strictly chemical-based ones, may be easier or faster, but they often don't work as well when it comes to getting good yields and purity from hard-to-work-with samples like bee tissues. This hybrid protocol combines the best parts of the enzymatic, CTAB, and Chelex methods to create a balanced solution that preserves rare biological materials while getting the best quality DNA. This integrated approach is a strong and flexible method for researchers in conservation, forensic science, and genomics (Ali et al., 2017; Greenslit, 2018), if it is carefully optimized, validated, and adapted to high-throughput formats.

This model protocol can be changed even more to meet the needs of the user, thanks to the evidence and detailed methods found in the literature. Researchers who use bee tissues should pay close attention to the size of the first sample, how evenly the tissue is broken up, and when each step should happen. Pilot experiments could be used to fine-tune each parameter even more to make sure that the final DNA preparation is of the best quality possible for uses like species barcoding and advanced genomic sequencing (Adema, 2021; Jangra & Ghosh, 2022).

This integrated method can be used as a starting point for creating next-generation protocols that not only increase yield and purity but also make it easier to reproduce and scale in different lab settings by systematically dealing with the problems that arise when extracting DNA from small, inhibitor-rich samples. The careful mix of enzyme, chemical, and resin purification steps makes sure that even when there is very little starting material, every effort is made to keep and enhance the detectable nucleic acid content, which allows for dependable molecular analyses that are important in conservation and ecological research (Ali et al., 2017; Greenslit, 2018).

In summary, using a single DNA extraction process that combines enzymatic digestion, CTAB-based precipitation, and Chelex-based cleanup is an effective way to obtain very pure DNA from small bee tissue samples. This method solves the problems of conserving limited natural resources while still providing high-quality nucleic acids for strong downstream applications (Ali et al., 2017; Greenslit, 2018).

Chapter 9

9. Automating Traditional Nucleic Acid Extraction: Integrating CTAB and Enzymatic Methods for High-Throughput Consistency

Automation in nucleic acid extraction has become a critical area of active development for high-throughput applications. This is because labs all over the world need standardized workflows that deliver consistent and strong results. Researchers are revisiting older techniques like cetyltrimethylammonium bromide (CTAB) extraction and enzymatic lysis to find ways to use them in automated systems, which would minimize differences caused by human handling and allow more samples to be processed at once (Benčič et al., 2024). These methods have long been valued for their cost-effectiveness and ability to handle complicated clinical and environmental matrices. The traditional methods of extracting CTAB and enzymes by hand rely on several steps that must be done in order, such as breaking down cells, digesting proteins, separating phases, precipitating, washing, and resuspending. Changes in incubation times, pipetting volumes, and the way technicians do their jobs can cause differences in the amount and quality of DNA produced from one batch to the next. Such variations can affect downstream applications like high-throughput sequencing and quantitative PCR (Dalla-Costa et al., 2017). To deal with these problems, robotic extraction systems are being made to automate these time-consuming steps so that every step of the extraction process, from homogenizing the sample to the final elution, is done with a level of consistency that can't be achieved by hand (Bogožalec Košir et al., 2023).

Robotic extraction platforms can help make important steps in CTAB-based and enzymatic methods more consistent by precisely managing the amounts of chemicals used, the incubation times, and the mixing settings. Combining these traditional methods with automated liquid handling and robotic arms significantly reduces the variability resulting from human actions. An automated system can ensure that all samples get the same reagents and are incubated at the same temps. Such an approach makes extraction processes more repeatable and reproducible across large groups of samples (Benčič et al., 2024). Such accuracy is especially important when working with clinically diverse samples like sputum or blood, where the presence of inhibitors and the complexity of the matrix can make any mistakes in the procedure even worse.

In this case, adding CTAB-based protocols to standardized robotic platforms could speed up workflows and improve the quality of the isolated DNA, both in terms of yield and integrity (Benčič et al., 2024).

Many labs have already shown that automated extraction is possible with commercial kit systems that use magnetic bead-based chemistries. These systems use closed, cartridge-based workflows to break down samples, bind nucleic acids to paramagnetic beads, wash them, and elute them all under very strict conditions (Wallinger et al., 2017). These kit-based methods usually produce more consistent results and less variation between operators than traditional manual methods. Furthermore, combining robotic platforms with kit-based methods lets you process many samples at once, which is very important for high-throughput diagnostics and research (Bogožalec Košir et al., 2023). Automated systems like the GXT NA/Arrow kit or the Quick Pick genomic DNA extraction kit make results more reliable because they help ensure that the amount of nucleic acid produced is more consistent and reduce the chances of DNA damage that can happen during manual CTAB or enzymatic extractions (Benčič et al., 2024; Bogožalec Košir et al., 2023).

Along with the move toward automation to improve throughput and reproducibility, recent studies have also stressed the need for real-time quality checks during the extraction process. Spectrophotometers are being adapted for real-time monitoring of nucleic acid purity and concentration, which lets you get immediate feedback on how well the extraction is going (Kuang et al., 2018; Raju et al., 2024). These tools evaluate absorbance ratios like A260/A280 and A260/A230, which quickly indicate if there is protein contamination, organic solvents, or other impurities that might affect molecular tests later. When combined with automated platforms, spectrophotometric monitoring establishes an adaptive feedback loop capable of instantly altering extraction parameters, such as reagent volumes or incubation times, if real-time data indicates that the extraction is not performing optimally (Benčič et al., 2024; Kuang et al., 2018). This feature not only cuts down on wasted samples, but it also makes sure that the nucleic acids obtained are always of the same high quality. This further cuts down on the differences between people that come up in manual protocols.

When you consider high-throughput applications, the importance of these changes becomes even clearer. For example, small differences in extraction performance can make a big difference in the sensitivity and specificity of assays used for clinical diagnostics, like those that look for antimicrobial resistance in sputum samples or viral RNA in COVID-19 diagnostics (Dalla-Costa et al., 2017; Joseph et al., 2022). Automated extraction systems assure that a dataset is more reliable and comparable by reducing the overall effect of small changes made at each manual step when hundreds or thousands of samples need to be processed. Robotic platforms not only make the extraction process more consistent, but they also let labs do more work without making more mistakes or having more operator variability (Beall et al., 2019; Bogožalec Košir et al., 2023).

CTAB-based methods usually involve several manual steps that are very sensitive to small changes in timing or temperature. These methods also require meticulous pipetting and manual centrifugation to prevent contamination and ensure complete phase separation. With automation, a robotic system can do each step with a high level of accuracy. For example, it can keep the incubation times at 65 °C during cell lysis or make sure that reagents are moved between wells or tubes exactly as they should be. This level of accuracy leads to more consistent extraction efficiencies and less variation in DNA yield and fragment integrity across different sample types. Such precision is important for the reproducibility of later analyses like targeted high-throughput sequencing (Benčič et al., 2024; Wallinger et al., 2017). Automating these steps has two main effects: it makes things more standardized, and it makes it possible to get almost the same extraction results regardless of who is doing it or where the lab is located.

Automation is also helping enzymatic extraction protocols, which use enzymes like lysozyme or proteinase K to break down cell walls and proteins. For these protocols to function effectively, they require precise timing of enzyme incubation, optimal mixing, and optimal temperature conditions. Robotic systems can coordinate these parameters in a single workflow, making sure that enzymatic reactions happen the same way on all samples (Benčič et al., 2024; Dalla-Costa et al., 2017). When different technicians use slightly different enzyme concentrations or incubation periods, the results can be different. By automating the process, labs can avoid this. Additionally, robotic platforms can ensure that potential inhibitors are consistently removed by using standardized enzymatic extraction routines. This makes downstream applications like digital PCR and next-generation sequencing more reliable (Bogožalec Košir et al., 2023; Dalla-Costa et al., 2017).

Automation enhances not only the conventional CTAB-based and enzymatic protocols, but also numerous commercial kits designed specifically for robotic systems. For example, magnetic bead-based DNA and RNA extraction kits have become popular because they work well with automated liquid handlers and magnetic separation systems. These kits make the extraction process easier by getting rid of the need for centrifugation and manual transfers between containers. Instead, they capture the nucleic acids directly on paramagnetic beads during lysis and then move them through a series of automated wash and elution steps (Beall et al., 2019; Wallinger et al., 2017). For high-throughput applications, where being able to process tens or hundreds of samples at once can greatly speed up turn around times and lower overall labor costs while keeping extraction uniformity high (Benčič et al., 2024; Bogožalec Košir et al., 2023), the consistency these platforms offer is very useful.

Combining real-time spectrophotometric monitoring with automated extraction systems is one of the most promising new developments in the field. Normally, people use spectrophotometers or fluorimeters to check the quality and concentration of nucleic acids after they have been eluted. However, putting these quality checks directly into the extraction pipeline is a big step forward in process control (Kuang et al., 2018; Quero et al., 2024). Real-time monitoring makes it possible to find things like suboptimal absorbance ratios right away. Such a ratio is a clear sign that the extraction process may need to be changed or that some contaminants are present. This type of tracking can be very advantageous in protocols that are sensitive to the presence of fragmented DNA or enzymatic inhibitors. A low A260/A230 ratio might mean that extraction buffers or wash steps aren't working as they should (Kuang et al., 2018).

For instance, if the automated system sees an absorbance ratio that is too high or too low, it might trigger an adaptive response like lengthening the washing step, changing the elution volume, or even notifying an operator in real time so that corrective action can be taken right away. This ability to adapt is very important for working with samples that are not all the same, since different tissues or biofluids can have completely unique amounts of proteins, polysaccharides, and other contaminants (García-Giménez et al., 2017). The ability to change on the fly based on real-time quality metrics not only improves the quality of the extracted nucleic acids, but it also makes the protocol more robust overall and cuts down on the need for expensive re-extractions. This saves valuable sample material and speeds up the overall processing time (Kuang et al., 2018; Raju et al., 2024).

Robotic extraction systems have more benefits than just better consistency and throughput. In high-throughput diagnostic labs, where hundreds of samples have to be processed every day, even small changes made by hand can cause significant differences in the results of the tests. For example, the sensitivity and specificity of downstream tests depend a lot on the quality of the nucleic acids that were extracted. This is true for tests that find pathogens or genotyping assays. Automated systems ensure uniform treatment of every sample, including mix times, reagent exposure, and incubation periods (Beall et al., 2019; Dalla-Costa et al., 2017). This reduces the amount of variation between people. This lower variability is important when diagnostic decisions are based on threshold values or when comparing results from different patient samples. It ensures that biological variability, not sample processing issues, is the cause of differences in assay results.

Robotic systems also lower the risk of cross-contamination, which is a common worry with manual extraction protocols, especially when working with many samples. Robotic systems work on closed platforms with little human interaction, which makes it much less likely that samples will spill or aerosolize, which can lead to contamination (Bogožalec Košir et al., 2023; Wallinger et al., 2017). These systems often incorporate built-in decontamination routines and adhere to strict sample handling sequencing. Such functionality helps keep nucleic acid preparations safe during high-throughput workflows.

The ability to combine different extraction chemistries into one platform makes automated systems even more flexible. For example, you could change the program settings and reagent cartridges on a robotic platform so that it could do both magnetic bead-based extractions and modified CTAB- or enzymatic-based protocols. Because of this flexibility, a single automated tool could be used in many different types of labs, such as those that do clinical diagnostics, research studies with environmental samples, agricultural tissues, or even forensic evidence (Benčič et al., 2024; Dalla-Costa et al., 2017). As high-throughput labs improve, changing extraction protocols for different types of samples without losing consistency becomes increasingly important.

Automated extraction workflows not only have technological benefits, but they also have big operational benefits. Laboratories that use robotic systems can often cut down on turnaround time by a lot, which frees up skilled workers to do other important tasks and lowers the cost per sample processed. This feature is especially important in places where resources are limited or when infectious diseases are spreading quickly, since quick and consistent nucleic acid extraction is necessary for making timely diagnosis and treatment decisions (R. K. Jain et al., 2024; Joseph et al., 2022). Automated workflows also let you monitor and control things from a distance using networked interfaces. Such versatility means that operators don't have to be there all the time, and it makes it possible to manage quality across several labs from one place.

Also, being able to expand these automated extraction platforms helps meet the increasing demand for processing a lot of samples quickly in today's genomic and transcriptomic research. It is important for large-scale epidemiological studies, population genomics projects, and extensive clinical trials to be able to process thousands of samples with little variation. In these situations, automated extraction systems give us the repeatability we need to make sure that data comparability stays strong across many samples processed over a long time (Beall et al., 2019; Bogožalec Košir et al., 2023).

Even though there are clear benefits, it is still difficult to fully automate extraction protocols for all sample types and extraction chemistries. For instance Adapting traditional CTAB-based techniques, for example, necessitates considerable adjustments to the way laboratories typically operate. We must carefully optimize reagent formulations, incubation times, and the physical mechanics of mixing and separation to transform these manual tasks into automated routines. Recent research, on the other hand, has shown that these "open" extraction methods can be automated successfully while keeping the cost benefits that made them appealing in the first place (Benčič et al., 2024; Dalla-Costa et al., 2017).

Researchers are actively looking into how to use microfluidic devices and custom-built robotic systems to maintain the delicate balance of chemical interactions needed for effective CTAB-based extraction. Ultimately, these efforts will lead to a future where choosing a method for extraction is more focused on the specific characteristics of the sample and target nucleic acid rather than just how quickly it can be done (Nikulin et al., 2024).

Another new area of research is how to combine machine learning and artificial intelligence (AI) with automated extraction processes. AI-driven algorithms can adjust extraction settings in real-time by analyzing data like absorbance ratios and fluorescence signals. These kinds of systems can use past extraction data to figure out when a protocol might not be meeting its performance goals and automatically change the length of steps, the volume of reagents, or even start additional washing cycles to get the best nucleic acid purity and yield (Kuang et al., 2018; Raju et al., 2024). This smart feedback loop makes it possible to adapt to an unprecedented level, allowing extraction protocols to be fine-tuned to the specific characteristics of a sample or the desired downstream applications. This process consistently reduces variability and improves data quality. The combination of robotic automation, real-time quality monitoring, and AI-driven control is a big step toward self-optimizing laboratory workflows.

These advances have effects that go far beyond the lab bench. Better reproducibility and less variability in nucleic acid extraction mean better outcomes for patients in clinical settings. Highly reliable extraction protocols make sure that diagnostic tests for infectious diseases, cancer biomarkers, or genetic disorders are based on high-quality input material. Such consistency makes the tests more sensitive, specific, and accurate overall (Dalla-Costa et al., 2017; Joseph et al., 2022). In research settings, processing many samples with very little human error makes studies more statistically powerful and helps scientists find small biological phenomena that technical noise might otherwise hide. This capability is especially important in areas like population genomics and metagenomics, where getting nucleic acids out consistently and with high fidelity is essential for accurately capturing biological diversity and the structure of microbial communities (Benčič et al., 2024; Bogožalec Košir et al., 2023).

Furthermore, the creation of automated extraction protocols has big effects on the economy. Laboratories can save money on labor and work more efficiently by cutting down on the time it takes to isolate nucleic acids. The fact that automated systems are consistent and can be used again and again means that there is less of a chance of having to do expensive re-runs or repeat extractions. This lowers the costs of large-scale screening programs even more. This benefit is especially important in places where resources are limited, like where there isn't enough money or people to do all the tests that are needed (R. K. Jain et al., 2024; Joseph et al., 2022).

Automated extraction is likely to become even more connected to downstream analytical platforms in the future. In a fully automated diagnostic workflow, receiving samples, extracting nucleic acids, checking quality, amplifying, and analyzing data all happen in one smooth process. Robotic extraction systems are crucial in these pipelines because they make sure that high-quality nucleic acids are always sent to amplification platforms like digital PCR or next-generation sequencing systems (Beall et al., 2019; Bogožalec Košir et al., 2023). If robotic handling and real-time quality assessment get better, these pipelines could become "plug-and-play" systems that don't need much operator input and can be set up quickly to meet new public health needs or research priorities.

In summary, automating nucleic acid extraction protocols significantly advances high-throughput laboratory applications. Robotic systems standardize every step of sample processing and cut down on the differences between people by turning traditional manual methods, like CTAB-based and enzymatic extractions, into automated processes (Benčič et al., 2024; Dalla-Costa et al., 2017). Combining automated, kit-based platforms with magnetic bead chemistries makes it even easier to reproduce results and process more samples at once. Real-time quality monitoring with spectrophotometric techniques also lets you change the protocol as needed (Kuang et al., 2018; Quero et al., 2024). These changes not only make the workflow in diagnostic and research labs more efficient, but they also reduce the variability that has made it difficult to understand nucleic acid-based assays for a long time. Ultimately, the integration of automation, real-time quality feedback, and intelligent data-driven process control will fundamentally transform nucleic acid extraction, enhancing the reliability, scalability, and cost-effectiveness of high-throughput molecular diagnostics (Nikulin et al., 2024; Raju et al., 2024). In the future, ongoing research is expected to make these automated systems even better by combining them with cutting-edge microfluidic devices and AI algorithms to make them even more consistent and accurate. As these technologies get better and more labs start using them, they will be able to use standardized nucleic acid extraction protocols that give the same results no matter how complicated the matrix is or who is doing the extraction. The next generation of robotic extraction systems will probably be more adaptable, which will make it easier to quickly optimize protocols to deal with the specific problems that different sample types or clinical situations present. This will lead to more accurate molecular diagnostics and stronger genomic research (Offornedo et al., 2022).

In conclusion, the field of automating extraction protocols is growing quickly and has a lot of potential for high-throughput uses. Researchers and doctors can get much more reliable and repeatable results across many samples by using robotic extraction systems to standardize complicated manual processes like CTAB-based and enzymatic methods. Adding real-time spectrophotometric quality checks to these automated workflows makes them even more flexible and useful. This process makes sure that nucleic acid preparations meet the strict standards of downstream molecular assays. These improvements set the stage for a future in which molecular diagnostics and genomic research are not only more efficient and cost-effective but also have a level of precision and standardization that has never been seen before (Benčič et al., 2024; Dalla-Costa et al., 2017).

Overall, combining robotic platforms with traditional extraction methods and real-time quality control is a big step toward getting around the inherent variability of manual nucleic acid isolation. Extensive studies comparing kit-based methods and robotic extraction platforms have shown that switching from traditional, labor-intensive workflows to automated systems cuts down on human error and offers a scalable solution that is necessary for using high-throughput methods in animal diagnostics, environmental screening, and large-scale genomic studies (Beall et al., 2019; Bogožalec Košir et al., 2023). We anticipate further advancements in this field to enhance extraction uniformity and boost operational efficiency. Such advances will make sure that molecular data from different labs can be compared directly and that diagnostic results are delivered quickly and with high confidence.

In the end, the ongoing work to automate nucleic acid extraction protocols means that labs will have to change the way they prepare samples. These standardized, high-throughput processes not only simplify the replication of results and reduce variability, but also pave the way for intelligent, integrated laboratory systems that will shape the future of molecular diagnostics and research. The next generation of extraction systems will be able to self-optimize and adapt to different sample conditions as real-time monitoring technologies continue to improve and be used together. These innovations will make sure that every step, from extraction to analysis, is done with the highest level of accuracy and consistency (R. K. Jain et al., 2024; Joseph et al., 2022).

In short, automating nucleic acid extraction is a crucial new development for high-throughput applications, whether it's done with CTAB-based methods, enzymatic protocols, or modern kit-based methods. Robotic extraction systems improve the consistency and reliability of samples by removing the variability that comes with manual processes and adding real-time quality controls. This leads to more accurate and robust molecular diagnostics. This field of research is constantly changing and promises to produce extraction workflows that are more efficient, scalable, and adaptable to meet the needs of modern biomedical and genomic research (Benčič et al., 2024; Raju et al., 2024).

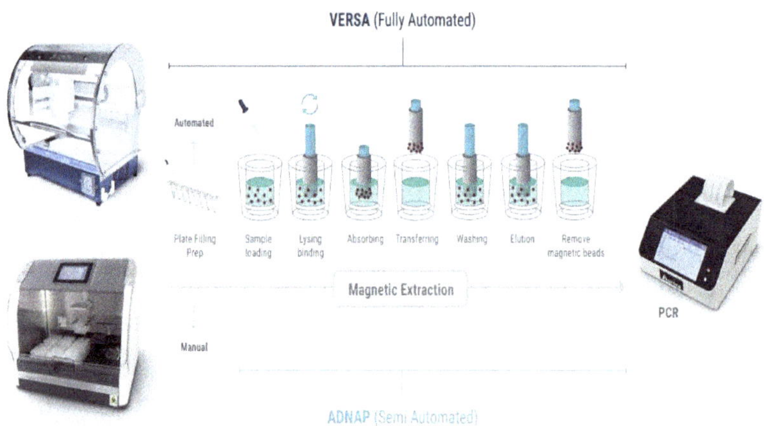

Fig. 12. VERSA™

96 purifications in 1.5 hours <15 min hands-on prep, Installation-and-play setup adjustable/modifiable protocol libraries

ADNAP:

96 purification in 30 minutes <30 min hands-on prep, Plug-and-play setup adjustable protocol libraries

Automated Purification Kits & Consumables

Aurora's extraction systems, including the VERSA™ and ADNAP series, utilize magnetic bead-based RNA/DNA extraction, providing efficient and high-yield results. Aurora offers magnetic bead nucleic acid extraction kits, and various vendors use small paramagnetic beads that strongly yet reversibly bind to RNA/DNA, ensuring consistent yields and purity. Magnetic beads offer flexibility, working with various sample types such as whole blood and cells.

Both the VERSA™ and ADNAP systems are compatible with leading commercial kits, including (NucleoMag kitsMacherey-NAGEL), (MagPure kitsAngen Biotech), and (MagMAX kitsThermo Fisher Scientific), ensuring seamless integration and versatility in your lab workflow Nucleic Acid (Extraction Systems), (https://www.aurorabiomed.com/solution/nucleic-acid-extraction-systems/).

Chapter 10

10. PCR Amplification and Primer Design Strategies for Mitochondrial DNA Markers in Phylogenetic Analysis

This chapter goes into excellent detail about the polymerase chain reaction (PCR) methods used to amplify mitochondrial DNA (mtDNA) regions for phylogenetic analyses. It explains how to create and improve primers that focus on key markers like cytochrome oxidase subunit I (COI), the space between cytochrome oxidase I and II (COI–COII), cytochrome b (cytb), parts of NADH dehydrogenase, and a specific area between tRNA^Leu and COII. The discussion goes over why these areas were chosen, how to design primers that don't amplify other things, and the optimization methods used to verify that the amplicons produced deliver clear diagnostic signals and strong phylogenetic signals.

To study mtDNA, it's important to get high-quality DNA and then use PCR to make copies of it. This is a key step in targeting specific parts of the mitochondrial genome. Researchers use PCR to make millions of copies of certain mtDNA fragments once they have obtained pure, intact DNA through strict extraction and purification protocols. The process solves problems with low template concentrations and makes it possible to do sequencing and comparative analyses (F. Chen et al., 2016).

The evolutionary properties and diagnostic usefulness of the target mtDNA regions help us choose them. For example, the COI gene is well-known in DNA barcoding studies because its structure is fairly stable across species but varies between species, making it a good choice for telling closely related species apart (Grass et al., 2016). The space between COI and COII is chosen because its differences can reliably identify various evolutionary groups (Grass et al., 2016). Additionally, regions like cytochrome b (cytb) and parts of NADH dehydrogenase are important for understanding evolutionary relationships because they change at rates that are not too fast or too slow (Grass et al., 2016).

Many studies look at the section between tRNA^Leu and COII because it usually has many changes and differences that are common when species evolve separately. This area is known for having different secondary structures in the tRNA and for being useful for figuring out phylogenetic relationships when other mitochondrial protein-coding genes don't have enough variation (Grass et al., 2016).

Primer design is a key part of amplifying mtDNA. This process must consider both the need for high specificity and the requirement to obtain sufficient phylogenetic information. Researchers usually pick sequences from conserved areas around the variable markers when making these primers. The goal is to get the most amplification possible without binding to host nuclear mitochondrial pseudogenes (NUMTs) or other areas that aren't related to the target (Pacheco et al., 2018). This careful choice prevents errors from happening, which could make the copied mtDNA look different from the real mitochondrial genome of the target organism, confusing future studies on evolutionary relationships (Pacheco et al., 2018).

We usually choose primer lengths that are between 17 and 28 base pairs long and have a GC content that is between 40% and 60% to make sure that the melting temperatures (Tm) are high and the binding stays stable during the annealing phase of PCR (Grass et al., 2016). The annealing temperature during PCR is carefully set to match the Tm of the primer pairs. This step helps the primers bind to the target template more specifically and lowers the chance of primer-dimer formation, which is a major cause of nonspecific amplification (Grass et al., 2016).

After the primers are made, the PCR amplification process usually follows a standard thermal cycling protocol. This includes starting with denaturation at a high temperature and then switching between denaturation, annealing, and extension cycles. The first step of most protocols is denaturation, which happens at about 94–95 °C for long enough to melt the double-stranded DNA template. This process makes sure that the single-stranded regions that serve as binding sites for the primers are accessible (F. Chen et al., 2016). Annealing temperatures are set based on the specific primers and can range from 50 °C to 62 °C, depending on the design of the primer and the area it is meant to target. Extension steps usually happen at 72 °C, which is the best temperature for DNA polymerases like Taq and their high-fidelity derivatives that make new DNA strands (Jiang et al., 2020).

Using overlapping PCR amplicons to create complete mitochondrial genomes is a popular technique because the overlapping pieces provide extra sequence information, which improves the accuracy of putting the pieces together and fixing mistakes (J. Ma et al., 2017). Fragments usually have overlaps of 50 to several hundred base pairs. This procedure makes sure that the sequence is continuous and lets you check its accuracy by confirming that the overlapping nucleotides are correct (Yamauchi et al., 2003).

Using both universal and species-specific primers is a key part of these methods. Universal primers are made to bind to conserved areas in a wide range of taxa, which makes them useful for studies that compare different species (Grass et al., 2016). On the other hand, species-specific primers are made from reference sequences from the target organism or a closely related species. This makes them more specific for binding, less likely to cross-amplify, and more likely to work well in PCR reactions even when the DNA samples are old or small (Hafeez et al., 2015).

Furthermore, some people use degenerate bases in primer sequences to allow for small differences in the primer-binding region between species. But it is important to find the right balance when adding these unclear bases because too much degeneracy can make the test less specific and make nonspecific amplification more likely (Pacheco et al., 2018). As a result, a lot of research focuses on reducing the number of degenerate positions while making sure that the primer can still bind to a wide range of sequences. Such an approach is the best way to strike a balance between specificity and broad utility (Pacheco et al., 2018).

When designing primers for areas like COI, cytb, and NADH dehydrogenase components, it's important to be able to tell closely related species apart and to ensure that the amplified regions have enough phylogenetic signal. The evolutionary rate of a marker determines how useful it is for phylogenetics. Markers that are too conserved may not show differences between recent divergences, while markers that are too variable may gather homoplasies that make it difficult to accurately reconstruct a tree (Grass et al., 2016). So, cytochrome oxidase I and other areas like it have become standard markers in DNA barcoding because their evolutionary rate is balanced, which means they can tell species apart while still being able to align across taxa (F. Chen et al., 2016; Grass et al., 2016).

In studies that need to tell the difference between evolutionary lineages, the area between tRNA^Leu and COII is crucial. The structure of this area changes a lot, with insertions, deletions, and sometimes different secondary structures. All of these changes make it more useful for diagnosis (Grass et al., 2016). When primers are made to go around this area, they pick up not only the variable nucleotide sequence but also the structural motifs that might help figure out how populations or closely related species are related in terms of their evolutionary history. In fact, a lot of research has shown how important it is for figuring out complicated phylogenetic patterns, especially in insects and vertebrates, where mitochondrial DNA has been used a lot for taxonomy and population genetics (Grass et al., 2016).

Taking all of these variables into account, a well-optimized PCR protocol for amplifying mtDNA markers usually has several steps for optimization. During the first step, the amounts of magnesium chloride ($MgCl_2$), dNTPs, primers, and template DNA are all carefully measured. Too much template DNA can make nonspecific amplification more likely, while not enough template concentrations can make it impossible to make amplifiable product. For this reason, many protocols suggest making a series of preliminary dilutions of the extracted DNA to find the best concentration for each reaction (Pacheco et al., 2018).

The conditions for thermal cycling also need to be set precisely. Touchdown PCR is a common method for improving specificity. It starts with cycles that use an annealing temperature higher than the Tm of the primers, and then the temperature gradually drops over the next cycles. This method has been proven effective at reducing unwanted binding and primer-dimer formation, particularly when the target area is close to sequences that are very similar to non-target areas (Grass et al., 2016). The length of the expected amplicon usually determines how long the extension phase will last. Longer fragments need longer extension phases to make sure the DNA template is fully synthesized (Jiang et al., 2020).

Another crucial consideration is the use of high-fidelity DNA polymerases, particularly for phylogenetic studies involving amplified mtDNA sequences. Enzymes like Q5® high-fidelity polymerase are more accurate because they lower the number of times nucleotides are incorrectly incorporated. This procedure is important because even small mistakes can lead to wrong phylogenetic interpretations (Grass et al., 2016). These enzymes are less likely to make chimeric PCR products because they can proofread and work quickly

(Grass et al., 2016). Chimeric PCR products can happen if the enzyme leaves the template too soon during amplification.

Studies that have successfully amplified complete mitochondrial genomes using overlapping PCR approaches are excellent examples of how these methods can be used in real life. For example, one study used 17 pairs of primers from different species to create overlapping PCR fragments that covered the whole mtDNA. The technique made it easier to put together the correct sequence and cover all of the regions, including COI, the COI–COII intergenic spacer, cytb, and NADH dehydrogenase genes (F. Chen et al., 2016). Overlap strategies not only improve sequence confirmation, but they also help reduce the loss of data that can happen when some segments are not amplified as well as they could be (J. Ma et al., 2017).

In addition, mitochondrial markers have several benefits when used in phylogenetic studies. Mitochondrial genes usually change more quickly than nuclear genes. This means that mutations build up more quickly in mitochondrial genes. We can use these mutations to correct recent differences, and their conservation allows alignment across a wider range of taxonomic groups (Grass et al., 2016). This dual trait makes them especially useful for research that involves identifying species, DNA barcoding, and making inferences about evolution. Another good thing about mtDNA is that it is usually present in large amounts in cells, which makes it more likely that it will be successfully amplified even from DNA samples that are damaged or low in quantity (Grass et al., 2016).

Case studies demonstrating methods for finding parasites further illustrate the usefulness of targeted mitochondrial PCR amplification. Researchers have created species-specific primers that target mtDNA COI in studies of avian haemosporidians. Researchers can use these primers to directly identify and amplify parasites such as Plasmodium, Haemoproteus, and Leucocytozoon from mixed infections (Hafeez et al., 2015; Pacheco et al., 2018). Researchers have successfully utilized nested multiplex PCR methods that differentiate between parasite genera in a single reaction by creating primer sets that produce different amplicon sizes. Such technology has made diagnostics and later phylogenetic studies easier (Hafeez et al., 2015).

The rules for making primers in these situations work for all types of animals, from vertebrates to invertebrates. For instance, in studies of fish and insects, primers that focus on mtDNA areas, particularly around the COI–COII intergenic spacer and the tRNA^Leu-COII region, are very useful for understanding how different species and populations are connected (Offornedo et al., 2022; Straeuli, 2022). The unique secondary structures of tRNA molecules and the different lengths of intergenic regions make these systems very useful for diagnosis. They assist researchers in identifying subtle phylogenetic differences that may go unnoticed when solely focusing on protein-coding genes (Offornedo et al., 2022).

Even though PCR-based mtDNA amplification has clear benefits, there are still several problems that need to be solved to get the best results. Non-specific amplification is still a big problem, especially when primers might attach to nuclear copies of mitochondrial genes (NUMTs) or when the target region has many repetitive sequences that make the reaction more likely to form primer-dimers. To avoid these issues, strict in silico analyses are done before primer synthesis. These analyses use software tools to ensure that the primers will only attach to the right DNA, check for any unwanted structures that might form, and look at how the primer's building blocks are arranged (Pacheco et al., 2018). Primers are only made after these computer tests indicate that they will work for the intended purpose (Pacheco et al., 2018).

Using special protocols for old or damaged DNA samples is also part of optimizing PCR conditions. These protocols favor the amplification of shorter amplicons usually less than (200 base pairs) to make it more likely that amplification will work. This method is especially helpful in studies where the only sources of genetic material are museum specimens or field samples that aren't excellent (Grass et al., 2016). In these situations, scientists often modify or create new primers to produce shorter overlapping pieces that still include key areas for studying evolutionary relationships, such as COI, cytb, and the tRNA Leu-COII intergenic spacer (Grass et al., 2016).

The careful planning and improvement of PCR methods have big effects on studies of phylogenetics and population genetics. When strong PCR protocols are in place, the mtDNA sequences can be put together to make big datasets that give better resolution at both the macroevolutionary and microevolutionary levels (Grass et al., 2016). Having detailed datasets that include protein-coding genes, rRNA genes, tRNA intergenic regions, and control regions makes it easier to accurately build evolutionary trees, stabilize connections, and identify evolutionary events (Grass et al., 2016). These benefits show how important it is to customize PCR methods and primer design strategies to the needs of the study, whether the goal is to identify species, create DNA barcodes, or figure out deep phylogenetic relationships.

Recent developments have made it possible to combine next-generation sequencing (NGS) technologies with single-tube long-range PCR amplification, in addition to traditional PCR methods. This combination has made it possible to amplify whole mitochondrial genomes at the same time using long overlapping fragments. We then sequence these fragments all at once to create a dataset with extensive coverage and high resolution (Yamauchi et al., 2003; R. Zhang et al., 2023). Using overlapping fragments created by long-range PCR reduces the likelihood of missing parts in the sequence and minimizes interference from NUMTs. This procedure makes the data more reliable for phylogenetic and evolutionary studies (Yamauchi et al., 2003).

Many studies on different types of organisms show that these methods work well. For example, we created overlapping long PCR pieces using special primers designed from partial sequences of COI, cytb, and srRNA to copy the whole mitochondrial genome of the swimming crab Portunus trituberculatus (Yamauchi et al., 2003). Using universal primers that target the COI gene in insect studies, along with extra, species-specific primers for areas like the tRNA Leu–COII intergenic spacer, has produced large datasets that are the basis for detailed phylogenetic trees and help figure out how closely related taxa are (Offornedo et al., 2022; Straeuli, 2022). These case studies indicate that the basic rules for designing primers and optimizing PCR are the same regardless of what organism you are trying to amplify. These rules are crucial for the success of the mtDNA amplification process as a whole.

In addition, the development of new PCR methods keeps phylogenetic studies more reliable and accurate. Researchers can now determine and separate multiple target regions within a single reaction setup thanks to the development of nested and multiplex PCR protocols. This approach saves money, time, and the risk of contamination that comes with running separate PCR assays for each target marker (Hafeez et al., 2015; Pacheco et al., 2018). Nested PCR, for example, amplifies a certain part of DNA after an initial round of amplification. This approach makes it more specific when there are low-abundance targets, like parasite mtDNA in host tissues. Multiplex PCR enhances this method by utilizing multiple primer pairs in a single reaction. This approach makes it easier to amplify different mitochondrial markers at the same time, like COI, cytb, and NADH dehydrogenase components (Hafeez et al., 2015).

When making multiplex PCR tests, it's vital to ensure that each primer pair works well under the same cycling conditions so that one target doesn't get amplified more than another. This needs a lot of real-world testing and often involves using computer programs to model how primers interact and guess the best conditions for the reaction. To get a balance that gives strong, specific amplification across all targets (Pacheco et al., 2018), changes are often made to the levels of magnesium, dNTPs, and the choice of polymerase.

Adding strict quality control steps, such as using negative controls for all PCR amplifications and regularly checking the expected amplicon size with gel electrophoresis, is another effective way to prevent unwanted amplification. These steps are essential when working with complex sample matrices where host DNA or contaminants may be present at levels that are much higher than the target mtDNA (Grass et al., 2016; Pacheco et al., 2018). Cleaning up PCR products before sequencing ensures that fake amplification products or primer-dimers don't interfere with downstream analyses. This makes the phylogenetic reconstructions that come after them more reliable.

One of the most important things that PCR has done is make it possible to amplify mitochondrial markers, which has led to a lot of research that marks taxonomic boundaries and provide us information about how evolution works and how populations are structured. Using mtDNA with strong PCR methods has changed fields like wildlife forensics, conservation genetics, and biogeography. Researchers have used mtDNA sequences to find hidden species, chart historical migration routes, and even find hybridization events within and between populations. All of this depends on the high accuracy of the PCR amplification process (F. Chen et al., 2016; Grass et al., 2016).

Looking ahead, combining PCR-based mtDNA amplification with high-throughput sequencing platforms should make it even easier for us to figure out complicated phylogenetic relationships. Deep sequencing of long-range PCR products can detect rare variations and low-frequency differences in the mitochondrial genome. Such findings can help us learn more about how mitochondria work, how they relate to disease, and how they evolved (Bayerri, 2019). Also, as bioinformatics tools improve for putting together and aligning sequences, combining complete mitochondrial genome data with many nuclear markers will enable more in-depth studies that examine both organellar and nuclear genomes (Pacheco et al., 2018).

In conclusion, PCR amplification of mtDNA regions is still a vital and reliable way to get phylogenetically informative sequences that help us learn more about how species are related to each other. The careful separation of high-quality DNA sets the stage for the next steps in PCR, where creating the right primers is essential for focusing on specific areas like COI, the COI–COII intergenic spacer, cytb, NADH dehydrogenase components, and the region between tRNA^Leu and COII that is useful for diagnosis (F. Chen et al., 2016; Grass et al., 2016). Researchers can improve the usefulness of mtDNA markers and avoid unwanted results by adjusting the reaction conditions, like changing the length of the primers, the GC content, and the annealing temperatures, and using better polymerases (Grass et al., 2016; Pacheco et al., 2018). These improved methods not only make it possible to accurately identify species and classify them phylogenetically across a wide range of taxa, but they also make it possible to do large-scale genomic studies that help us understand how evolution works and how diverse life is. The PCR methods discussed here will remain crucial as the field expands and introduces new sequencing technologies. These improvements will make sure that mtDNA stays a powerful tool in molecular phylogenetics and population genetics (Hafeez et al., 2015; Pacheco et al., 2018).

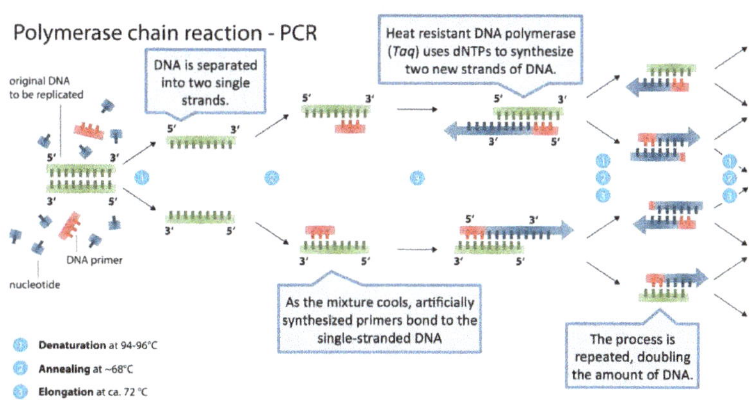

Fig. 13. Enzoklop CC BY-SA 3.0 (https://creativecommons.org/licenses/by-sa/3.0)

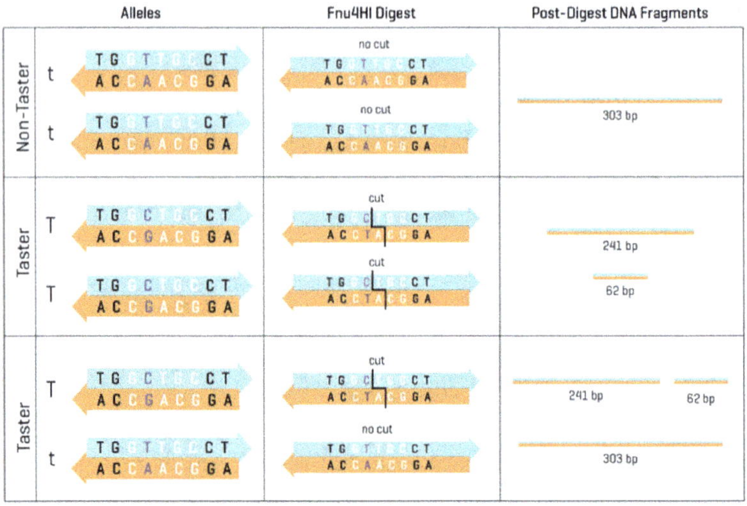

Fig. 14 Fred Hutchinson Science Education Partnership. (http://libguides.fredhutch.org/ld.php?content_id=36813729).

Chapter 11

11. Sequencing Methods for Mitochondrial DNA Analysis: Traditional Sanger Versus Next-Generation Approaches

11.1. Abstract

Mitochondrial DNA (mtDNA) analysis has been crucial for research on genetic variation, haplotype diversity, population structure, and evolutionary history. After amplifying mtDNA fragments, the next step is to sequence these amplicons to find nucleotide variants and correctly assign haplotypes. Sanger sequencing has long been the most accurate and reliable method for single-locus analyses, even though it has a lower throughput than other methods. Recent advances in technology have made next-generation sequencing (NGS) platforms like Illumina very popular. These platforms can sequence whole mitochondrial genomes quickly and easily. In this chapter, we look at the main sequencing methods used after mtDNA amplification and compare and contrast the old Sanger method with newer NGS methods. We will discuss the pros and cons of each method and how high-throughput sequencing has changed mtDNA research by giving us a more profound understanding of the structure and evolutionary history of mitochondrial populations.

11.2. Introduction

For a long time, studying mitochondrial DNA has been a key part of genetic research in fields like forensic science and evolutionary biology. mtDNA is a valuable marker for measuring nucleotide variation and haplotype diversity because it is passed down from mother to child, doesn't recombine, and has a relatively high mutation rate. After using polymerase chain reaction (PCR) or long-range PCR to make copies of specific mtDNA regions, sequencing is the most important step for understanding the genetic information in these areas. Sanger sequencing has been widely used in the past because it gives accurate sequence data for individual loci, making it useful for confirmatory analyses and small-scale studies (Carroll et al., 2014). As the need for in-depth analysis of whole mitochondrial genomes grew, next-generation sequencing (NGS) platforms slowly took over Sanger methods for everyday use. NGS platforms like Illumina let you look at the whole genome quickly and cheaply while also giving you the detailed information needed to accurately measure levels of heteroplasmy, rare variant alleles, and small differences in haplotypes (R. Chen et al., 2021). This chaper goes into excellent detail about these sequencing methods, starting with the traditional ones and then moving on to the new ones made possible by NGS. These new methods have entirely changed how we can collect large amounts of mitochondrial sequence data for population and evolutionary studies (Legati et al., 2021).

11.3. Traditional Sanger Sequencing for mtDNA Analysis

11.3.1. Fundamental Principles and Workflow

Sanger sequencing, which was created in the 1970s, is still one of the most reliable ways to sequence DNA because of its chain-termination method. This method uses dideoxynucleotides (ddNTPs) during DNA synthesis to stop extension at certain nucleotide positions. This procedure makes sequence-specific fragments that can be separated by capillary electrophoresis (Lyons et al., 2013). After copying mtDNA, specific parts or nearby areas such as the hyper variable regions in the control region display changes in nucleotides at the single base level (Carroll et al., 2014).

11.3.2. Pros and Cons

The main benefits of Sanger sequencing for mtDNA studies are that it has well-known methods, is very accurate, and can make long, continuous reads. These qualities make them ideal for confirmatory analyses where validating the sequence is crucial (Dames et al., 2013). Sanger sequencing is reliable, but it has some drawbacks when used for large-scale or whole mitochondrial genome analyses. These drawbacks include low throughput, high cost per base, and the significant effort required to perform the analysis. This method also usually only looks at one gene region or the mtDNA control region, which means it only gives a partial picture of the overall diversity of the mitochondrial sequence (Desalle et al., 2017).

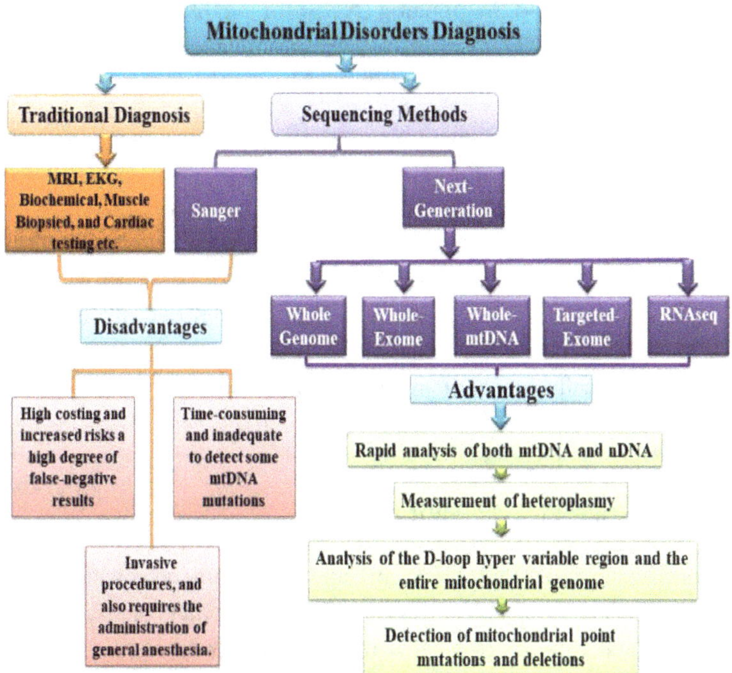

Diagrammatic representations of traditional and next-generation sequencing for identifying mitochondrial disorders (Mahmud et al., 2022)

11.3.4. Applications in Forensic and Clinical Contexts

Forensic applications have traditionally relied on Sanger sequencing, which is important for making high-quality mtDNA profiles for identifying individuals and tracing maternal lineages. It is excellent at finding known single nucleotide polymorphisms (SNPs) and mutations, which makes sure that haplotype assignments stay strong (Lyons et al., 2013). Additionally, Sanger sequencing creates reference mtDNA sequences for comparing new variants discovered by other methods (Carroll et al., 2014).

11.3.5. Next-Generation Sequencing (NGS) for mtDNA Analysis

11.3.5.1. An Overview of NGS Technologies

Illumina and other companies have made next-generation sequencing platforms that can generate millions of sequence reads at once. This technology makes it possible to sequence whole mitochondrial genomes in one experiment, which lets researchers capture a lot of nucleotide variation and look at haplotype diversity on a scale that has never been seen before (R. Chen et al., 2021). Usually, NGS methods break up the amplified mtDNA and add sequencing adapters, then clone the DNA and use sequencing-by-synthesis to get accurate base calls over short-read lengths (Huang, 2011).

11.3.5.2. Workflow Post-Amplification

After amplification, mtDNA fragments are usually mixed together in equal amounts and used to make a library. Most researchers use the Illumina workflow, which cuts the fragments into pieces between 150 and 300 base pairs long. This procedure can be done with enzymes or machines. Each fragment is linked to an adapter molecule with a unique barcode index, which lets you sequence many samples at once (Huang, 2011). Then, the prepared libraries are amplified on the flow cell, and sequencing begins using cycles of reversible terminator chemistry. Paired-end sequencing is often used to help align and put together the mitochondrial genome better. This approach makes it possible to do a full analysis of nucleotide diversity (Huang, 2011).

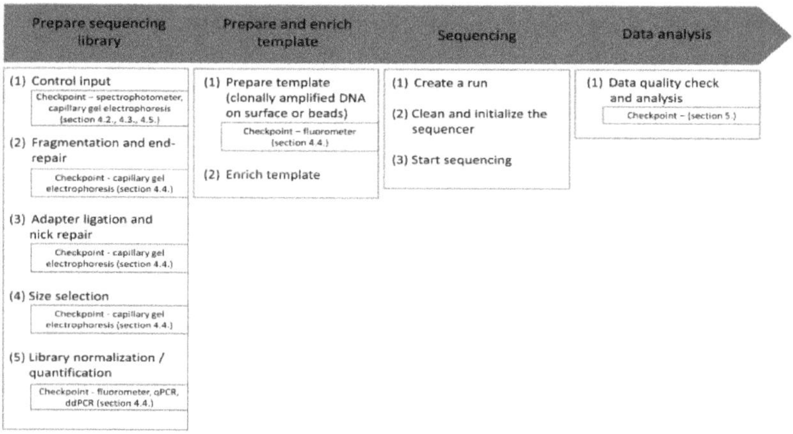

Fig. 1. Overview of the general NGS workflow.

The main steps library and template preparation, enrichment, sequencing and data analysis are divided into substeps containing recommendations for checkpoints which are proposed for QC(Endrullat et al., 2016).

11.3.5.3. Advantages of NGS Approaches

There are several clear benefits to using NGS methods instead of traditional Sanger sequencing. First, they can sequence complete mitochondrial genomes from hundreds of samples at the same time, which is crucial for studies that look at whole populations (R. Chen et al., 2021). NGS excels in identifying common variants due to its extensive coverage. It can also identify low-level heteroplasmic variants that may only be present at 5% or even lower frequencies. This capability is very important for bees, clinical and forensic investigations (Harvey et al., 2019; Huang, 2011). Furthermore, NGS covers the whole mitochondrial genome, which reduces amplification biases that are common in single-locus analyses. The result gives a more complete picture of haplotype diversity and population structure (Mahmud et al., 2022).

11.3.5.4. Impact on Population Genetic Studies

NGS has changed the way we view population genetics and evolutionary biology by changing how we look at mtDNA. The ability to make large-scale sequence datasets makes it possible to find small nucleotide changes and rare haplotypes that tell us about the demographic history and phylogeography of bee populations (K. Ma et al., 2016; C. R. Taylor et al., 2020). Researchers can now map out entire haplotype networks and figure out mutation rates and migration patterns with much more statistical power than they could with analyses that only looked at control region sequences made by Sanger sequencing (Ghasemi et al., 2025).

Lyophilized NGS Library Preparation Kit

Lyophilized reactions in tubes, packed in sealed pouches for ambient temperature stability.

Fig. 15. Three tubes consisting of the lyophilized components for end-repair and A-tailing, ligation, and library amplification. These can be stored at ambient temperature and are ready to use, just add DNA solution.

Workflow

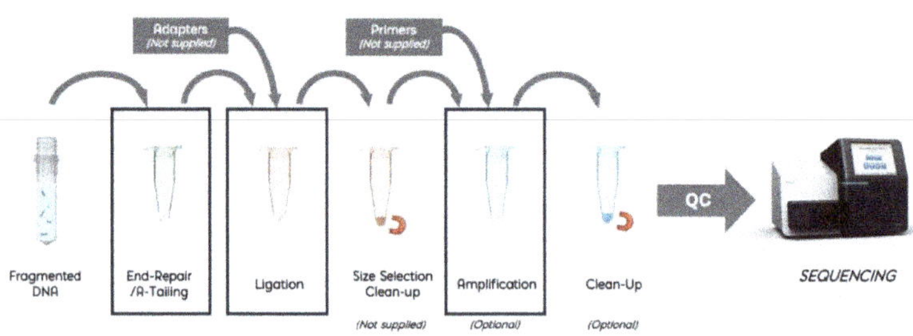

Fig.16. The straightforward design ensures that the NGS library preparation process is fast and convenient, contributing to cost-effective, sustainable, and accessible NGS technology. (https://www.meridianbioscience.com/lifescience/ngs-enzymes/lyophilized-ngs-library-preparation-kit/).

10.3.5.5. High-Throughput Sequencing in Forensics and Clinical Diagnostics

Forensic science is one of the most important uses of NGS in mtDNA research. High-throughput sequencing platforms can analyze damaged samples with little DNA. They can also identify mixtures that may come from more than one source. NGS is very sensitive, so it can isolate low-level variants and accurately measure heteroplasmy. This makes forensic profiles clearer than those made by Sanger sequencing (Avila et al., 2019; Harvey et al., 2019). Using NGS in clinical diagnostics has made it possible to screen for mitochondrial disorders faster and at a lower cost. By sequencing the entire mitochondrial genome with high coverage, doctors can identify pathogenic mutations that targeted approaches might miss. This improves the diagnostic yields for patients suspected of having mitochondrial dysfunction (Mahmud et al., 2022).

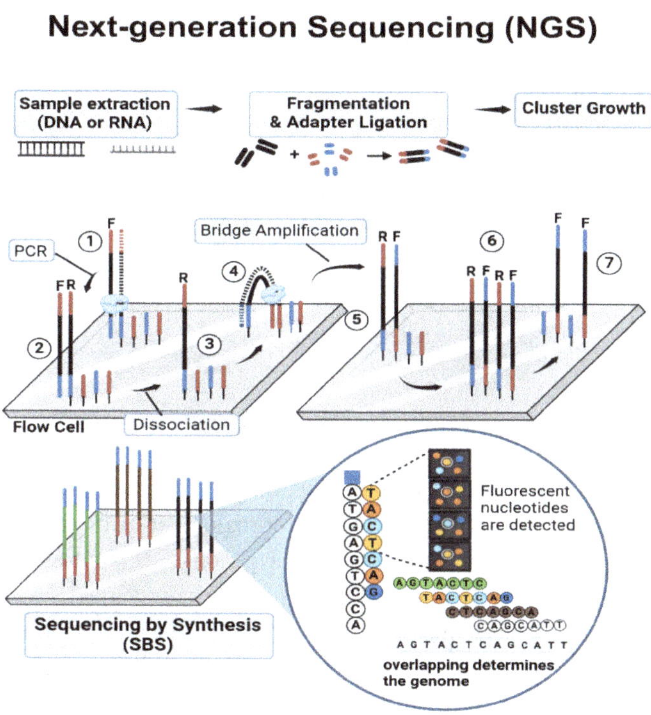

Description

This figure template illustrates a workflow of NGS, and clearly shows the formation of clusters: Oligonucleotides attached to flowcell match adapters, allowing PCR generation with DNA polymerase. The binding between DNA and florescent nucleotide stimulates different colors. In a completed DNA sequence, overlapping nucleotides are utilised to confirming the location of DNA fragments to determine the entire DNA (Next-generation Sequencing (Workflow) | BioRender Science Templates).

11.3.5.6. Comparative Analysis: Sanger Sequencing versus NGS

11.3.5.6.1. Accuracy, Throughput, and Sensitivity

Sanger sequencing is very accurate and can read long sequences, but it doesn't have a very high throughput and isn't as good at finding low-frequency variants because it isn't very sensitive when it comes to measuring heteroplasmy (Carroll et al., 2014; Lyons et al., 2013) . On the other hand, NGS, especially Illumina-based platforms, offers massively parallel sequencing with high depth. This increases throughput and makes it possible to identify variants that are present at very low frequencies. This higher sensitivity is crucial for accurately measuring heteroplasmy levels, which can have big effects on health (R. Chen et al., 2021; Huang, 2011).

11.3.5.6.2. Coverage and Genome-Wide Perspective

Sanger sequencing usually only works on one gene or a small area, like the hypervariable regions in the mtDNA control region. NGS, on the other hand, lets you sequence the whole mitochondrial genome. This all-encompassing method lets scientists find both coding and non-coding variants, which gives them a better idea of how mitochondrial variation affects haplotype diversity (R. Chen et al., 2021; C. R. Taylor et al., 2020). When trying to answer questions about evolutionary history and population genetics, it's especially important to have such complete genomic coverage. This is because small differences across the whole genome can provide us with information about demographic events and selective pressures (Lindberg et al., 2016).

11.3.5.6.3. Cost-Effectiveness and Scalability

The capacity to scale and the cost per base are two additional factors that render NGS a viable option in the present research environment. The initial costs of NGS platforms such as Illumina can be substantial; however, the cost per sample decreases significantly when processing large groupings of samples due to the platform's capacity to multiplex. Conversely, Sanger sequencing is unsuitable for large-scale population studies due to its high cost and lengthy scaling process (Mahmud et al., 2022; Peck et al., 2018). The process is rendered even more cost-effective by bioinformatic pipelines that can rapidly process and analyze substantial quantities of sequencing data with minimal human assistance (Bris et al., 2018).

11.3.5.6.4. Data Analysis and Bioinformatics Integration

NGS creates a lot of data, so we need strong bioinformatics tools to put reads together correctly, locate variants, and add annotations to haplotypes. It is now common to use pipelines that combine sequence alignment algorithms, variant callers, and haplogroup assignment tools. These tools are essential for getting biologically useful information from raw sequencing data (García-Olivares et al., 2021; Lindberg et al., 2016). These tools help tell the difference between real biological variants and sequencing artifacts, which makes sure that the data is high quality and can be reproduced. For these analyses, the Burrows-Wheeler AlignerBWA) is essential for read mapping, and variant calling platforms such as the Genome Analysis ToolkitGATK) are also very important (Harvey et al., 2019; Huang, 2011).

11.3.5.7. Advances in Sequencing Technologies and Future Perspectives

11.3.5.7.1. Enhancements in Library Preparation and Amplification

Recent advances in how libraries are made have made NGS platforms even more powerful. New methods such as PCR-free library construction and improved long-range PCR techniques greatly reduce the chances of biased amplification and the simultaneous amplification of nuclear mitochondrial pseudogenes (NUMTs). These new methods have improved mitochondrial genome data overall and have made it easier to find nucleotide variation and heteroplasmy (Abicht et al., 2018; Dames et al., 2013).

11.3.5.7.2. Emerging Long-Read Sequencing Technologies

Illumina and other short-read NGS platforms have been the most popular in the field, but new long-read sequencing technologies from companies like Oxford Nanopore and Pacific Biosciences are starting to offer useful features that work well with short-read NGS. These platforms can create sequence reads that cover whole mitochondrial genomes. The capability lets you phase variants over long genomic distances and learn more about structural changes that short-read methods might miss. Long-read technologies, on the other hand, have higher per-base error rates than Illumina platforms and are usually used with short-read data for full analyses (Harvey et al., 2019; Lindberg et al., 2016).

11.3.5.7.3. Integration of Multi-Omics Data

Sequencing the genome will be important for the future of mitochondrial genetics, and combining genomic data with transcriptomic and epigenetic data will also play a crucial role. RNA sequencing (RNA-seq) can be used to examine patterns of gene expression in mitochondria. Third-generation sequencing methods may one day let us determine epigenetic changes and nucleotide changes at the same time. These combined multi-omics methods should give us a deeper understanding of how mitochondria work, what causes mitochondrial diseases at the molecular level, and how they affect how cells use energy (Macken et al., 2021).

11.3.5.7.4. Clinical and Diagnostic Applications

In clinical settings, the ability to quickly and easily capture the entire mitochondrial genome has greatly improved the molecular diagnosis of mitochondrial disorders. Many mitochondrial diseases show heteroplasmy, which means that the amount of mutated mtDNA can be entirely unique in different tissues. Deep NGS coverage is necessary to get an accurate count of these rare variants. Such information helps with genotype-phenotype correlations and the diagnosis of primary mitochondrial disorders (Macken et al., 2021; Mahmud et al., 2022). The shift toward using NGS for full genome-wide assessments is changing how doctors work, allowing for earlier and more accurate diagnoses, as well as better-informed prognoses and treatment plans.

11.3.5.8. Discussion

One of the most significant changes in mitochondrial genetics technology is the switch from traditional Sanger sequencing to modern NGS platforms. Sanger sequencing has been around for a long time and has been shown to be accurate, which makes it necessary for targeted analyses and validation. Its ability to make long, clear reads has made it a standard for finding mtDNA variants (Carroll et al., 2014; Lyons et al., 2013). But modern research needs a deeper understanding of how mtDNA changes across whole genomes and large groups of bees and animal and humans, which Sanger sequencing can't do. NGS platforms like Illumina have filled this gap by providing highly sensitive, scalable, and affordable sequencing that can look at nucleotide variation and haplotype diversity across a whole population (R. Chen et al., 2021; Huang, 2011).

NGS has also made it possible to find low-level heteroplasmy, which can be crucial for diagnosing mitochondrial diseases and for forensic work when mixed samples are present. Deep sequencing protocols have made it possible to find heteroplasmic variants at rates as low as 5% or less, with a lot of confidence (Harvey et al., 2019; Huang, 2011). NGS also has a high throughput that lets it sequence the entire mitochondrial genomes of hundreds of bees or people at the same time. This ability to analyze a lot of data at once makes it possible to make big mtDNA databases, which lets researchers study things like population structure, migration patterns, and evolutionary dynamics in detail (Desalle et al., 2017; C. R. Taylor et al., 2020).

Bioinformatics pipelines have changed along with sequencing technologies to accommodate the problems that arise when working with a lot of NGS data. Using advanced algorithms and reference databases, integrated software platforms now simplify the processes of checking data quality, aligning reads, identifying variants, and assigning haplogroups (García-Olivares et al., 2021; Lindberg et al., 2016). These computer programs are necessary to tell the difference between real mtDNA variants and sequencing artifacts and to accurately put together mitochondrial haplotypes, which then provide us with information about population structure and evolutionary history.

The quick growth of NGS technologies has also made people more interested in long-read sequencing. Oxford Nanopore and other platforms are having trouble with raw read accuracy right now, but ongoing improvements promise to make them work better and eventually allow the sequencing of entire mitochondrial genomes in single, contiguous reads. Such information would be a huge help for phasing and structural variant analysis (Harvey et al., 2019; Lindberg et al., 2016).

In the future, combining genomic, transcriptomic, and epigenetic data may present us a better overall picture of mitochondrial biology. We expect that these multi-omic approaches will help us learn more about how mitochondria work in both health and disease, which will help us create targeted treatment plans. In this case, sequencing methods are still important tools, and NGS platforms are the best at giving us a full picture of the genome (Legati et al., 2021; Macken et al., 2021).

11.3.5.9. Conclusion

The next step after amplifying mtDNA is sequencing, which is crucial for finding differences in nucleotides and haplotypes. For many years, mtDNA studies have relied on traditional Sanger sequencing, which is very accurate and reliable for single-locus analyses (Carroll et al., 2014; Lyons et al., 2013). Next-generation sequencing platforms, like Illumina, have changed the field of mitochondrial genomics. These NGS platforms allow us to quickly and cheaply sequence a lot of data, making it easier to study complete mitochondrial genomes and detect small variations in genetic material (R. Chen et al., 2021; Huang, 2011). The large-scale datasets that we now have have changed how we study mitochondrial population structure, evolution, and the genetic causes of mitochondrial disorders.

NGS methods have become even better thanks to improvements in library preparation, amplification strategies, and bioinformatics pipelines. These improvements have fixed many of the problems that come with traditional sequencing methods. Forensic settings still use Sanger sequencing for confirmatory analyses and strict validation requirements. NGS is now the most popular method for research and clinical diagnostics due to its scalability and resolution. As sequencing technologies get better, combining long-read sequencing with multi-omics will probably give us even more information about mtDNA variation, which will help us learn more about how mitochondria work and how they evolved (Avila et al., 2019; Ghasemi et al., 2025; Harvey et al., 2019).

In short, mitochondrial genomics has changed from using Sanger sequencing to using NGS technologies, which provide depth, accuracy, and throughput that have never been seen before. This change not only makes it possible to fully describe mitochondrial nucleotide variants and haplotype structures, but it also provides us important information about population genetics and evolutionary biology. As these technologies get better, they will definitely help us figure out more about mitochondrial DNA and how it affects bees, and human health and disease.

Chapter 12

12. Quality Control Programs for Next-Generation Sequencing in Mitochondrial Genome Reconstruction

This chapter goes into detail about the strict quality control (QC) steps that must be followed when using next-generation sequencing (NGS) to put together full mitochondrial genomes. In modern genomics, the huge amount of data produced by NGS platforms requires rules that make sure only high-fidelity reads are kept for later assembly and analysis. These steps begin with checking the raw sequences for issues, then cleaning them up using computer programs, followed by putting the sequences together with tools like SPAdes, and finally confirming the results by comparing them to known sequences and analyzing the data. This combined method is needed to reduce issues caused by sequencing errors, contamination from adapters, and poor-quality base calls, ensuring that the final genomic assembly correctly reflects the target mitochondrial genome (Yamauchi et al., 2003).

The first step in an NGS workflow is to examine the quality of the raw sequences to observe problems like adapter contamination, low-quality base calls, overrepresented sequences, and differences in read length. Fast QC (https://www.bioinformatics.babraham.ac.uk/projects/fastqc/), and other tools are often used to make detailed graphical reports that show the quality of each base, the GC content, and the distribution of quality scores. However, the next step, which is just as important, is to systematically get rid of unwanted sequences from the raw data. In this case, programs like Trimmomatic (http://www.usadellab.org/cms/?page=trimmomatic), are often used to get rid of adapter sequences and cut off low-quality bases from short reads. This approach makes the assembly processmore accurate (Razzaq, 2020). This trimming process uses adjustable settings, like sliding window quality thresholds and minimum read length requirements, to make sure that only reads with high-confidence base calls are sent on to the assembly pipeline (Sewe et al., 2022).

After the raw reads have been cleaned, the next step in the QC pipeline is to put together the mitochondrial genome. **De novo** assembly algorithms like SPAdes are excellent for this job because they can handle short reads well and use **de Bruijn graph strategies** to figure out complicated genomic regions. SPAdes (https://ablab.github.io/spades/getting-started.html), uses more than one k-mer size to make a more contiguous assembly. It also has built-in error correction steps that make the genomic contigs even better (Fisher et al., 2020). We chose **SPAdes** to assemble mitochondrial genomes because it does an impressive job of making contiguous sequences that accurately reflect the underlying mitochondrial DNA. This is especially true when comparing SPAdes to other assemblers, which may struggle with regions that are less complex or contain a significant amount of repetitive content (Forouzan et al., 2017).

Validation is another important part of the quality control workflow. Researchers usually compare the de novo assembled mitochondrial genome to reference sequences that can be found in public databases like **GenBank** (https://www.ncbi.nlm.nih.gov/genbank/), after it has been put together. The main goals of this comparison are to ensure the assembly is accurate by checking that the mitochondrial gene content, gene order, and total sequence length are the same as those in known reference genomes, and to identify any mistakes or mixed-up sections that might have occurred during the assembly process (H. Li et al., 2011). You can find any differences between the assembled sequence and reference sequences by doing a **BLAST** or similar sequence similarity search. If you need to, you can then introduce changes to the assembly parameters (Hayes et al., 2024).

Another important step in making sure data is accurate is **validation based on mapping**. At this point, the original reads that have been trimmed for quality are realigned to the assembled mitochondrial genome using rapid aligners like **Bowtie.** Researchers can check the uniformity and depth of coverage across the whole genome during the realignment process. This mapping step is vital for finding areas of low or uneven coverage, which could mean gaps in the assembly or biased sequencing. Additionally, it's crucial to ensure that sufficient high-quality reads support the assembled sequence (Razzaq, 2020). Furthermore, having uniform coverage across the mitochondrial genome is a good way to make sure that any possible misassemblies or structural artifacts are either not there or have been fixed properly during the assembly process (Ariede et al., 2018).

The overall QC pipeline includes several iterative procedures in addition to the ones already mentioned. For instance, people often **use statistical measures**

like N50, L50, and total assembly length to judge how excellent the assembly is. These metrics provide us with information about how connected and complete the assembly is, which is important for later analyses like phylogenetic studies or population genetics studies. After that, the quality metrics from these statistics are compared to mapping-based coverage assessments. Any parts of the assembly that have strange coverage are marked for further study. In many cases, making small changes to the trimming and assembly parameters based on these evaluations helps make the genomic dataset even better (Pratas et al., 2020).

Finding the right balance between strict filtering and maintaining sufficient sequence data is crucial for effective quality control in NGS workflows. If trimming parameters are too aggressive, they can get rid of low-quality bases and adapter contamination, but they can also lose important genomic information if they aren't set up correctly. On the other hand, settings that are too lenient may keep wrong sequences, which can make it harder to put things together correctly and cause mistakes in downstream analyses like variant calling or gene annotation. So, to find the best balance between sensitivity and specificity in the quality control process, we need to use an iterative and carefully bench marked method (Razzaq, 2020).

Another important thing to think about is **how barcode and multiplexing sequences affect pooled NGS runs**. In a lot of high-throughput sequencing projects, several samples are sequenced at once. This means that the QC process needs an extra step to separate the pooled reads and assign them to the right sample. This step of demultiplexing is often built into quality control software and needs to be done very carefully so that the next analyses are done on datasets that are correctly labeled. If you can't reliably demultiplex reads, samples could get mixed up, which could draw any biological conclusions drawn from the research less reliable (Yamauchi et al., 2003).

The QC steps that were just described are not only necessary for putting together mitochondrial genomes; they are also necessary for all types of NGS analyses. The analytical workflow for assembling a viral genome or sequencing a whole bacterial genome begins with Trimmomatic, which removes low-quality sequences and adapter contamination. Then, SPAdes assembles the genome from scratch, and alignment to trusted reference databases checks that the assembly is correct. These guidelines also apply when creating phylogenomic datasets or conducting metagenomic analyses, as issues like contaminants, sequencing mistakes, or incorrectly assembled pieces can significantly affect the conclusions drawn later (H. Li et al., 2011).

The final assembly and validation steps are not the end of quality control. **It is important to examine the quality of the assembly after it is done**, for example, by using coverage plotting tools to confirm that no parts of the mitochondrial genome were unintentionally left out because of sequencing biases or technical problems. For example, visualization tools that combine read depth and GC-content profiles provide us more information about how reliable the genome assembly is by showing us any problems that might need more attention. Some of the time, these visualization techniques are not necessary, but they are a useful part of thorough quality control procedures (Jenke & Gruenstaeudl, 2022), because they help confirm that there are no systematic biases.

In addition, **using multiple compatible analysis methods greatly aids in accurately reconstructing mitochondrial genomes**. Trimmomatic is a very important tool for trimming the quality of the first reads, but you can also use other QC tools, like Fastp or BBDuk, at the same time to double-check the quality of the trimmed reads. These different tools supply separate evaluations of read quality and can often show more about the types of sequencing errors or regional quality differences that a single software package might not be able to fully capture (Razzaq, 2020). Using more than one quality control tool adds redundancy to the process, making it more reliable. This ensures that another tool will identify any issues that one tool might overlook.

Researchers have also come to realize how important it is to include quality control (QC) steps in **automated pipelines**. Automation reduces the chance of human error and makes it easier to reproduce results by using the same QC process on different datasets or projects. Many genomics labs now **use pipelines** that **combine trimming, assembly, mapping, statistical evaluation, and visualization.** These pipelines are always getting better as new technologies and experiences come along. Automated workflows like these not only improve the accuracy of the data, but they also speed up genomic discovery by quickly creating high-quality genome assemblies that can be used for more research (Razzaq, 2020).

One of the most important steps in the QC process is still checking against well-known databases like **GenBank.** After making a temporary assembly, researchers usually check the mitochondrial genome's identity by doing sequence similarity searches against GenBank entries. They also check that the gene order and content match what is known for the species being studied. This last step of comparing the results to known standards, helps make sure that any new discoveries are really due to biological differences and not just technical errors. Routine validation against GenBank not only makes the assembly more reliable, but it also makes it easier to put new data into the larger field of genomic research by linking it to a lot of already available sequence information (Hayes et al., 2024).

People don't always realize how important it is to check the complexity and duplication levels of a library as part of quality control. When making libraries for NGS, the PCR amplification steps often create copies that can affect later analyses, especially when looking at coverage depth or identifying genetic variants. In these situations, deduplication steps, which are often done by tools that mark and remove PCR duplicates, are essential to assure that the final assembly and any analyses that come after it are based on unique and separate sequence reads. This process of removing duplicates makes coverage assessments even more reliable by making sure that areas of high coverage really show biological replication and not just technical redundancy (Keane, 2020).

Furthermore, consider how the **quality of the sample affects the QC results**. The success of all the steps that follow depends heavily on the integrity and purity of the first DNA sample. To get reliable sequence data, you need high-quality DNA extraction protocols that keep degradation and contamination to a minimum. Poor DNA quality can lead to broken assemblies, uneven coverage, and a higher chance of sequencing errors. All of these things can make it harder to analyze and understand the mitochondrial genome later on. So, strict quality control (QC) doesn't just start with the computer work; it also includes the lab work done to extract DNA and prepare the library (Yamauchi et al., 2003).

Along with the strategies mentioned above, **new developments in sequencing technologies and computational methods have made QC procedures even better.** To deal with the problems caused by longer reads and different sequencing chemistries, new versions of trimming algorithms and assembly software are always being made. One example of these improvements is the iterative improvements seen in tools like SPAdes, which use feedback from real-world quality assessments to improve assembly performance over time.

Also, the **creation of hybrid assembly pipelines** that mix short-read and long-read data has made it possible to put together genomic regions that were hard to put together before. But these kinds of hybrid methods still depend on the basic QC rules of removing adapters, trimming quality, and checking coverage to make sure that both types of data are properly combined (Razzaq, 2020).

Strict quality control procedures have effects that go beyond the technical world; they are necessary for correctly understanding biological data. The quality of the input sequence data directly impacts how sensitive and accurate later analyses are, such as finding variants, reconstructing haplotypes, and making phylogenetic inferences. This is true in population genetics studies, for example. When QC procedures are followed closely, the biological conclusions that are drawn from the data are much more reliable. This, in turn, supports strong evolutionary and ecological conclusions that could have big effects on fields like conservation genetics and disease research (Ariede et al., 2018).

The iterative QC paradigm is also crucial for finding technical artifacts that might look like real biological variation. If quality control fails to identify and correct misassemblies, chimeric contigs, and sequencing biases, it can lead to incorrect data interpretation. Researchers can find areas where the coverage is unusually low or uneven by using tools like **Bowtie to map truncated reads** back to the assembled mitochondrial genome. Then, these errors can be looked at more closely and fixed by putting the pieces back together by hand, which makes the genome assembly more reliable overall (Pratas et al., 2020).

Additionally, **multi-sample studies** demonstrate the importance of using strict quality control methods when consistency across datasets is **crucial.** In these situations, **automated QC pipelines** assuret hat every sample is processed using the same set of filtering and threshold criteria. This reduces the differences that come from processing samples differently. This uniformity not only makes it easier to compare results, but it also helps with meta-analyses that combine data from several labs, experiments, or even consortium-based projects. The fact that sequence assemblies and variant calls can be repeated in different studies is a direct result of using strong quality control protocols in a systematic way (Orsini et al., 2016).

The technical details in this chapter are mostly about putting together mitochondrial genomes, but many of **the same quality control methods can be used** on nuclear genomes, transcriptomes, and metagenomic datasets as well. For example, when putting together a nuclear genome, the amount of data may be orders of magnitude larger, but the basic ideas behind removing adapters, trimming quality, fixing errors, building from scratch, and validating based on mapping stay mostly the same. In these situations, the computational demands are higher, but making sure that the input data is accurate is even more important because the genome is more complex and there is a greater chance of encountering repetitive elements or structural variants that make assembly more difficult (Forouzan et al., 2017).

Another important part of quality control is **checking for contamination**. In a lot of sequencing experiments, especially those that use environmental samples or samples with low biomass, DNA from outside sources like host DNA, (bacterial symbionts, or lab reagents) can mess up the analysis. Stringent quality control (QC) workflows include steps to find and remove these contaminant sequences. Such removal is usually done by mapping the reads against a large database of known contaminants or by using taxonomic classification tools. The procedure makes sure that the assembled mitochondrial genome comes only from the organism of interest and not from other species that were sequenced at the same time, which makes taxonomic assignment and biological interpretation more accurate (Linck, 2019).

In summary, there are **numerous strict quality control steps used in NGS analyses that work together effectively**. They include the following: first, using tools like **Trimmomatic** to get rid of adapter sequences and trim low-quality bases from raw reads to verify that only high-quality data are used for further analysis; second, **using sequence assembly algorithms like SPAdes** to put together the whole mitochondrial genome from these high-fidelity reads; third, **using high-confidence comparisons with reference sequences** in public databases like **GenBank** to confirm the assembled mitochondrial genome; and finally, **using aligners like Bowtie** to map the original reads back to the assembled genome to make certain that there is uniform coverage and to identify any possible misassemblies. Each of these parts is very important for reducing technical artifacts, addressing the issues associated with short-read data, and ensuring that downstream analyses rely on accurate and reliable genomic assemblies (Fisher et al., 2020; Razzaq, 2020; Yamauchi et al., 2003).

The steps outlined here show how important it is to take a whole-system approach to quality control in NGS workflows. Researchers are always coming up with new ways to ensure that the data they collect is as accurate and reproducible as possible. Researchers can make genomic datasets that improve our understanding of mitochondrial biology and set the standard for high-quality, reproducible, and biologically meaningful genomic research (Keane, 2020; H. Li et al., 2011), by using strict QC methods at every step, from trimming and assembly to validation and coverage analysis.

This careful method of quality control is a key part of modern genomics and shows how closely data must be checked in a field where data integrity is crucial. Researchers can confidently draw biological conclusions from their NGS data by making sure that every step, from getting the raw data to validating the final assembly, is done with the highest level of care. Such an approach helps both scientific discovery and practical uses in fields like population genetics and clinical diagnostics (Hayes et al., 2024; J¨unemann, 2022).

In conclusion, strict quality control steps in NGS workflows are necessary to make sure that only high-quality sequence data are used in genome assembly and later analyses. The assembled mitochondrial genome is not only accurate, but so are any applications that come after it. This is because it was put together using tools like Trimmomatic, SPAdes, and Bowtie, and it was thoroughly checked against trusted reference databases like GenBank. New technologies and the evolving needs of genomic research constantly drive the improvement of these protocols. The study shows that strict quality control is an essential part of modern bioinformatics pipelines (Ariede et al., 2018).

Chapter 13

13. A Step-by-Step Protocol for High-Resolution Melting (HRM) Analysis Coupled with Real-Time PCR to Detect Mitochondrial DNA Variation in Honey Bees for Conservation Genetics

This chapter provides a step-by-step guide on how to use high-resolution melting (HRM) analysis with real-time PCR to determine small changes in mitochondrial DNA (mtDNA) sequences. This method has become an important tool for telling apart closely related mitochondrial haplotypes in honeybees, especially in conservation programs that want to keep track of introgression and protect genetic integrity (Galuszynski & Potts, 2020a).

13.1. Introduction

High-resolution melting analysis is a technique that comes after PCR. It uses sensitive detection of differences in how amplified double-stranded DNA fragments melt to find even single nucleotide differences (Zhao et al., 2025). The technique is becoming more and more popular for finding mtDNA sequence variations because mitochondrial genomes are passed down from mother to child, are small, and have a relatively high mutation rate, which makes them perfect for haplotype discrimination in honeybees (Soares et al., 2018). Fast screening of mtDNA variation is especially useful for conservation programs that need to find introgression from non-native lineages and keep an eye on the genetic integrity of native honey bee subspecies (Porrini et al., 2021).

13.2. Background and Rationale

Because they are haploid, evolve quickly, and are passed down from mother to child, mitochondrial markers have been used in population genetics for a long time. This makes recombination less likely to change haplotype structure (Soares et al., 2019). In honeybees, finding small differences in mtDNA sequences is very important for finding different maternal lineages and keeping an eye on introgression events that could hurt the **genetic purity** of native subspecies (Kahya, 2020). HRM analysis and real-time PCR together offer a quick, cheap, and high-throughput way to screen large sample sets for small nucleotide differences (Galuszynski & Potts, 2020a). is a better option than traditional sequencing methods.

13.3. Materials and Methods

13.3.1. Sample Collection and DNA Extraction

The first step is to get high-quality biological samples, usually bee tissues like thoraces or spermathecal contents from queens, and ensure that they don't break down too much over time (Kahya, 2020). We extract DNA using methods specifically designed for insect tissues. For example, commercial insect DNA extraction kits or CTAB methods can give you DNA that is very clean and intact (Nikolikj, 2017). We use UV spectrophotometry and agarose gel electrophoresis to **measure the amount of DNA** we have and **confirm its quality**. We then set the concentrations to a working level for example, (5–7 ng/μL), which is important for PCR amplification that can be repeated (Mohamadzade Namin et al., 2022).

13.3.2. Primer Design and Target Selection

To do a good HRM analysis, you need to make primers that make a short piece of target mtDNA usually (75–200 bp) with useful single nucleotide polymorphisms (SNPs) or other sequence changes (Kahya, 2020). Primers are made in conserved flanking regions to ensure that amplification is strong. The length of the amplicon is also crucial because shorter fragments often show bigger differences in melting profiles (Galuszynski & Potts, 2020b). **Cytochrome oxidase subunit I (COI) and cytochrome (b)** are two parts of the mitochondria that scientists often study in honeybees because they contain specific SNPs that can **help distinguish different lineages** (Porrini et al., 2021).

13.3.3. Reaction Setup—Real-Time PCR with HRM-Compatible Mix

To save valuable template and reagents and make sure that the cycling conditions are the same for all samples, the **PCR reaction** is usually **set up in small volumes** (10–20 μL) (Mohamadzade Namin et al., 2023). The reaction mix has the DNA that was taken out, a pair of specific primers, and a high-resolution melting master mix that has a saturating fluorescent dye usually (EvaGreen or SYBR Green I) and other things like DNA polymerase, dNTPs, $MgCl_2$, and buffer (Kahya, 2020). The saturating dye is very important because it binds to double-stranded DNA without stopping amplification and gives off fluorescence that can be repeated and is related to the amount of double-stranded DNA present (Galuszynski & Potts, 2020a).

13.3.4. Thermal Cycling Conditions

The initial phase of real-time PCR heat cycling is denaturation, typically occurring at (95 °C for 2 to 10 minutes), to fully dissociate the DNA strands (Borland & Kading, 2021). Subsequently, (40 to 50), cycles of denaturation occur at (95 °C for 10 to 15 seconds), followed by annealing at the primers' melting temperatures, typically between (55 to 60 °C for 15 to 30 seconds), and extension at (72 °C for 20 to 30 seconds) (Kahya, 2020). We monitor each cycle in real time, and upon completion of the extension phase, we quantify the fluorescence to ensure accurate amplicon enumeration (Galuszynski & Potts, 2020a).

13.3.5. High-Resolution Melting (HRM) Analysis

After PCR amplification, the instrument goes straight to the melting analysis phase without opening the reaction tubes. This lowers the risk of contamination. Starting at a set temperature like (65 °C), the temperature is slowly raised in small steps like (0.1–0.2 °C per step) until it reaches a maximum of about 95 °C . When the amplicons melt, they release the fluorescent dye, causing the fluorescence to decrease. This change is recorded to make a melting curve (Soares et al., 2018).

13.3.6. Data Acquisition and Analysis

The HRM software that comes with the real-time PCR instrument automatically normalizes the melting curves and makes difference plots, which show small differences in melting temperatures between samples (Porrini et al., 2021). The melting profiles are grouped together based on the shape and temperature at which the fluorescence drops sharply, which is the melting temperature (Tm) of the amplicons (Galuszynski & Potts, 2020a). These groups are thought to be haplotypes. For example, samples that melt at slightly different Tms show that there are single nucleotide differences or small indels in the targeted mtDNA region (Kahya, 2020).

13.3.7. Confirmation and Validation of HRM-Detected Variants

HRM analysis is very sensitive and specific, but it is common to use Sanger sequencing of the amplicons to confirm a subset of the detected clusters and check the nucleotide variation that caused them (Galuszynski & Potts, 2020b). In most cases, only a few samples from each HRM cluster are sequenced. This cuts down on the need for full sequencing, which saves time and money (Nikolikj, 2017). This two-step approach first checking with HRM and then doing confirmatory sequencing ensures that we can process many samples quickly while being very sure about identifying haplotypes (Porrini et al., 2021).

Table. 11. HRM Assay and Reagent Optimization Tip: Shopping List for HRM Analysis Suggested Reagents

Item	Size/Description	Part Number
Forward and reverse PCR primers optional: (HPLC purified)	Custom	-
Appropriate HRM products for your instrument	-	-
MeltDoctor™ HRM Master Mix	5 mL	4415440
MeltDoctor™ HRM Master Mix	5 x 5 mL	4415452
MeltDoctor™ HRM Master Mix	10 x 5 mL	4415450
MeltDoctor™ HRM Master Mix	50 mL	4409535
MeltDoctor™ HRM Positive Control Kit	1 kit	4410126
MeltDoctor™ HRM Reagent Kit	1 kit	4425557
MeltDoctor™ HRM Calibration Plate (384-well)	1 plate	4425559

MeltDoctor™ HRM Calibration PlateFast (96-well)	1 plate	4425618
MeltDoctor™ HRM Calibration Standard	1 tube, 1 mL	4425562
MicroAmp® Optical 384-Well Reaction Plate with Barcode	0.1 mL, 50 plates	4309849
MicroAmp® Optical 384-Well Reaction Plate with Barcode	0.1 mL, 1,000 plates	4343814
MicroAmp® Optical 96-Well Reaction Plate with Barcode	0.2 mL, 20 plates	4306737
MicroAmp® Optical Adhesive Film	-	4360954
MicroAmp® Optical Adhesive Film	-	4311971

Step 1: Collect samples and get DNA out of them.

- Get tissue samples from honeybees, like drone thoraces and queen spermathecal contents, in a way that keeps the DNA intact (Kahya, 2020).
- Use a protocol that works best for insects to get genomic DNA, and then measure the DNA to make sure it is of good quality and concentration (Nikolikj, 2017).

Step 2: Design the Primer

- Choose a mitochondrial target region like (COI or cytochrome b) that is known to separate different maternal lineages of honeybees (Porrini et al., 2021).
- Make primers that go around areas with known SNPs or indels. The amplicon size should be (75–200 bp) for the best sensitivity in HRM analysis (Kahya, 2020).

Step 3: Setting up the Real-Time PCR Reaction

- Make a reaction mixture that has the standardized DNA template, species-specific primers, and a master mix that works with HRM and has a fluorescent dye that saturates it (Mohamadzade Namin et al., 2022).
- Put the reaction mix into 96-well plates to make it easier to process many samples at once (Galuszynski & Potts, 2020a).

Step 4: Thermal Cycling

- Set the thermal cycler to start with a denaturation step at (95 °C for 2 to 10 minutes, then do 40 to 50 cycles) of denaturation (10 to 15 seconds at 95 °C), annealing15 to 30 seconds at (55 to 60 °C), and extension (20 to 30 seconds at 72 °C) (Kahya, 2020).

Step 5: Melting Analysis with High Resolution

- After the PCR amplification is done, start the HRM phase by slowly raising the temperature for example, from 65 °C to 95 °C in (0.1–0.2 °C increments) while measuring fluorescence all the time (Borland & Kading, 2021).
- Make melting curves that show how the fluorescence goes down as the DNA melts, and make sure the instrument can pick up on small changes in how the DNA melts (Galuszynski & Potts, 2020a).

Step 6: Analyzing the data and grouping haplotypes

- Use HRM analysis software to standardize the raw melting data and make difference plots that show how the samples differ from each other (Soares et al., 2018).
- Group samples together automatically or by hand based on their melting profiles; these groups are different mtDNA haplotypes (Porrini et al., 2021).

Step 7: Verify by sequencing.

- Choose a few samples from each HRM-defined cluster to do confirmatory sequencing on. This process will validate the nucleotide differences responsible for the observed melting variations (Galuszynski & Potts, 2020b).
- Use the confirmed sequences to give each haplotype a unique identity, which makes it possible to tell haplotypes apart and do population genetic analyses (Nikolikj, 2017).

13.3.8. Applications in Conservation Genetics

The HRM method is particularly advantageous in conservation genetics due to its ability to rapidly and economically analyze genetic variation in substantial sample sizes (Soares et al., 2019). Researchers can track maternal lineages and identify introgression events that may happen when migratory beekeeping practices or non-native queen bees are brought in by focusing on mtDNA regions (Kahya, 2020). The HRM technique is very sensitive, which means it can determine even low-frequency haplotypes. This attribute is crucial when checking the genetic health of native honey bee populations (Y. Z. Zhang et al., 2019). In conservation programs, these quick screening methods make it easier to make proactive management decisions, like stopping the movement of bees between areas with different haplotype profiles or starting breeding programs that keep the genetic purity of the bees (Soares et al., 2019).

Furthermore, HRM analysis is a useful way to find small changes in haplotype frequencies that could mean that non-native mitochondrial genomes are coming in, which can act as an early warning system for genetic introgression (Porrini et al., 2021). Longitudinal studies, which require monitoring population changes over time to inform conservation strategies and policies, greatly benefit from this high-throughput capacity (Mohamadzade Namin et al., 2023). Furthermore, combining HRM data with morphometric and ecological analyses makes population structure assessments even more accurate, which allows for a multidimensional approach to conservation genetics (Soares et al., 2018).

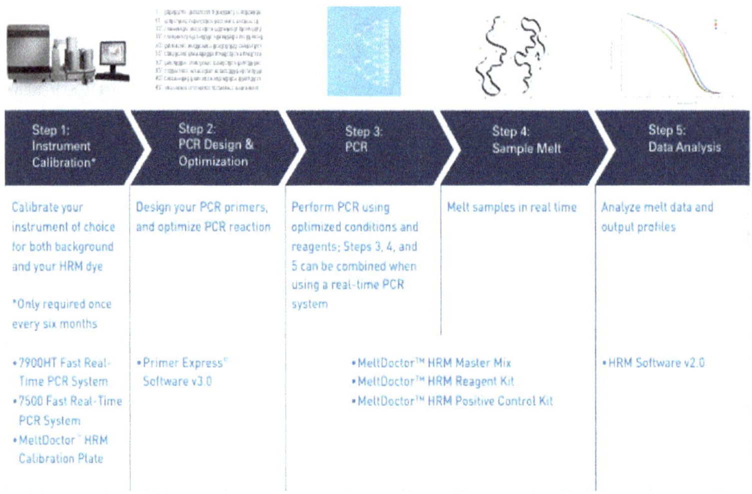

Fig. 17. The HRM Analysis Workflow. (https://assets.thermofisher.com/TFS-Assets/LSG/manuals/cms_070283.pdf)

13.3.9. Pros and cons

There are many benefits to HRM analysis, but the most important one is that it is quick and cheap. Traditional sequencing can take a long time and cost a lot of money. HRM, on the other hand, gives you feedback on genetic variation almost right away and only needs confirmation sequencing for a small number of samples (Galuszynski & Potts, 2020a). The fact that HRM is in a closed tube makes it less likely to get contaminated and makes the workflow much easier (Borland & Kading, 2021). HRM is also very sensitive to even small changes in nucleotides, which lets it tell apart haplotypes that are very similar and might not be able to be told apart by gel electrophoresis or regular PCR methods (Kahya, 2020).

But the method does have some drawbacks. **The type of nucleotide change can affect how well HRM works**. For example, transitions like (A↔G) are usually easier to figure out than transversions like (A↔T) because they affect the melting temperature in different ways (Galuszynski & Potts, 2020b). Sometimes, samples with very few sequence differences can have melting profiles that overlap, which could lead to false negatives that need to be fixed by sequencing (Soares et al., 2018). To ensure consistent results, the best assay conditions, such as primer amount, amplicon length, and dye saturation, must be adjusted (Mohamadzade Namin et al., 2022). HRM is still one of the best ways to quickly screen mtDNA haplotypes, even with these problems (Kahya, 2020).

13.3.10. Case Studies and Comparative Insights

Several studies have utilized HRM analysis to examine honey bee research, monitor mtDNA variation, and identify instances of introgression. For example, research on distinguishing mitochondrial haplotypes in different subspecies of *Apis mellifera*, showed that HRM analysis can reliably differentiate between Africanized and European lineages based on small differences in melting temperature (Porrini et al., 2021). These studies indicate that the HRM method not only finds the natural genetic diversity in honeybee populations, but it can also find small mixes that might show recent introgression events (Y. Z. Zhang et al., 2019). Another example is when HRM was used to examine mitochondrial SNP markers in the sperm storage of honeybee queens that were artificially inseminated with mixed semen from different subspecies. The presence of heteroduplex peaks in the melting curves confirmed the existence of mixed mitochondrial haplotypes (Kahya, 2020). These kinds of uses show how flexible HRM analysis is for studying both simple populations and complicated mating systems with polyandry, which is common in honeybees (Kahya, 2020).

Furthermore, studies that compare HRM to other molecular techniques like PCR-RFLP or sequencing have shown that these other methods are more accurate but require a lot more work and are not as good for high-throughput screening (Galuszynski & Potts, 2020a). HRM analysis, on the other hand, strikes a satisfactory balance between sensitivity, specificity, and operational efficiency. This combination makes it perfect for quick field and lab screening in conservation settings (Borland & Kading, 2021; Soares et al., 2019).

13.3.11. Integration with Broader Conservation Strategies

Larger conservation genetics frameworks can utilize HRM analysis for mtDNA haplotyping. Sampling plans, breeding program design, and monitoring the effects of human-caused pressures such as habitat fragmentation or the introduction of non-native genetic material can benefit from the use of genetic data from HRM screening (Soares et al., 2019). HRM-based genetic profiles, when used with ecological data and morphometric analysis, can give us a lot of information about the structure of honey bee populations, how they move around, and how they evolve over time (Moya-Moraga & Pérez-Ruíz, 2022). This multi-pronged approach is necessary for coming up with effective ways to keep genetic diversity and control introgression, which will help native honey bee subspecies survive and keep their genetics intact in the long run (Soares et al., 2018).

HRM analysis has a rapid turnaround time, which makes it possible to monitor the genetic status of a population almost in real time. This is especially important in areas where honeybee populations are at risk from diseases, climate change, and invasive species (Mohamadzade Namin et al., 2023). In conservation, its use means that HRM data can be used to quickly change management plans, like by limiting the movement of queens between areas or choosing breeding stock based on confirmed haplotype purity (Adolph, 2016). HRM analysis helps manage honeybee populations in a proactive way by giving a sensitive and reliable measure of genetic variation. This helps protect their evolutionary history and ecological roles (D. S. M. Henriques, 2018).

13.3.12. Future Directions and Technological Advancements

As real-time PCR instruments and melting curve analysis software get better, so do the resolution and throughput of HRM analysis. In the future, it is likely that more work will be done to make reaction volumes even smaller, automate data analysis even more, and combine these results with those from next-generation sequencing to create complete haplotype databases (Galuszynski & Potts, 2020b). HRM may be more sensitive with more stable fluorescent dyes that can hold more color, allowing it to detect subtler sequence variations (Kahya, 2020). Also, multiplex HRM assays that examine more than one mtDNA region at a time could simplify workflows and provide a more detailed picture of haplotype variation in a single reaction (Porrini et al., 2021).

As conservation genetics shifts to more comprehensive approaches that integrate genetic, environmental, and physical data, HRM analysis is well-prepared to be a crucial tool for fast, on-site molecular testing (Soares et al.,

2019). HRM is an important part of modern conservation toolkits because it can determine low-level introgression and keep track of changes in haplotype frequencies in almost real time. This capability is especially important for protecting native honeybee subspecies that are threatened by globalization and climate change (Soares et al., 2018; Y. Z. Zhang et al., 2019).

13.3.13. Conclusion

High-resolution melting analysis and real-time PCR together make a strong, sensitive, and quick way to observe small differences in mtDNA sequence. This method provides an effective means to distinguish between closely related mitochondrial haplotypes in honeybees (Galuszynski & Potts, 2020a). The protocol calls for careful sample collection, optimized DNA extraction and quantification, precise primer design targeting polymorphic mtDNA regions, and careful PCR and HRM cycling conditions that create unique melting profiles (Kahya, 2020). After amplification, the gradual heating of PCR products and the analysis of the melting curve make it possible to distinguish haplotypes apart based on small differences in how they melt (Porrini et al., 2021). Sanger sequencing can confirm the accuracy of the detected variations, which lowers the cost and time it takes to sequence them (Nikolikj, 2017). This simplified method has many benefits for conservation programs. It makes it possible to quickly screen many samples to check the genetic integrity of native honeybee populations and find introgression events (Borland & Kading, 2021; Soares et al., 2019). HRM analysis is an integral component of broader conservation plans designed to develop effective management practices that safeguard genetic diversity against environmental and human-made challenges (Moya-Moraga & Pérez-Ruíz, 2022).

HRM analysis is a revolutionary tool in honey bee genetics that helps make quick decisions to protect and preserve native genetic resources. It is both technically sound and useful in real-world conservation situations (Adolph, 2016; D. S. M. Henriques, 2018). The assay conditions are expected to improve and be combined with other genomic methods, making it even more useful in conservation genetics. These improvements will strengthen its position as an important part of the modern molecular toolkit for managing honey bee populations (Galuszynski & Potts, 2020a).

This detailed protocol explains the step-by-step process for performing HRM analysis, which involves collecting samples, extracting DNA, creating primers, conducting real-time PCR, performing incremental melting analysis, and sequencing to verify the results (Kahya, 2020; Porrini et al., 2021). HRM is an essential tool for finding mtDNA haplotype variation because it can quickly and accurately tell the difference between tiny nucleotide differences. The technique allows conservation programs to keep an eye on introgression and protect the genetic integrity of native honey bee subspecies (Soares et al., 2019; Y. Z. Zhang et al., 2019).

In short, the detailed procedure described here shows that HRM analysis combined with real-time PCR is not only a useful and effective way to determine small changes in mtDNA sequences, but it is also an important tool for protecting and managing honey bee genetic resources in a way that is beneficial for the environment (Galuszynski & Potts, 2020a; Kahya, 2020). When used in the field, it could change the way genetic monitoring is done and help protect and strengthen native honey bee populations as they face global environmental challenges (Borland & Kading, 2021; Soares et al., 2018).

Chapter 14

14. PCR-RFLP Protocol for Mitochondrial DNA-Based Identification of Honeybee Haplotypes and Subspecies

The PCR-RFLP (Polymerase Chain Reaction–Restriction Fragment Length Polymorphism) method is a well-known molecular technique that uses nucleotide polymorphisms in mitochondrial DNA (mtDNA) to identify different honeybee haplotypes and subspecies apart (Alburaki et al., 2023). The first step in this method is to pick a specific part of the mtDNA, like the COI–COII intergenic segment, that has differences connected to certain lineages and haplotypes. This versatility makes it a great target for phylogeographic and diversity studies (Alburaki et al., 2023). Some studies have also looked at other mitochondrial genes, such as COI, 16S rDNA, ND5, and CYT B, in addition to the COI–COII region. Studies have demonstrated that these genes possess the capacity to display unique diagnostic restriction sites across various evolutionary lineages (Kekeçoğlu et al., 2024). The first step in an in silico analysis of the chosen mtDNA region is to align reference sequences from databases like GenBank and use tools for restriction site analysis to find enzymes like DraI, EcoRI, or HinfI that cut at places where haplotypes have different nucleotides (Jenke & Gruenstaeudl, 2022). When choosing the best restriction enzyme, you should look for one that can create a unique and repeatable banding pattern when it digests the PCR-amplified product (Jenke & Gruenstaeudl, 2022).

The first step in the experiment is to collect tissue samples from honeybees, usually from worker bees. We then preserve the tissues, typically the legs, thoraxes, or heads, in ethanol to ensure the DNA remains intact (Alburaki et al., 2023). After we collect the samples, we extract all the DNA using methods like Chelex-based extraction or other common techniques to make a good quality template for PCR amplification. Attia 2023 Mitochondrial DNA Variation, pages 1–4, illustrates the use of a CTAB-based extraction method. It is critical that the DNA that was taken out **is free** of things like **proteins and polysaccharides** that could stop the PCR enzyme from working. Therefore, we often check DNA purity using spectrophotometric analysis and agarose gel electrophoresis before proceeding (Linck, 2019). A polymerase chain reaction is set up using a master mix that has Taq DNA polymerase, dNTPs, specific primers which can be either previously validated primer pairs or newly designed ones, $MgCl_2$, and a suitable reaction buffer (Rahimi et al., 2016). The primers are made to surround the target mtDNA region, making sure that the whole segment of interest is amplified. They are usually tested in earlier studies to make sure they are specific to honeybee mtDNA (Rahimi et al., 2018). The **volume of a typical PCR** reaction can be between **25 and 40 µL**. The reaction should have about 20–50 ng of template DNA, 0.5 µM of each primer, and enough $MgCl_2$ according to the recommended conditions for the chosen polymerase (Kekeçoğlu et al., 2024). The thermal cycling parameters are carefully fine-tuned. The first step is denaturation for (2 to 5 minutes at 92 to 95 °C). This stage is followed by 30 to 35 cycles of denaturation (92 to 95 °C for 30 to 40 seconds), annealingthe temperature depends on primer T_m, which is usually between (47 °C and 55 °C for 1 minute), and extension at (63 to 72 °C for 1 to 2 minutes). Finally, there is a final extension at (63 to 72 °C for about 10 minutes) to make sure that all strands are fully extended (Alburaki et al., 2023; Rahimi et al., 2016).

The next important step after PCR amplification is to assess the size of the product and the specificity of the amplification. To check this, a part of the PCR reaction usually (5–10 µL) is run on an agarose gel usually (1–2% concentration) that has been stained with a DNA-binding dye like ethidium bromide or GelRed™ (Jenke & Gruenstaeudl, 2022). To verify that the amplification is correct, you can compare the observed band to a molecular weight marker, like a 100 bp ladder. The ladder makes sure that the amplicon size is the same as the expected length, which can change depending on the region being targeted (Alburaki et al., 2023). A single, strong band means that the amplification worked and there aren't many non-specific products. If multiple bands appear, adjust the annealing temperature or primer concentration to improve the PCR conditions (Kekeçoğlu et al., 2024). Once researchers confirm that the PCR amplicon is correct and clean, they **digest** a **small amount of the PCR product** with **restriction enzymes** to discover **polymorphisms**.

The main idea behind the restriction enzyme digestion step is that **the amplified mtDNA product has sites that certain enzymes can recognize and cut**. The **presence or absence** of these **sites will be different in different haplotypes.** For example, the enzyme DraI will cut the PCR product at a specific nucleotide sequence. If there is a **mutation that removes the enzyme** site, the samples will either not be cut or will produce a different pattern of fragments (Porrini et al., 2021). A typical digestion reaction mixes a certain amount of restriction enzyme (measured in units according to the manufacturer's instructions), the recommended 10× reaction buffer, and sterile, nuclease-free water with the PCR product (usually 2–10 μL) to make a final volume of about 20–30 μL (Jenke & Gruenstaeudl, 2022). The mixture is then kept at the enzyme's ideal temperature (usually 37°C) for 1 to 3 hours, though some protocols suggest leaving it overnight to make sure the cutting is complete (Alburaki et al., 2023). It is critical that the enzyme reaction happen in a way that makes it possible for complete cleavage to happen. **If the digestion is incomplete, it could cause unclear banding patterns** that make it hard to understand haplotypes (Rahimi et al., 2018).

Once the digest is complete, we need to solve and observe the restriction fragments. Usually, this involves transferring the entire digestion reaction or a portion of it onto an agarose gel. A gel with a concentration of 1–1.5% is suitable for large fragments, while a gel with a concentration of 2–3% agarose may be needed for smaller fragments to get the best resolution (Kekeçoğlu et al., 2024). A buffer like 1× TBE is used to make the gel, and the electrophoresis is run at a steady voltage (usually between 80 and 120 V) until the bands are clearly visible. Electrophoresis takes 1–3 hours, depending on gel concentration and fragment size distribution (Wu et al., 2017). After the run, the gel is stained with a dye that only works on DNA. If you have already stained the gel, you can view the banding pattern under UV or blue light (Khalid et al., 2022).

The pattern of bands seen on the gel is a DNA **"fingerprint"** that can tell you which mtDNA lineages or haplotypes are present. This is because different honeybee subspecies will have different numbers of restriction sites in the amplified region, which will cause the fragments to be different sizes (Tanasković et al., 2021). For example, if you cut a COI–COII PCR product with DraI and get two pieces, one 350 bp and one 200 bp, for one haplotype, and one whole piece, 550 bp, for another haplotype that doesn't have a DraI site, these different results help you figure out each sample's mtDNA lineage (Alburaki et al., 2023; Wu et al., 2017). Some studies use multiple restriction enzymes to create a composite RFLP profile. The result makes the test more accurate (Jenke & Gruenstaeudl, 2022; Rahimi et al., 2016). Researchers can then compare the resulting profiles to reference patterns from earlier studies or experimental databases. This process lets them put the honeybee sample into a specific subspecies group or phylogenetic cluster (Kekeçoğlu et al., 2024).

After the banding patterns have been recorded, the next step in the analysis is to do a detailed interpretation that will help assign haplotypes and make phylogenetic inferences. To make this interpretation, you need to compare the sizes and numbers of bands from each sample with those found in published studies that define the standard RFLP profiles for different honeybee lineages (Rahimi et al., 2018). Many cases may require a computer-assisted analysis. This means that the banding patterns are converted into digital files, and software is used to calculate genetic differences, create diagrams like dendrograms using methods such as UPGMA, and build phylogenetic trees using neighbor-joining or maximum likelihood methods (Porrini et al., 2021; Rahimi et al., 2016). This kind of analysis not only confirms the haplotype's identity, but it also shows the genetic diversity that exists within and between populations, which is useful information for breeding and conservation programs (Alburaki et al., 2023).

There are many benefits to PCR-RFLP. It's fast, cheap, and doesn't need any special sequencing equipment, which makes it very useful for large-scale population surveys where hundreds or thousands of samples have to be screened (Alburaki et al., 2023; Kekeçoğlu et al., 2024). Furthermore, the method works well even with partially degraded DNA samples from field collections, as long as the targeted amplicon is small enough to be reliably amplified. Furthermore, since mitochondrial DNA comes from the mother and doesn't mix with other DNA, the RFLP patterns show a direct matrilineal ancestry. This makes it easier to understand population structure and phylogeographic patterns (Tanasković et al., 2021).

Even though PCR-RFLP techniques have some benefits, they also have some drawbacks. For example, **incomplete digestion** or intra-individual heteroplasmy in the mitochondrial genome may result in **unclear banding patterns.** In these cases, a clear haplotype identification may require direct sequencing or other methods (Rahimi et al., 2018). In addition, the method depends a lot on knowing how the mtDNA sequence varies in the target population and having accurate restriction maps on hand so that enzymes can be chosen that make diagnostic fragments (Jenke & Gruenstaeudl, 2022). Because of this, it is essential to carefully do the first sequence analysis and primer design, often using data from past studies to make sure that the chosen primers and enzymes are specific to the honeybee groups being studied (Kowalczyk et al., 2021; Rahimi et al., 2018).

Here is a summary of the whole PCR-RFLP process, broken down into steps: First, gather and store samples of honeybees while making sure that their DNA doesn't break down too much (Alburaki et al., 2023). Second, use a method like Chelex, CTAB, or standard phenol-chloroform extraction to get genomic DNA. Then, check the quality of the DNA using spectrophotometry and agarose gel electrophoresis (Linck, 2019). Third, either design or use validated primers that flank the mtDNA region of interest. Make sure that they only make one clean amplification product of the right size (Rahimi et al., 2016). Fourth, make the PCR reaction mix with all the parts it needs, and fine-tune the cycling conditions to get strong amplification of the target region (Kekeçoğlu et al., 2024). Fifth, use the right DNA ladders to check the size of the PCR product with agarose gel electrophoresis (Jenke & Gruenstaeudl, 2022). Sixth, mix a specific amount of the PCR product with a restriction enzyme chosen based on computer predictions of where the DNA can be cut, and set up the right conditions for the reaction to happen (Alburaki et al., 2023). Seventh, place the digestion products on a gel made with a high percentage of agarose (usually 2–3% if you expect small pieces) in a 1× TBE buffer and stain the gel to see the bands (Wu et al., 2017). Eighth, take pictures or otherwise record the banding patterns and compare them to known reference patterns or a pre-existing database to figure out the haplotypes and sort the honeybee samples (Jenke & Gruenstaeudl, 2022). Lastly, if you need to, do more analyses like cluster analysis or phylogenetic tree construction using numerical data from the RFLP profiles to figure out how different populations are related to each other (Rahimi et al., 2016).

The PCR-RFLP method is especially useful for large-scale surveys in different parts of the world because it is quick and cheap; hundreds of samples can be processed at once without the need for expensive sequencing reagents or equipment (Alburaki et al., 2023; Atoui & El Khoury, 2017). This method is good for samples collected in the field and for retrospective studies where archival material might be the only source available (Khalid et al., 2022). This is because it is easy to use and can work with small amounts of degraded DNA. PCR-RFLP can be used initially to check samples and identify which ones require more detailed analysis using high-resolution melting analysis or direct sequencing (Porrini et al., 2021; Rahimi et al., 2018). Researchers can get a broad view of genetic diversity and a detailed look at haplotype structure within honeybee populations by combining PCR-RFLP with other molecular methods (Syromyatnikov et al., 2018).

Furthermore, it's important to carefully plan experiments so that technical artifacts don't lead to wrong conclusions. If the enzymes in the DNA sample aren't working well or if there are inhibitors, the digestion may not be complete, which can cause bands to fade, show up unexpectedly, or add bands that don't match real haplotypes (Gajić et al., 2016). To avoid these problems, reactions usually include both a positive control (a sample with a known restriction pattern) and a negative control to make sure that the digestion went as planned (Garrick et al., 2015). If the results are unclear, it may be necessary to re-digest or amplify the PCR amplicon using different primer sets. Confirmatory sequencing of the PCR amplicon adds another level of validation (Rahimi et al., 2018; Vesterlund et al., 2014).

The interpretive process means closely looking at the sizes of the fragments we see and comparing them to the sizes we expect based on computer models of known mtDNA haplotype sequences. This comparison is usually done by hand by looking at the gel images or by using special software that can measure the sizes and intensities of the bands (Garrick et al., 2015). Some studies go even further and code the patterns into binary matrices to make it easier to do statistical analyses of genetic distance and population structure. Dendrograms or principal coordinate analyses can display these patterns (Rahimi et al., 2016). These analyses provide us with information about how genetically different populations are and how genes have moved between them over time. This information is very useful for making decisions about breeding programs and conservation strategies (Göze & Özdil, 2023). These kinds of studies have the extra benefit of being able to determine possible introgression from non-native populations. This is because different RFLP patterns can sometimes show that haplotypes are present that are not common in that area (Alburaki et al., 2023).

The PCR-RFLP method is also easy to use and can be used on a wide range of species, including Varroa mites, bumblebees, and even vertebrates, showing that it is a flexible molecular marker technique (Vesterlund et al., 2014). No matter what kind of organism it is, the basic steps are the same: amplify a target region, digest it with one or more restriction enzymes that show nucleotide polymorphisms, and then use electrophoresis to separate the resulting fragments (Xu et al., 2016). In each situation, choosing the right primers and enzymes ensures that the RFLP patterns produced are reliable indicators for identifying species, haplotypes, or subspecies (Kim, Kim, et al., 2019).

A standard PCR-RFLP process for looking at honeybee mtDNA can be broken down into separate steps, each of which needs careful attention to detail. During the sample preparation phase, samples are handled in a way that reduces the risk of cross-contamination. Pre-PCR procedures, for instance, use specialized pipettes and distinct work areas. This procedure keeps the DNA in each sample pure and high enough quality for further analysis (Rahimi et al., 2018). Following the established procedures and checking the DNA quality with agarose gel electrophoresis while extracting DNA helps ensure that we don't overlook any positive results during PCR amplification (Szalanski et al., 2016). During the PCR amplification phase, following the thermal cycling conditions exactly and using the right amounts of primers and $MgCl_2$ ensures that each sample makes only one specific amplicon (Rahimi et al., 2018). This step is crucial because any non-specific amplification can make the pattern seen after restriction digestion more complicated (Kim, Kim, et al., 2019).

Once the PCR product is confirmed to be the right length, the next step, restriction digestion, needs cautious pipetting and incubation. It is important to use fresh enzymes and follow the manufacturer's instructions for handling them. Carefully adjust the reaction conditions for buffer concentration, temperature, and time (Atoui & El Khoury, 2017). This step is the most important part of the method because it turns small differences in nucleotides into significant differences in fragment length, which gives each haplotype a digital fingerprint (Porrini et al., 2021). When you do a control digestion with a known template, you can be sure that the enzyme is working properly. This process gives the banding patterns you see more credibility (Garrick et al., 2015).

The next important step is to use electrophoresis on the restriction fragments. The expected sizes of the fragments determine the agarose concentration. Smaller fragments do better with higher concentrations, while larger fragments do better with lower concentrations (Wu et al., 2017). It is important to load a DNA ladder with known fragment sizes with each sample to use as a molecular weight marker. This is because accurately measuring fragment sizes is necessary for assigning haplotypes (Khalid et al., 2022). The gel is usually run under a constant voltage until the bands are far enough apart. This process is watched in real time to find the best end point (Garrick et al., 2015).

After electrophoresis, we expose the gel to UV light and record the resulting banding pattern. To find out what evolutionary lineage the sample belongs to, the pattern is then compared to known RFLP profiles that have been made in earlier studies (Alburaki et al., 2023). Often, the pattern may consist of a series of bands whose sizes, when taken together, are diagnostic for a particular haplotype; for example, a profile might exhibit bands of 246 bp and 782 bp indicative of one lineage, whereas a different profile might show an intact, uncut fragment representing another haplotype (Wu et al., 2017). If new or unclear patterns are found, DNA sequencing can be used to confirm the mutation (s) that caused the unique RFLP profile (Rahimi et al., 2016).

Statistical methods are also used to understand RFLP patterns. For instance, researchers can create binary matrices that indicate the presence or absence of a specific band using a "1" or "0" symbol. These matrices are then used as inputs for clustering methods, like UPGMA or neighbor-joining algorithms, to show how samples are genetically related (Rahimi et al., 2016). Researchers can make dendrograms that show how honeybee populations are related and find genetic clusters that correspond to different subspecies or lineages (Kowalczyk et al., 2021). This analytical phase is crucial because it turns the qualitative observations from the gel into quantitative data. This data can then be statistically checked for patterns of diversity and structure (Gajić et al., 2016).

Because PCR-RFLP is fast and cheap, many studies have used it to look at genetic diversity in different parts of the world. This has made it possible to do large-scale screening programs that can help with breeding and conservation strategies (Alburaki et al., 2023). The method is especially important when populations of *Apis mellifera*, in a certain area have low mtDNA haplotype diversity or when there is a risk of introgression from imported stock (Alburaki et al., 2023). In these cases, PCR-RFLP is an easy first step that can be followed by more detailed high-resolution methods, like direct sequencing or high-resolution melting analysis, when fine-scale resolution is needed (Porrini et al., 2021).

For PCR-RFLP to work well for analyzing honeybee mtDNA, it needs to follow strict lab rules and have a full analysis and interpretation of the data. Researchers need to keep very detailed records of the reaction conditions, such as enzyme lot numbers, buffer conditions, and thermal cycling settings. These factors can significantly impact the reproducibility of the RFLP patterns (Syromyatnikov et al., 2018). To make sure that unexpected results are due to real genetic differences and not technical problems, it is helpful to include both positive and negative controls in every run (Garrick et al., 2015), Maintaining consistency in sample handling, reaction setup, and documentation is crucial for the method to function effectively. This is especially true when the goal is to compare genetic diversity across large datasets or several populations (Xu et al., 2016).

To sum up, the PCR-RFLP method for analyzing mtDNA in honeybees is a multi-step process that starts with collecting the best samples and extracting the DNA. Then, specific PCR amplification is performed using specially designed primers, and finally, the PCR products are cut with specific restriction enzymes. Agarose gel electrophoresis is used to sort the DNA fragments, and then the patterns are compared to known samples to identify the haplotypes and subspecies. The method is great for large-scale surveys of honeybee populations in different parts of the world because it is quick, cheap, and fairly easy to use. This procedure helps researcher make smart choices about how to protect, breed, and manage this species, which is important for the environment and the economy (Alburaki et al., 2023; Kekeçoğlu et al., 2024).

Previous research has shown that PCR-RFLP is a reliable and reproducible molecular marker technique in honeybees and other organisms at every step of the process. This shows how useful it is as a basic tool in molecular ecology (Rahimi et al., 2016; Vesterlund et al., 2014). When done and understood correctly, PCR-RFLP is a strong molecular test that turns tiny changes in mitochondrial DNA into patterns that are easy to understand. This technique lets researchers build maternal lineage, look at genetic structure, and keep an eye on changes in honeybee populations over time (Tanasković et al., 2021; Wu et al., 2017).

Overall, incorporating PCR-RFLP into honeybee genetic research has significantly enhanced our understanding of the species' evolutionary history and population structure. It has also provided a cheaper alternative to sequencing-based methods for large-scale screening and monitoring projects (Rahimi et al., 2018; Syromyatnikov et al., 2018). This strong method is useful for research and is also necessary to keep genetic diversity and sustainability in managed and wild honeybee populations around the world (Göze & Özdil, 2023).

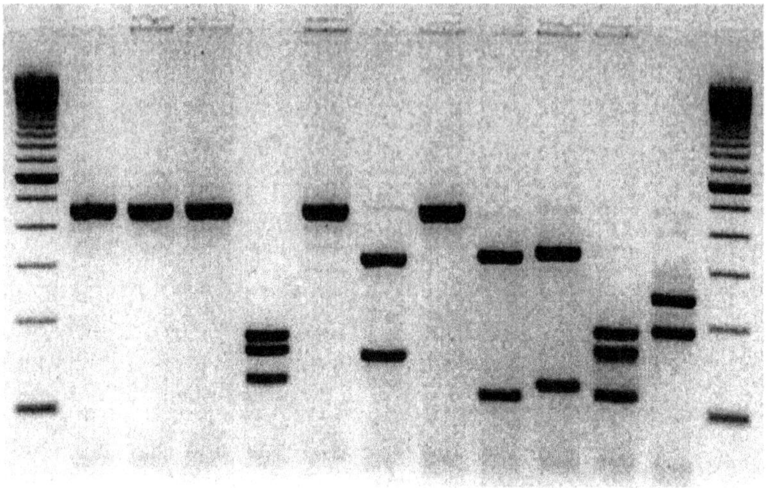

Fig. 18. Restriction analysis of the of 464 bp cyt b amplicon of red deer (CW94) with 11 enzymes (Wolf et al., 1999)

Evolutionary lineages and sub-lineages identification by COI-COII haplotypes

DNA amplification and sequencing of the COI-COII intergenic region was successful for all analysed individual (sn = 1184). Three evolutionary lineages were detected: the African A lineage (P_0Q n = 647, P_0QQ n = 73, P_1QQ n = 9 and P_1QQQ n = 3); the European C lineage (Q n = 413); and the European M lineage (PQ n = 5, PQQ n = 28 and PQQQ n = 6). Additional file 1: Table (S2). Three African sub-lineages A_I *(DraI* profiles A1, A4, A6, A64, A65, A66 and A67), A_{III} *(DraI* profiles A11, A14, A16) and Z (*DraI* profiles Z2, Z7), were identified from sequencing data coupled to (*DraI* restriction profilesFig. 2) (Techer, Clémencet, Simiand, Preeaduth, et al., 2017).

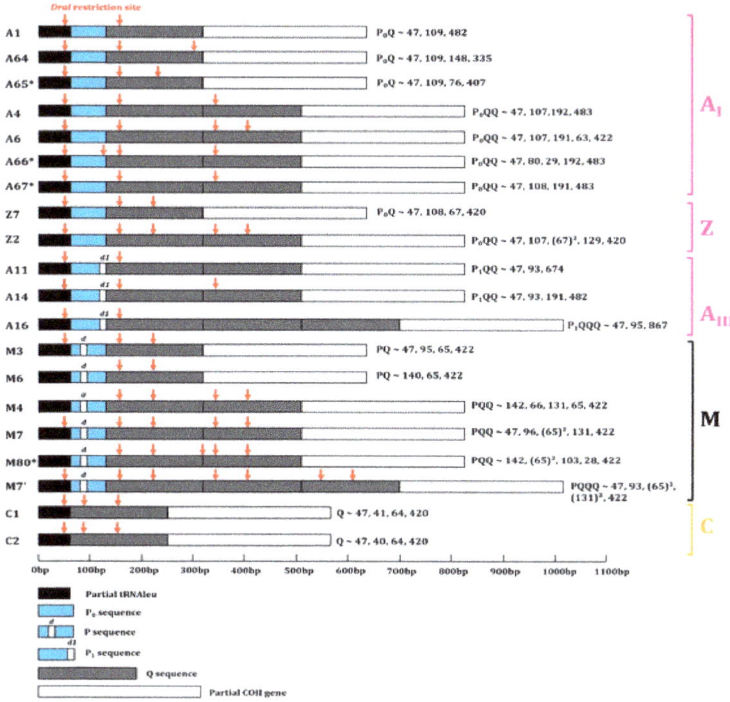

Fig. 19. COI-COII honey bee sequences composition and *DraI* restriction patterns detected as a tool to discriminate evolutionary lineages and sub-lineages.

For each restriction pattern, the composition in (PP_0, P or P_1) and Q unitsfrom (one to three units observed) is indicated. "d" and "d1" are characteristic deletions of the M lineage and Atlantic A lineage, respectively. Red arrows indicate restriction sites, numbers and restriction fragments sizes are also presented~ indicates that size could slightly varied from a few bp due to insertion/deletion within sequence not modifying restriction sites. Profiles with * were newly described (Techer, Clémencet, Simiand, Preeaduth, et al., 2017).

Table 12. Polymorphisms analyzed for determination of A, M, C, Z, Y, O, and H mtDNA haplogroups in *Apis mellifera*, including all requested details

HAPLOGROUP	POLYMORPHISM	PRIMER NAME & SEQUENCE(5'-3')	PCR PRODUCT SIZE (BP)	DIGESTION ENZYME	PATTERN/LINEAGE	POSITIVE FRAGMENT (S)(BP)	NEGATIVE FRAGMENT (S)(BP)
A	A663G	L614: 6FAM-ACCTCCTCAAAGCAATACACTGAAA H774: GGCTAAGCGTTTTGAGCTGC	204	HaeIII	African (A)	60	72
M	C4715T	L4680: 6FAM CCACCAACTAATCTCACCCG H4918: GGGTAGGTGGTAGTTTGTGCT	285	MseI	West European (M)	97	285
C	A13263G	L13088: 6FAM-GCCCTACTCCACTCAAGCAC H13547: AGATAGGGCTCAGGCGTTTG	498	AluI	East European (C)	172	443
Z	T1843C	L1800: 6FAM-CCACCTACTAATCTCACCCG H2030: GGTAGGTGGTAGTTTGTGCT	275	NlaIII	Mediterranean (Z)	132	275

Primer Sequence
TACCTACTTATTCTCACCCG
GGTAGGTGGTAGTTTGTGCT
ACCTCCTAATAGCAATACACTG
GGCTAAGCGTTTTGAGCTGC
GCCTGACTGGCATTGTATTAGC
GTCTAGGGTGTAGCCTGAGAATAG

Table Legend:

Haplogroup: *mtDNA* haplogroup designation.
Polymorphism: SNP or indel used for discrimination.
Primer Name & Sequence: Forward (6FAM-labeled) and reverse primer sequences.
PCR Product Size: Expected amplicon size.
Digestion Enzyme: Restriction enzyme used to detect the polymorphism.
Pattern/Lineage: Evolutionary lineage or region.
Positive Fragment (s): Fragment size (s) when the enzyme cuts (polymorphism present).
Negative Fragment (s): Fragment size (s) when the enzyme does not cut (polymorphism absent)

Key Notes

Primer Labels:

All forward primers (L-series) are labeled with 6 (FAM at the 5’ end for fluorescence detection.

Reverse primers H-series) are unlabeled.

Digestion Patterns:

Positive Peak: Represents fragment size (s) when the polymorphism is present (enzyme cuts the site).

Negative Peak: Represents fragment size (s) when the polymorphism is absent (no cut).

Electrophoresis:

Fragments analyzed via capillary electrophoresise.g., (ABI 3500xL).

Peak positions correspond to labeled fragment sizes in base pairs (bp).

Lineage-Specific Diagnostics:

Haplogroup	**Evolutionary Lineage**	**Common Subspecies**
A	African	*A. m. scutellata, A. m. Capensis*
M	West European	*A. m. Mellifera*
C	East European	*A. m. ligustica, A. m. Carnica*
Z	Mediterranean	*A. m. Cypria*
Y	Yemenite	*A. m. Yemenitica*
O	Oriental	*A. m. syriaca, A. m. Anatoliaca*
H	Balkan	*A. m. adami*Crete)

References

Primer Design: Adapted from (Arias et al.1996) *Molecular Phylogenetics and Evolution*.

Restriction Enzymes: Selected based on lineage-specific SNP databasese.g., (Whitfield et al., 2006 *Science*).

Fragment Sizes: Calculated using *in silico* digest toolse.g., (NEBcutter).

For regional subspecies validation, combine this protocol with the PCR-RFLP method for COI-COII analysis.

Steps Involved in Restriction Fragment Length Polymorphism (RFLP), (https://microbenotes.com/restriction-fragment-length-polymorphism-rflp/).

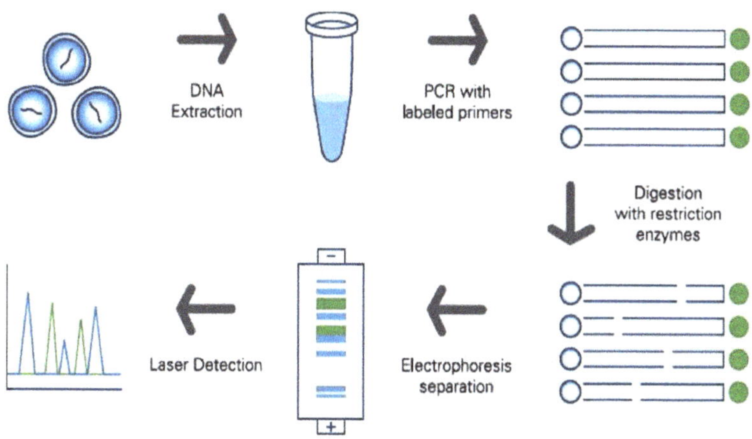

Chapter 15

15. Bioinformatics Pipeline for Mitochondrial DNA-Based Phylogenetic Analysis in *Apis mellifera* Subspecies

By looking at mitochondrial DNA (mtDNA) sequence data from different populations and subspecies, bioinformatics methods have become essential for figuring out the evolutionary history of organisms like *Apis mellifera* (Sarhaddi Avval et al., 2024). These methods include a series of connected steps, starting with getting and checking the quality of sequence data and ending with making phylogenetic trees that show how species are related through evolution (Jacques et al., 2023). We go into excellent detail about each step of the bioinformatics pipeline and the ideas behind the methods used in this chapter. We pay special attention to alignment tools like MUSCLE, MAFFT, and ClustalW, as well as phylogenetic inference platforms like MEGA and RAxML (Jacques et al., 2023).

The first step in this process is getting raw sequence data, which is usually taken from well-maintained public databases like NCBI or GenBank (Sarhaddi Avval et al., 2024). Before moving on to the alignment step (Tihelka et al., 2020), **quality control** steps must be taken. These include checking for sequencing errors and making sure that all of the mtDNA regions are present. The next step in the research is **to align the sequence data** to make sure that homologous positions are compared correctly (Jacques et al., 2023).

Multiple sequence alignment (MSA) is an important step in bioinformatics analyses because it lays the ground work for all the evolutionary conclusions that follow. The positions in the alignment show nucleotide or amino acid residues that are thought to be related through evolution (Sarhaddi Avval et al., 2024). There are many algorithms that can help you obtain accurate alignments. MUSCLE, MAFFT, and ClustalW are three of the most popular ones because they are fast and accurate (Jacques et al., 2023). These tools use different methods: MUSCLE uses iterative refinement and profile alignment to improve similarity scores (Tihelka et al., 2020), MAFFT uses fast Fourier transformation to accelerate the alignment process while keeping it accurate, and ClustalW uses a hierarchical clustering method that compares pairs of sequences and aligns them progressively (Jacques et al., 2023).

The main idea behind multiple sequence alignment is to put the sequences in a matrix so that each column represents the same evolutionary position among the sequences (Jacques et al., 2023). This helps identify stable areas that are important for function or structure and changing areas that might be influenced by evolution or differences between species (Sarhaddi Avval et al., 2024). Researchers pay special attention to AT-rich noncoding regions when studying mtDNA because they are very variable and can create gaps that are difficult to align (Sarhaddi Avval et al., 2024).

The next step after making the first alignments with MUSCLE, MAFFT, or ClustalW is to **identify and eliminate any unclear areas** that could change the way we approach evolutionary relationships (Jacques et al., 2023). When it comes to filtering out unclear segments, labs often choose to either look at these areas by hand using visualization tools or use specialized software to automatically trim poorly aligned areas. Examples of computational methods used for this purpose are trimAl, BMGE, and Guidance2 (Sarhaddi Avval et al., 2024). The reason for omitting unclear areas is that positions that are misaligned or unclear can lead to the wrong identification of polymorphisms and possibly throw off the phylogenetic reconstruction (Jacques et al., 2023). The curated set of aligned sequences, **after alignment and trimming, forms a dataset for polymorphism detection** (Tihelka et al., 2020). Finding polymorphisms means searching for places in the alignment where the sequences have different nucleotides, like single nucleotide polymorphisms (SNPs), insertions, or deletions (Jacques et al., 2023). It is important to find these variable sites because they are molecular markers that show how different subspecies are genetically and how they have evolved under different pressures (Moškrič et al., 2022). MEGA and other software offer a wide range of tools for not only visualizing alignments but also for calculating genetic distances, estimating nucleotide diversity indices, and finding polymorphic loci (Tihelka et al., 2020).

The polymorphism detection phase often lists the variable sites along with estimates of genetic diversity, like haplotype diversity, nucleotide diversity (π), and the ratio of nonsynonymous to synonymous substitutionsd (N/dS). These numbers can show how strong and what kind of natural selection is happening (Sarhaddi Avval et al., 2024). Researchers calculate these indices across the mtDNA sequences, providing a numerical measure of variability that aids in subsequent analyses of evolutionary relationships (Sarhaddi Avval et al., 2024). Researchers can also use the polymorphic sites they found to split the dataset into groups that might be related to different phylogenetic clades or subspecies (Sarhaddi Avval et al., 2024).

After the alignment has been improved and polymorphisms have been found, the next important step is **to build a phylogenetic tree**, which shows how the sequences have evolved (Jacques et al., 2023). There are two main types of phylogenetic inference methods: distance-based methods and character-based methods. Maximum parsimony and maximum likelihood are two examples of character-based methods. Maximum likelihood (ML) methods are preferred because they are more statistically sound (Jacques et al., 2023). When looking at *Apis mellifera* subspecies, ML-based methods are especially helpful because they can handle complicated models of nucleotide substitution and rate differences between sites (Tihelka et al., 2020).

Software platforms like MEGA and RAxML are very important for doing ML phylogenetic analyses (Jacques et al., 2023). MEGA is a user-friendly, all-in-one program that lets you do sequence alignment, model testing, and build trees using different algorithms like Neighbor-Joining, Maximum Parsimony, and Maximum Likelihood (Sarhaddi Avval et al., 2024). **RAxML**, on the other hand, is **better at working with big datasets** and complicated models. It uses rapid bootstrapping methods to statistically back up the inferred tree topology (Jacques et al., 2023). When looking at mtDNA sequences (Attardo et al., 2019), it's important to choose the right model of nucleotide substitution. GTR (General Time Reversible) models, for instance, are excellent options since they account for potential rate heterogeneity and fluctuating substitution rates.

It usually takes software tools like Model Test or Partition Finder to test different evolutionary models and see which ones are the best. These tools use statistical criteria like the Akaike Information Criterion (AIC) or Bayesian Information Criterion (BIC) to do this (Tihelka et al., 2020). Once the best model is chosen, the phylogenetic inference software uses it to figure out branch lengths, build the tree's shape, and find support values using bootstrapping or Bayesian posterior probabilities (Sarhaddi Avval et al., 2024). The phylogenetic trees show how species are related to each other. The lengths of the branches often show how many changes happened at each site, and the clade support values show how reliable the inferred relationships are (Sarhaddi Avval et al., 2024).

To check how statistically reliable the tree nodes are, **bootstrapping** is a common method that involves resampling the alignment data many times to make copies of the trees (Jacques et al., 2023). High bootstrap values, usually over 70–80%, mean that the branches are strongly supported. On the other hand, low values may mean that there is uncertainty or that the branches are evolving quickly (Tihelka et al., 2020). When the phylogeny is unclear, complementary methods like Bayesian inference (BI) use Markov chain Monte Carlo (MCMC) methods to figure out the posterior probability of tree topologies (Jacques et al., 2023). BI can use complicated models like the CAT-GTR+G mixture model, which is effective at dealing with sequence heterogeneity and compositional bias. This gives BI a different way to look at things than maximum likelihood methods (Tihelka et al., 2020).

In some studies, researchers create multiple datasets from the aligned sequences to avoid issues related to sequence saturation, particularly at the third codon position (Sarhaddi Avval et al., 2024). For instance, separate analyses could be done on datasets that only include the first and second codon positions (P12) and omit the third codon positions, which are more likely to become saturated. Alternatively, researchers could conduct analyses on datasets containing ribosomal RNA genes to integrate variable and conserved markers (Tihelka et al., 2020). It is possible to learn more about how strong the inferred relationships are by comparing phylogenetic trees made from these different datasets. Such an approach also helps figure out if some evolutionary patterns stay the same regardless of what dataset is used (Tihelka et al., 2020).

Another important part of bioinformatics analysis (Sarhaddi Avval et al., 2024), involves finding and understanding polymorphisms linked to the evolutionary history of populations. To find polymorphisms, you have to look at the aligned mtDNA sequences and find nucleotide substitutions, insertions, and deletions that are different between the subspecies (Jacques et al., 2023). To measure these differences, scientists often count the number of polymorphic sites and look at diversity indices like nucleotide diversity (π) and haplotype diversity (Sarhaddi Avval et al., 2024). The distribution and frequency of these polymorphisms can tell us about past demographic events like population growth, bottlenecks, or gene flow between subspecies (Sarhaddi Avval et al., 2024).

The dN/dS ratio is an important part of figuring out what selective pressures are acting on mitochondrial genes (Sarhaddi Avval et al., 2024). It compares the rate of nonsynonymous substitutions (dN) to the rate of synonymous substitutions (dS). A dN/dS ratio of less than one usually means that purifying selection is happening, which means that undesirable mutations are being removed from the population (Sarhaddi Avval et al., 2024). These kinds of studies are very useful for protein-coding areas like the cytochrome oxidase I (COI) gene, which is often used as a molecular marker in studies of honey bee phylogeny (Sarhaddi Avval et al., 2024).

Following the alignment and polymorphism discovery steps, we create a phylogenetic tree to illustrate the evolutionary history of subspecies (Jacques et al., 2023). During this step, the improved alignment, which is now free of unclear areas and full of known polymorphisms, is fed into tree-building software (Tihelka et al., 2020). Researchers new to phylogenetic analysis often use MEGA because it has a full set of tools for molecular evolutionary genetics analysis and is easy to use (Sarhaddi Avval et al., 2024). RAxML is more suitable when working with larger datasets or when the analysis requires the robust statistics provided by maximum likelihood methods. It has rapid computation and reliable bootstrapping methods for checking node support (Jacques et al., 2023).

These tools can make phylogenetic trees, and they also let researchers improve these trees by using statistical tests to make sure that the evolutionary relationships they see aren't just the result of alignment noise or the wrong model (Jacques et al., 2023). For instance, if different methods produce tree topologies that don't match up, it might be worth looking into further, like going back over the alignment to get rid of more unclear areas or trying out different evolutionary models with Partition Finder (Tihelka et al., 2020). To accurately show the evolutionary history of *Apis mellifera subspecies* (Jacques et al., 2023), researchers need to make these kinds of changes repeatedly.

One problem with studying *Apis mellifera* mtDNA is that it has areas **with a lot of A+T**, which can make it difficult to line up and cause fake gaps or misalignments (Sarhaddi Avval et al., 2024). **To fix this**, the alignment tools mentioned above need to be set up carefully, which often means **changing the gap penalties or using amino acid translations when aligning protein-coding regions** to give the alignment a biologically sound basis (Jacques et al., 2023). Often, converting nucleotide sequences into amino acids can help with the alignment process by taking advantage of the fact that protein sequences are more stable than their underlying nucleotide sequences. This reduces the amount of ambiguity in the alignment (Tihelka et al., 2020).

After the phylogenetic tree is made and its statistics are looked at, the next step is to figure out **how the different subspecies** of *Apis mellifera* are **related to each other in terms of evolution**. Strong bootstrap support for clades means that they are thought to be separate evolutionary lineages. The tree's shape can also provide us with information about how genes have moved around in the past, how populations have split up, and where they live (Sarhaddi Avval et al., 2024). In many instances, phylogenies indicate that subspecies from various regions of the world can cluster together, potentially due to factors such as migration or hybridization (Sarhaddi Avval et al., 2024).

Bayesian inference methods are now widely used in phylogenetic studies because they can manage complex evolution models and provide direct estimates of the likelihood of different groups (clades) (Jacques et al., 2023). These methods are used along with traditional ML and distance methods. Bayesian methods, which are used in programs like MrBayes and PhyloBayes, work well with ML methods from RAxML because they provide different ways to check how reliable the tree topology is (Tihelka et al., 2020). Researchers can compare phylogenetic trees made with different assumptions about substitution rates and model parameters by combining these methods. Such evidence indicates that the inferred evolutionary relationships are strong (Jacques et al., 2023).

The entire bioinformatics pipeline is iterative and involves getting sequence data, checking its quality, aligning multiple sequences, dealing with ambiguous regions, finding polymorphisms, and building a phylogenetic tree (Jacques et al., 2023). Researchers might have to revisit earlier steps if later results, like weak bootstrap support or surprising tree shapes, indicate that the sequence alignment or model choice needs improvement (Jacques et al., 2023). It is important that these analyses are done multiple times to make sure that the bioinformatics results are strong and show real evolutionary patterns instead of just being the result of the analysis process (Jacques et al., 2023).

For example, after making a preliminary phylogenetic tree with MEGA or RAxML, one might notice that some clades have low support values. This would lead to a second look at the data for misaligned regions or recombination events (Miller et al., 2010). Researchers often use other programs to do extra analyses, like **network analyses or coalescent methods**, to confirm the phylogenetic signal in their mtDNA datasets (Moškrič et al., 2022). This multifaceted approach ensures that the phylogenetic inference is both statistically sound and biologically meaningful (Njeri, N & Njeri, 2024).

It's important to understand that the bioinformatics methods described here can be used on many other organisms besides *Apis mellifera*. For example, they can be used to figure out the evolutionary history of other organisms by looking at their mtDNA sequences (Schmidt et al., 2023). For instance, researchers have used similar methods to look at the evolutionary relationships between beetles, stingless bees, and other insect groups. The evidence shows that there is a universal framework that combines high-throughput sequencing data, strict alignment protocols, and strong phylogenetic inference (Tihelka et al., 2020). Because these methods can be used in many different fields of systematics and evolutionary biology, they help make evolutionary studies more consistent (Tihelka et al., 2020).

High-Performance Computing (HPC) clusters and science gateways like CIPRES have made it possible for researchers to run resource-intensive analyses like maximum likelihood phylogeny reconstructions on enormous datasets (Miller et al., 2010). These tools have also made a lot of progress recently. These platforms have easy-to-use graphical interfaces that make it easier to do complex evolutionary analyses, even for people who don't know much about computers (Jacques et al., 2023). HPC and cloud-based solutions have made tools like RAxML much faster and more scalable. Such technology has made it easier to do more in-depth phylogenetic studies of species that are found all over the world, like the Western honeybee (Jacques et al., 2023).

Researchers have also started using integrative bioinformatics strategies more and more. These strategies combine sequence data with other types of biological information, like geographical distribution, ecological data, and morphometric characteristics, to learn more about how subspecies evolve (Sarhaddi Avval et al., 2024). These kinds of approaches from different fields add to the phylogenetic framework by giving molecular data a context in which to understand it in terms of ecological and evolutionary processes (Jacques et al., 2023). Scientists can figure out where *Apis mellifera subspecies* came from and how they got there by combining sequence data with field observations and ecological niche modeling (Sarhaddi Avval et al., 2024).

In short, the bioinformatics process for figuring out how different subspecies of *Apis mellifera* are related to each other starts with carefully collecting and checking the quality of mtDNA sequences. Next, scientists use tools such as MUSCLE, MAFFT, or ClustalW to align the sequences and identify homologous regions (Sarhaddi Avval et al., 2024). We carefully construct this alignment to eliminate difficult-to-read areas commonly found in AT-rich noncoding regions. The process makes sure that the data is of high quality for later analyses (Sarhaddi Avval et al., 2024). After establishing a robust alignment, polymorphism detection searches for variable positions that serve as markers for genetic differences among populations (Jacques et al., 2023). Lastly, we create phylogenetic trees using platforms such as MEGA and RAxML. These trees show how species have evolved over time and give us information about subspecies boundaries and their historical relationships (Jacques et al., 2023).

Every step is crucial in the process of studying evolution, and it's important to carefully consider every detail, like adjusting the alignment settings and selecting the appropriate substitution models, to ensure that the conclusions about evolution are both statistically reliable and biologically relevant (Tihelka et al., 2020). Combining these bioinformatics methods has revolutionized the process of deciphering complex evolutionary histories. It has also helped us learn a lot more about how species like *Apis mellifera* have changed and become more diverse over time (Sarhaddi Avval et al., 2024).

Furthermore, as sequencing technologies get better and make bigger, more detailed datasets, the bioinformatics tools and methods talked about in this chapter will definitely change to keep up with these changes (Jacques et al., 2023). Better computer algorithms and more powerful computers will make it easier to do even more precise alignments, identify subtle polymorphisms more accurately, and make phylogenetic inferences that are more reliable (Jacques et al., 2023). As a result, these improvements will make it possible to learn more about the evolutionary processes that lead to speciation and subspecies divergence. Such advances will allow researchers to keep using molecular data to answer basic questions in evolutionary biology (Jacques et al., 2023).

To sum up, the step-by-step bioinformatics method for analyzing mtDNA in *Apis mellifera* has several stages that depend on each other. First, established public databases are used to check the quality of the mtDNA sequences that were obtained. Next, we use MUSCLE, MAFFT, or ClustalW to perform a strict multiple sequence alignment, ensuring homologous comparisons (Sarhaddi Avval et al., 2024). Second, any alignment regions that aren't clear are either fixed or left out of further analysis to reduce the chance of making mistakes in later analyses (Jacques et al., 2023). Thirdly, polymorphism detection is done with tools like MEGA, which find differences in nucleotides and give estimates of genetic diversity indices that are important for understanding how evolution works (Tihelka et al., 2020). Finally, we use advanced machine learning methods in MEGA and RAxML to build a phylogenetic tree using the curated data. This graph shows how the different subspecies of *Apis mellifera* have evolved over time (Jacques et al., 2023).

This complete pipeline gives researchers a strong way to understand the genetic differences between and within honey bee populations. It also gives them a glimpse into the long-term processes of evolution that shape biodiversity around the world (Sarhaddi Avval et al., 2024). The bioinformatics community is well-prepared to deal with the problems that come with larger genomic datasets and to help us learn more about evolutionary biology (Jacques et al., 2023). They do this by constantly improving alignment methods, polymorphism detection algorithms, and phylogenetic inference software (Jacques et al., 2023).

Chapter 16

16. Bayesian Inference and Maximum Likelihood in Molecular Phylogenetics

16.1. Introduction: The Pillars of Molecular Phylogenetics

Bayesian inference and maximum likelihood (ML) methods are two of the most important tools in modern molecular phylogenetics. When used together, they give us a complete picture of the evolutionary history stored in mitochondrial genomes, like those found in honeybees (Nascimento et al., 2017). Many resesrchers use these methods because they each have their strengths. For example, Bayesian methods give posterior probability estimates that show how confident you are in each clade, and ML methods use bootstrap replicates to check how strong the inferred tree structure is (F. Song et al., 2016).

16.2. Bayesian Inference: Principles and Implementation

16.2.1. Bayesian Framework and Prior Knowledge

Bayesian inference is based on Bayes's theorem, which lets you combine what you already know with observed sequence data to find the posterior probability of a phylogenetic tree and its parameters (Yang & Rannala, 2012). This framework needs a prior distribution over tree topologies and evolutionary model parameters, which can be very flexible to include information from past studies or expert opinion (Nascimento et al., 2017). Bayesian methods look at the high-dimensional space of possible trees by creating a posterior distribution that shows how uncertain the data is and how complicated the evolutionary process is (Ronquist et al., 2012). They do this by using sampling techniques like Markov chain Monte Carlo (MCMC).

16.2.2. Modeling Evolutionary Heterogeneity

The Bayesian method is very good at including complicated evolutionary models that take into account different substitution rates among nucleotide sites, changing base frequencies, and different patterns of rate change over time (He et al., 2018). MrBayes and other programs use these models by letting users choose partitioned datasets and the substitution models that go with each partition. This makes the model fit better for different gene regions or codon positions in the mitochondrial genome (Ronquist et al., 2012). When working with mtDNA, it's important to be able to model evolutionary heterogeneity because different genes or even different parts of the same gene may evolve under different mutation rates and selective pressures (Condamine et al., 2018).

16.2.3. Probabilistic Clade Support and Interpretability

Bayesian inference is powerful because it can give direct probabilistic estimates of clade support through posterior probability values, which show how likely it is that a certain clade is correct given the model and data (D. Henriques, Chávez-Galarza, et al., 2019). If the model is correct, a clade with a posterior probability of **0.95** means that there is a 95% chance that it is real (Nascimento et al., 2017). This ability to understand results is particularly helpful for uncovering complex relationships in honeybee mitochondrial phylogenies, where slight changes in sequence can lead to confusing results if not modeled properly (Tihelka et al., 2020).

16.2.4. Model Averaging and Relaxed Clocks

One big benefit of Bayesian methods is that they can deal with model uncertainty by using techniques like reversible-jump (MCMC), which lets you compare and average over several substitution models instead of just one fixed model (Nascimento et al., 2017). This averaging of models lowers the chance of bias caused by model misspecification and helps make clade support measures stronger (Yang & Rannala, 2012). Bayesian frameworks can also handle relaxed molecular clock models that let rates change between lineages. This means that you can estimate the time of divergence even when mutation rates aren't the same across the phylogeny (Condamine et al., 2018).

16.3. Maximum Likelihood Approaches

16.3.1. Principles and Practice

16.3.1.1. Likelihood Optimization and Tree Search

Maximum likelihood (ML) methods, on the other hand, work by finding the tree topology and branch lengths that make the observed sequence data most likely to happen given a certain substitution model (Yang & Rannala, 2012). ML methods are based on well-known statistical rules and are quick to compute, especially when used in software that uses fast tree search algorithms and heuristics to find its way through the huge number of possible trees (F. Song et al., 2016). One important part of **ML** inference is checking how reliable each branch is by **using bootstrap replicates**. This means resampling the dataset over and over to see how often certain clades are found (Eimanifar, Kimball, et al., 2018).

16.3.1.2. Bootstrap Analysis and Support Values

In ML, bootstrap analysis means making a lot of fake replicate datasets (usually hundreds or thousands) and then using them to recreate the phylogeny for each one (Hassler et al., 2023). The percentage of replicates that show a clade gives a bootstrap support value, which is a frequentist way to measure confidence (García-Sandoval, 2014). ML bootstrap values and Bayesian posterior probabilities are usually understood in different ways. Bootstrap values are often considered more conservative than Bayesian support. However, both of these values help confirm that the inferred phylogenetic hypotheses are strong (He et al., 2018).

16.3.1.3. Assessing Stability and Robustness

The iterative process of bootstrap resampling lets ML methods show how stable certain branching patterns are when the data changes, which helps find nodes that might be sensitive to sampling error or model assumptions (Barrow et al., 2015). Researchers can be more sure that the inferred topology is not just a result of certain data partitions or model choices when different ML bootstrap analyses give the same results (Eimanifar, Kimball, et al., 2018). So, ML methods give us another way to check the robustness of phylogenetic trees that works with Bayesian posterior probabilities (Tihelka et al., 2020).

16.4. Application to Honey Bee Mitochondrial DNA Evolution

16.4.1. Complexity of Honey Bee mtDNA

Because mtDNA is so complicated and changes so quickly, both Bayesian inference and ML methods have been very useful for studying its evolution (He et al., 2018). Honey bee mtDNA has a lot of mutations, and different genes often have different substitution patterns, so we need to use complicated models that can accurately show how these changes happen over time (Hassler et al., 2023). In these situations, Bayesian methods not only guess the shape of the tree, but they also present you a lot of information about the model's parameters, like how branch lengths and substitution rates change. This information is crucial for understanding how evolution works (Nascimento et al., 2017).

16.4.2. Advanced Modeling and Temporal Frameworks

Bayesian inference tools like MrBayes can handle the complexity and randomness of molecular evolution in honey bee mitochondria by using advanced models like the general time-reversible model with gamma-distributed rate heterogeneity (Condamine et al., 2018). Researchers can also use fossil calibrations or other biological evidence to set priors, which can help narrow down estimates of when species diverged and provide a time frame for the evolutionary history of *Apis mellifera* (Tihelka et al., 2020). The Bayesian posterior probabilities that came out of this study not only confirm the clade structure in the phylogeny, but they also help us make new guesses about how biogeography and lineage diversification happened in the past (D. Henriques, Chávez-Galarza, et al., 2019).

16.4.3. ML Analyses and Empirical Support

At the same time, ML methods have been used successfully to look at concatenated datasets with multiple mitochondrial genes, giving strong bootstrap support values for nodes in the phylogenetic trees that come out of the analysis (Eimanifar, Kimball, et al., 2018). ML analyses are especially adept at finding the overall tree topology that best explains the observed patterns of nucleotide substitutions (F. Song et al., 2016), because they optimize a likelihood function over the whole dataset. This process of maximizing, along with many bootstrap replicates, makes sure that the ML-inferred tree has strong empirical support. This draws the conclusions drawn from the data more reliable (Ronquist et al., 2012).

16.4.4. Mitigating Methodological Risks

Using both Bayesian inference and ML together not only presents you more statistical evidence, but it also lowers the risks that come with using only one method (Hassler et al., 2023). For example, Bayesian posterior probabilities can sometimes overestimate clade support when certain model conditions are met. On the other hand, ML bootstrap values are usually more conservative, which is a useful way to double-check that you aren't too confident in the Bayesian estimates (He et al., 2018). In cases where model misspecification or data partitioning problems could cause topological differences, it is critical that the two approaches work well together (García-Sandoval, 2014).

Furthermore, combining both methods can help identify possible systematic errors, like long-branch attraction (LBA), which is a big problem in studies of mitochondrial genomes that change quickly (Talavera & Vila, 2011). Bayesian inference, particularly with site-heterogeneous models like the CAT model in PhyloBayes, has been proven to reduce long-branch attraction (LBA) more effectively than the standard homogeneous models typically used in maximum likelihood (ML) analyses (F. Song et al., 2016). In these cases, using advanced Bayesian models makes us more sure of our results because they clearly show the differences in substitutions and reduce the impact of compositional bias on the inferred phylogeny (Yang & Rannala, 2012).

Honey bee mitochondrial genomes are small (Eimanifar, Kimball, et al., 2018). Because mtDNA evolves so quickly, some areas can become saturated with substitutions. To obtain reliable phylogenetic reconstructions, it is important to choose the right datasets, such as leaving out third codon positions or certain non-coding regions (He et al., 2018). By letting you choose models and split data in ways that maximize the balance between resolution and accuracy (Condamine et al., 2018), both Bayesian and ML methods can be made to work with these data-specific problems.

Researchers often use both methods to do parallel analyses in real life to compare the topologies that come out and see how consistent the branching patterns are (Eimanifar, Kimball, et al., 2018). For example, if a certain clade has both high Bayesian posterior probabilities and high ML bootstrap values, this greatly increases confidence in the evolutionary relationships (Tihelka et al., 2020). If the methods disagree, it may mean that the model isn't adequate enough or that there are alignment issues, which would require more research and data refinement (Hassler et al., 2023).

When comparing Bayesian and ML trees, it is common to use statistical tests and model selection criteria. For example, in the Bayesian framework, Bayes factors are used, and in ML analyses, the Akaike Information Criterion (AIC) is used (Nascimento et al., 2017). These metrics supply a formal way to compare how well different evolutionary models work, and they are essential for making sure that the inferred phylogenetic relationships are strong even when the model is not certain (Yang & Rannala, 2012). In this way, both methods can be improved by using partitioning strategies that assign different models to different parts of the data based on things like gene identity or codon position. Such an approach makes the analysis even more accurate (Condamine et al., 2018).

Another benefit of using complementary methods is that they let researchers look at all the different types of phylogenetic uncertainty (Nascimento et al., 2017). Bayesian methods supply a natural estimate of uncertainty by making a distribution of plausible trees, each weighted by its posterior probability. ML methods, on the other hand, capture uncertainty through the variability seen in bootstrap replicates (He et al., 2018) . This detailed description of uncertainty is essential for later analyses like estimating divergence time, biogeographical reconstructions, and studies of evolutionary trait correlations (Hassler et al., 2023).

When it comes to the evolution of honey bee mtDNA, using both Bayesian and ML methods has given us important information about how subspecies are related to each other and how the mitochondrial genome changes over time (Eimanifar, Kimball, et al., 2018). For instance, research has shown that honeybee populations have many different haplotypes, but their nucleotides are not very different from each other. This pattern may be due to historical bottlenecks or selective sweeps that have affected the mitochondrial genome (D. Henriques, Chávez-Galarza, et al., 2019). Bayesian posterior probabilities and ML bootstrap replicates both provide strong statistical support to these results. Such evidence makes them more believable and makes sure that the inferred phylogenetic signal is not just a result of methodological bias (Tihelka et al., 2020).

Furthermore, these two methods have helped solve complicated phylogenetic problems where incomplete lineage sorting, introgression, or gene flow could make it difficult to see the real evolutionary history (He et al., 2018). In these situations, Bayesian methods can provide more detailed interpretations by taking into account the uncertainty in gene trees and model parameters. ML analyses, on the other hand, present a different view that helps confirm the inferred relationships (Bossert et al., 2020). This teamwork is particularly crucial in groups like honeybees, where mixing and sharing of mitochondrial haplotypes between populations can make it difficult to understand their evolutionary relationships (Eimanifar, Kimball, et al., 2018).

But there are still problems with making sure that both Bayesian and ML analyses are done in the best way possible. Bayesian MCMC analyses can be very demanding on computers, especially when working with large datasets or complicated models. To ensure proper convergence of the MCMC chains, it is often necessary to closely monitor them and allow them to run for an extended period (Ronquist et al., 2012). In the same way, the choice of resampling parameters in ML bootstrap analyses can affect the stability of the support values. If biases in the data, such as differences in composition, are not properly addressed during model selection and data splitting, they can negatively impact how well both methods work (Talavera & Vila, 2011). Recent methodological advances have focused on making models more accurate and computers more efficient to overcome around these problems. For instance, the addition of high-performance computing libraries like BEAGLE has sped up likelihood calculations by a significant amount, making it possible to use complex Bayesian models on whole mitochondrial genomes (Baele &

Lemey, 2013). The creation of better machine learning software that uses improved search methods and more precise support estimates has made phylogenetic reconstructions based on machine learning more trustworthy (Blair & Ané, 2019).

Additionally, modern phylogenomic studies are placing more and more importance on doing sensitivity analyses by changing the models and parameters used in the analyses (Nascimento et al., 2017). Researchers can confirm if the inferred phylogenetic signal is strong enough to hold up under the assumptions of each method by comparing results from different methodological frameworks. For honey bee mtDNA in particular, this kind of cross-validation has been crucial for confirming theories about how different subspecies have spread and changed over time, as well as for finding possible cases of introgression or mis-assignment of lineages (Condamine et al., 2018).

Combining Bayesian and ML methods has also made it possible to do more thorough analyses that combine data from different sources. For example, researchers can create more detailed family trees that show both maternal and biparental inheritance patterns by combining mitochondrial data with nuclear genome information in a comprehensive approach (He et al., 2018). Not only do these kinds of integrative methods draw the phylogenetic conclusions stronger, but they also help us understand how genomes evolve, like how selection, drift, and demographic events affect different parts of the genome (Hassler et al., 2023).

In short, combining Bayesian inference and maximum likelihood is a highly effective method for accurately supporting the reconstruction of mitochondrial phylogenies (Ronquist et al., 2012). Bayesian methods offer us detailed probabilistic measures of clade support through posterior probabilities. They also let us use complex evolutionary models that take into account the fact that different substitution processes can happen at different rates (He et al., 2018). At the same time, ML methods offer strong bootstrap support values that can be used as an independent measure of clade reliability and help confirm the overall tree structure (F. Song et al., 2016). These complementary methods lower the chance of getting the wrong idea because of model misspecification or data quirks. This means that the phylogenetic conclusions are both statistically sound and biologically meaningful (Talavera & Vila, 2011).

Researchers who are studying how honeybee mitochondrial DNA has changed over time have found that using both of these methods together has made it easier to see complicated patterns of diversification, adaptive evolution, and historical demographic processes that are difficult to understand with just one method (D. Henriques, Chávez-Galarza, et al., 2019). The high mutation rate and unique inheritance patterns of mtDNA make it important to be cautious when figuring out how temperate, tropical, and hybrid populations of *Apis mellifera* are related (Eimanifar, Kimball, et al., 2018). Advancements in computational phylogenetics have facilitated increasingly detailed analyses. Such information presents us a better understanding of the genetic basis of honeybee diversity and the evolutionary forces that shape it (Hassler et al., 2023).

Furthermore, the iterative process of model testing, sensitivity analysis, and cross-validation that comes with using both Bayesian and ML methods makes it less likely that artifacts like long-branch attraction will show up. These can lead to wrong conclusions about phylogenetic relationships in rapidly evolving mitochondrial genomes (F. Song et al., 2016). By carefully changing evolutionary models to fit the unique characteristics of mtDNA data, researchers can build phylogenies that truly reflect the actual history of organisms and show the small differences in branch length and rate between different groups (Yang & Rannala, 2012).

These methods present us more than just information about tree topology. We can expand Bayesian frameworks to include divergence time estimation under relaxed-clock models. Such analysis presents us chronological views of evolutionary events that happen at the same time as changes in the environment or biogeographical processes (Condamine et al., 2018). On the other hand, ML methods provide us strict likelihood-based estimates that make the inferred phylogenetic relationships more believable, especially when tested through many bootstrap simulations (Eimanifar, Kimball, et al., 2018).

In practice, the first step in a solid phylogenetic study is usually organizing the data, which means carefully aligning and splitting mitochondrial sequences into gene sections or codon positions to obtain the clearest phylogenetic signal (He et al., 2018). Next, a series of model selection procedures take place, often using tools like PartitionFinder to identify the best substitution models for each partition (Condamine et al., 2018). Researchers then do parallel Bayesian and ML analyses after getting the data ready and picking the right models. They make sure to compare the results from both methods and look into any differences (Nascimento et al., 2017).

The last step in these kinds of in-depth studies is to look at the phylogenetic trees. At each node, the reader sees both Bayesian posterior probabilities and ML bootstrap values, which provide a more complete picture of how reliable the tree is (Ronquist et al., 2012). When the two methods come up with a similar topology that has a lot of statistical support, the resulting phylogeny is considered strong. Such an outcome makes it easier to trust future biological interpretations, like finding monophyletic clades, ancestral relationships, or patterns of geographic dispersal (He et al., 2018).

In the end, using both Bayesian and ML methods together has entirely changed the field of molecular phylogenetics. This has made it possible for scientists to answer questions that were previously impossible to answer about the evolution of mitochondrial genomes in honeybees and other organisms (Hassler et al., 2023). The innovative approaches demonstrated by these methods are not merely superficial; they significantly impact conservation biology, taxonomy, and our overall understanding of how evolutionary processes influence biodiversity (Blair & Ané, 2019).

As computer tools acquire better and bigger genomic datasets become available, it is expected that combining different phylogenetic methods will reveal us even more information about when, how, and at what rate evolution happens in mitochondrial genomes (Baele & Lemey, 2013). This ongoing improvement in methods will help us put together a more detailed and accurate picture of the evolutionary history of honeybees. These findings will help us understand the genetic processes that drive adaptation, speciation, and population divergence over time (Talavera & Vila, 2011).

In summary, using both Bayesian inference and maximum likelihood methods together in studying the evolutionary relationships in honey bee mitochondrial genomes is a powerful and beneficial approach (Eimanifar, Kimball, et al., 2018). The Bayesian framework's ability to combine prior knowledge and complicated evolutionary models provides a probabilistic measure of support that is essential for nuanced interpretations. The ML approach's use of bootstrap replicates provides an independent, data-driven assessment of node stability (Ronquist et al., 2012). Together, these methods ensure that multiple lines of evidence support the branching patterns observed in mtDNA. Such an approach makes phylogenetic conclusions more reliable and strong overall and allows for a more in-depth study of mtDNA evolution in honeybees (He et al., 2018).

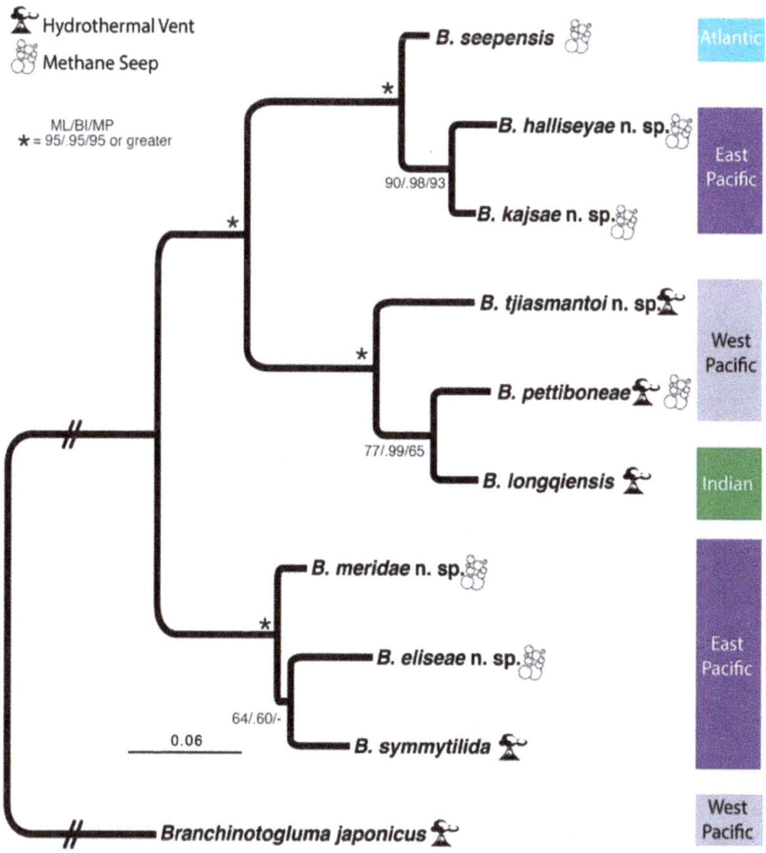

Fig. 20. Maximum likelihood (ML) and Bayesian inference tree from the analysis of the combined sequences from COI, 16S, and ITS.

The topology for the single maximum parsimony (MP) tree differed in showing *Branchipolynoe eliseae n. sp.* as sister to *B. meridae n. sp.* Numbers on relevant nodes are ML bootstrap BI posterior probability and MP jack knife below. * indicates 95% bootstrap/jackknife or greater and 0.95 posterior probability or greater; '-' indicates the node was not found (Lindgren et al., 2019).

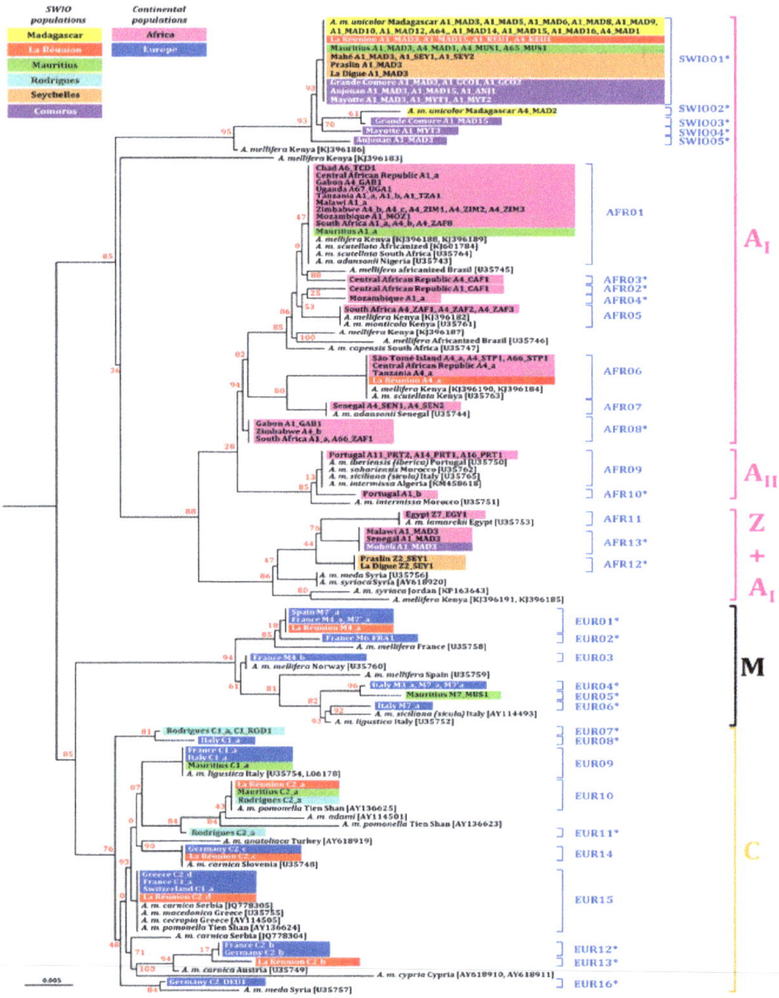

Fig. 21. Phylogenetic relations among ND2 haplotypes linked to COI-COII haplotypes and geographical origins support a South West Indian Ocean *A. mellifera* African sub-group.

The tree is an Approximately-Maximum-Likelihood phylogenetic tree based on ND2 haplotype and using GTR + CAT model. The tree was based on the 34 obtained ND2 sequences, including 49 references sequences from GenBank in (brackets). For each ND2 haplotype, the origin of the sample is indicated (South West Indian Ocean islands, African or European countries) and highlighted by a color. Additionally, the observed associated COI-COII haplotype is also specified for each island or country taxon. Only one individual of each ND2–COI-COII and location are represented in the tree for better visibilitysee also Additional file 1: Table (S4). Bootstraps are indicated in red near the node (Techer, Clémencet, Simiand, Preeaduth, et al., 2017).

Chapter 17

17. Advances in SNP-Based Mitochondrial DNA Analysis for Conservation and Genetic Monitoring of *Apis mellifera*

Researchers have been using methods like PCR-RFLP and targeted sequencing of mitochondrial regions to study mitochondrial variation in *Apis mellifera* for a long time. Single nucleotide polymorphism (SNP) analysis has become an important methodological advance recently. The resolution at which we characterize mitochondrial variation has greatly improved. This chapter explains how SNP analysis helps us learn more about the diversity of mitochondria, lets us tell the difference between strains and evolutionary lineages, and supports conservation management through high-throughput genotyping.

17.1. Introduction

Mitochondrial DNA (mtDNA) has been a popular genetic marker in honeybee studies because it is inherited from the queen bee, is haploid, and has a small genome size of about 16 kilobases. Researchers used specific areas between the COI and COII genes and methods like PCR-RFLP, such as the (DraI test) to distinguish different haplotypes in early studies. These methods provide us a starting point for tracking lineage and measuring diversity, but they weren't perfect because they didn't have very high resolution, and there was a chance that haplotypes could be shared between subspecies, as shown in regular mitochondrial studies (Buswell, 2022). SNP analysis, on the other hand, not only makes it easier to tell the difference between samples, but it also opens up more possibilities for mitochondrial variation studies by allowing researchers to identify small sequence differences like point mutations, small insertions, and deletions that happen all over the mitochondrial genome.

17.2. Methodological Advances Through Complete Mitochondrial Genome Comparisons

There have been improvements in research methods due to full mitochondrial genome comparisons. One of the best things about the new SNP-based approach is that it can do comparative sequence analyses on complete mitochondrial genomes from more than one person. Researchers can identify single-nucleotide polymorphisms that are unique to certain strains or evolutionary lineages by sequencing the whole mitochondrial genome. This full comparison goes beyond the problems that come with single- or few-locus approaches by looking at variation across the whole genome. We choose diagnostic SNP markers because they show many differences between evolutionary lineages. These markers exhibit statistically significant differences in allele frequency, which are associated with specific maternal lineages or adapted ecotypes (D. Henriques, Parejo, et al., 2019). These kinds of genome-wide assessments help find small-scale differences that other methods might miss. This, in turn, helps us understand the microevolutionary processes that shape mitochondrial diversity in natural populations.

The first step in the comparative approach is to sequence the full mitochondrial genome of many people from different parts of the world or subspecies. Advanced bioinformatics pipelines then make alignments that list the differences in SNPs. These studies indicate that certain evolutionary lineages have unique patterns of nucleotide substitution. For instance, some SNPs may always be the same in *Apis mellifera mellifera* but different in related subspecies like *A. m. ligustica* or *A. m. carnica.* By identifying foreign SNP signatures in an otherwise uniform mitochondrial background, this diagnostic potential also enables researchers to monitor the introduction of non-native alleles into native populations (D. Henriques, Parejo, et al., 2019).

17.3. High-Throughput Genotyping Platforms and Multiplex Analysis

The Agena BioScience iPLEX MassARRAY system and other high-throughput genotyping platforms have made it much easier to detect and study these mitochondrial SNP markers. These platforms enable multiplexed analysis of hundreds of SNPs simultaneously. Researchers can now confirm the genetic integrity of multiple genomes and see how much introgression there is between them in a cost-effective and accurate way by using both mitochondrial and nuclear markers in the same tests. The iPLEX MassARRAY system is a powerful tool that uses mass spectrometry to distinguish between alleles by measuring their mass differences after specific reactions. This system is excellent for standardizing conservation genetic monitoring in *Apis mellifera* because it has a high sample throughput and can be used in different labs (D. Henriques, Browne, et al., 2018).

To make multiplexed SNP assays, you first choose candidate SNPs that are very informative based on things like FST values, which show how different the populations are from each other. Once you select a set of diagnostic candidate markers, we combine these SNPs into a single panel. Then, comparative sequence analyses show that the SNPs that were found always separate evolutionary lineages or strains within full mitochondrial genomes. Because these platforms are all connected, they can genotype mitochondrial and nuclear markers at the same time. The result provides us a view of genetic integrity and gene flow in honeybee populations from multiple locations. This two-pronged approach is especially useful because it makes up for the shortcomings of single-locus mitochondrial analyses, which only look at maternal inheritance patterns (D. Henriques, Parejo, et al., 2019).

These high-throughput methods make it possible to quickly process hundreds to thousands of samples with lower costs per sample and little need for human input. This ability to scale up is crucial when monitoring large populations for conservation purposes, where thousands of colonies may need regular genetic screening to ensure that non-native gene flow is kept to a minimum and the genetic identity of endangered subspecies like *A. m. mellifera* is kept safe (Carina Ferreira Matias, 2021).

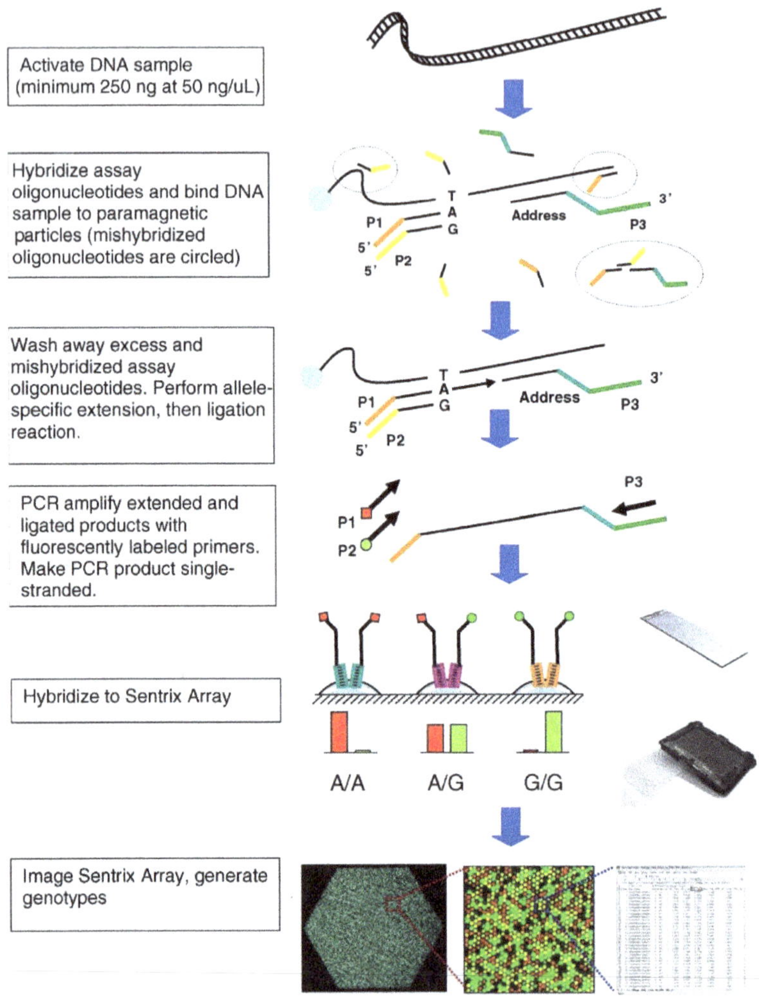

Fig. 22. Illustration of the Golden Gate genotyping assay process (Shen et al., 2005)

17.4. Diagnostic SNP Markers and Their Role in Conservation Management

Finding and describing diagnostic SNP markers is very important for efficient conservation management. In the case of *Apis mellifera*, these markers are very accurate molecular tools for finding non-native alleles in protected populations. For instance, conservation programs must constantly monitor colonies to safeguard the genetic integrity of European dark honeybees and to prevent the intermingling of popular C-lineage subspecies. Diagnostic SNP markers are a precise and high-resolution way to tell the difference between pure native genotypes and those that have been mixed with other genotypes through human-mediated introductions (Buswell et al., 2023).

Using a diagnostic panel provides a high level of certainty in identifying even small amounts of introgression. This is crucial because small amounts of genetic mixing that might not show up with regular mitochondrial or microsatellite markers can have big effects on how well native populations can adapt and how long they can live. Conservation managers can identify isolate, and eliminate colonies that show signs of introgression by regularly screening populations with these targeted SNP panels. This proactive management keeps local adaptations from getting weaker and makes sure that selective breeding programs keep the genetic integrity of native honeybee stocks (Dogantzis & Zayed, 2019).

SNP-based diagnostic markers also work well with machine learning classification methods, which is another benefit. Using advanced statistical methods like supervised classification, we can combine SNP genotypes to accurately place individual bees or whole colonies in their evolutionary lineage. This automated process reduces the likelihood of human error and allows conservation practitioners to make rapid, data-driven decisions regarding the management of honeybee populations. The overall impact of these approaches directly translates into more robust conservation policies that are adaptive, science-based, and economically sustainable (Momeni et al., 2021).

Studies that compare different methods have shown that smaller panels of informative SNPs can produce ancestry assignments that are very similar to those from full genome analyses. Although complete mitochondrial sequencing provides a breadth of genetic information, the ability to distill this information into a small set of highly discriminant markers is powerful. These panels not only make the genotyping process easier, but they also make it easier to compare results from different studies or labs. This makes conservation genetic initiatives more reproducible and interoperable (Muñoz et al., 2015, 2017).

Using these cytoplasmic (mitochondrial) and nuclear SNP markers together in an analytical framework also helps us develop a better picture of the genetic integrity of honeybees. Mitochondrial markers provide us with information about the female lineage, which is important for understanding how populations changed over time and how genes were passed down from maternal to offspring. Nuclear markers show how genes were passed down from both parents and how much male-mediated admixture and overall genomic introgression there was. High-throughput SNP genotyping has shown that it can identify cases of cytonuclear discordance in many honeybee populations. This condition happens when the mitochondrial haplotype does not match the nuclear genetic background because of hybridization events (D. Henriques, Parejo, et al., 2019). We need this level of diagnostic accuracy to protect genetically unique populations at risk of being overwhelmed by introgressive hybridization.

Sanger's method of gene sequencing

(https://www.onlinebiologynotes.com/sangers-method-gene-sequencing/).

Sanger's method of gene sequencing is also known as dideoxy chain termination method. It generates nested set of labelled fragments from a template strand of DNA to be sequenced by replicating that template strand and interrupting the replication process at one of the four bases.

Four different reaction mixtures are produced that terminates in A. T. G or C respectively.

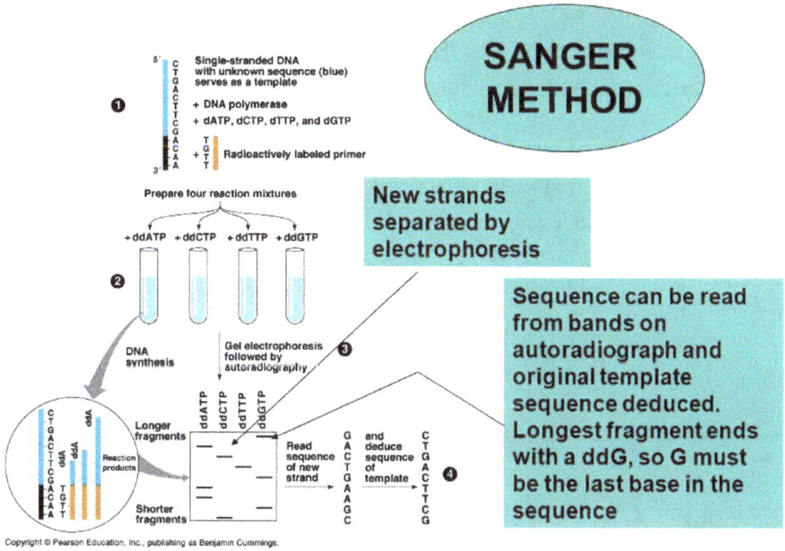

Fig. 23. Diagrammatic representation of Sanger sequencing

<u>Principle</u>

A DNA primer is attached by hybridization to the template strand and deoxynucleosides triphosphates (dNTPPs) are sequentially added to the primer strand by DNA polymerase.

The primer is designed for the known sequences at 3’ end of the template strand.

M13 sequences is generally attached to 3’ end and the primer of this M13 is made.

The reaction mixture also contains dideoxynucleoside triphosphate (ddNTPs) along with usual (dNTPs).

If during replication ddNTPs is incorporated instead of usual dNTPs in the growing DNA strand then the replication stops at that nucleotide.

The ddNTPs are analogue of dNTPs

ddNTPs lacks hydroxyl (group-OH) at c3 of ribose sugar, so it cannot make phosphodiester bond with nest nucleotide, thus terminates the nucleotide chain

Respective ddNTPs of dNTPs terminates chain at their respective site. For example ddATP terminates at A site. Similarly ddCTP, ddGTP and ddTTP terminates at C, G and T site respectively.

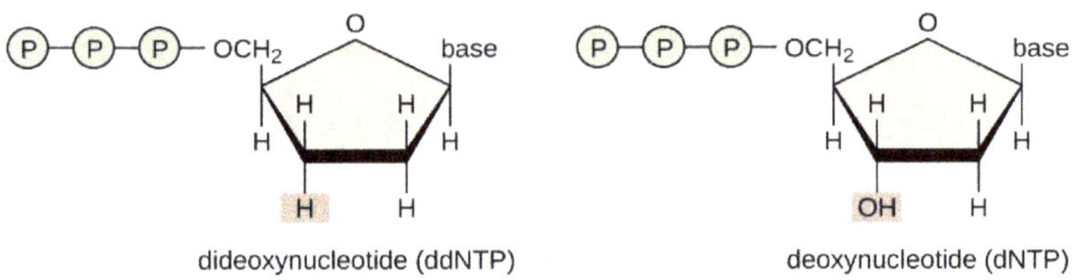

Procedure

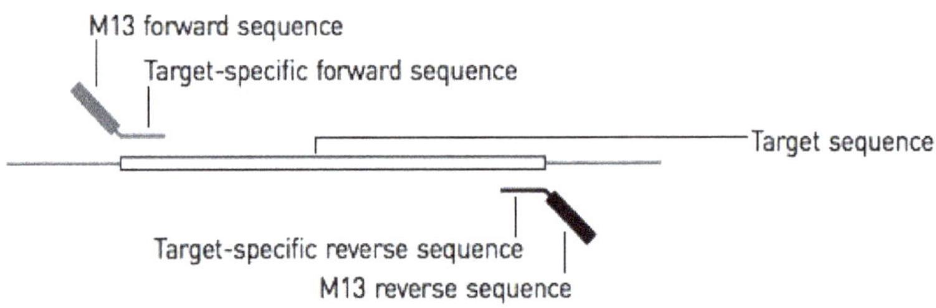

1. Template preparation:

m13-forward-sequence

Copies of template strand to be sequenced must be prepared with short known sequences at 3’ end of the template strand.

A DNA primere is essential to initiate replication of template, so primer preparation of known sequences at 3’end is always required.

For this purpose a single stranded cloning vector M13 is flanked with template strand at 3’end which serves as binding site for primer.

2. Generation of nested set of labelled fragments:

Copies of each template is divided into four batches and each batch is used for different replication reaction.

Copies of standard primer and DNA polymerase I are used in all four batches.

To synthesize fragments that terminates at A, ddATP is added to the reaction mixture on batch I along with dATP, dTTP,dCTP and dGTP, standard primer and DNA polymerase I.

Similarly, to generate, all fragments that terminates at C, G and T, the respective ddNTPs ie ddCTP,ddGTP and ddTTP are added respectively to different reaction mixture on different batch along with usual dNTPs.

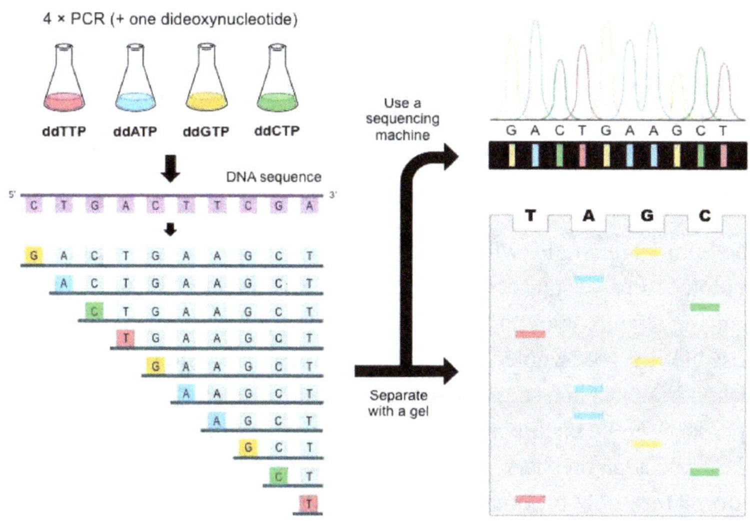

Figs. 23, 24, 25, 26. Sanger sequencing Process

17.5. Implications and Future Directions for *Apis mellifera* Conservation

The ability to use SNP analysis to show small differences in mitochondria is not just a small step forward; it is a giant change in the field of conservation genetics for *Apis mellifera*. Researchers can now create a detailed family tree that shows how *Apis mellifera* populations have been separated in the past and how they share genes today, thanks to the easier comparison of complete mitochondrial genomes at high resolution. For several reasons, this better phylogenetic resolution is critical. First, it provides us with information about how evolution has shaped the structures and adaptations of today's populations. For instance, their adaptability to local environmental conditions, pathogen pressures, and climatic changes is noteworthy. Second, being able to watch introgression events in near real-time helps with quick management actions that can fix or lessen the effects of unwanted hybridization caused by trading commercial queens and moving beekeeping practices (Buswell, 2022).

Additionally, using high-throughput SNP analysis in conservation efforts allows for "precision conservation," which means using genetic information to make very specific and informed management choices. This means that conservation programs can set threshold criteria based on diagnostic SNP profiles in real life. For instance, conservation programs can quarantine colonies or breeding stocks with excessive non-native alleles, selectively cull them, or conduct controlled breeding. In many areas of Europe, where laws protect native honeybee types like *Apis mellifera mellifera*, authorities can require the use of SNP-based genetic testing for certification and border control (Momeni et al., 2021).

In the future, the field will probably benefit from further refinements in technology. For instance, the ongoing development of next-generation sequencing technologies and the decline in the cost of each sample may make it possible to use whole-genome approaches that combine nuclear and mitochondrial data in a single analytical pipeline in even more situations. Such an approach would enable regular population monitoring and in-depth studies on the molecular mechanisms of selection and adaptation. Additionally, as bioinformatics tools become increasingly sophisticated, there is potential for integrating large-scale SNP databases with environmental meta-data, thereby enabling the mapping of adaptive traits to specific genomic regions. It is important to use these kinds of approaches to figure out how to protect native populations from quickly changing environments and pathogens that are always changing (Muñoz et al., 2015).

High-throughput SNP analysis also promises to contribute to international collaborative efforts aimed at safeguarding the genetic resources of *Apis mellifera*. Once they are tested, standardized SNP panels can be sent to labs all over the world. This will provies them a consistent way to measure genetic diversity and introgression. When these standardized panels are used, data from different parts of the world can be put into global databases. Meta-analyses and large-scale conservation plans can then utilize these databases. These efforts are especially important right now because people all over the world are starting to realize how important honeybees are to the economy, the environment, and culture (Dogantzis & Zayed, 2019).

17.6. Challenges and Considerations

Despite the numerous benefits of SNP-based methods for mitochondrial variation studies, there remain certain issues that require resolution. One problem is making sure that the SNP markers chosen cover all of the different types of mitochondria. Since many honeybee populations have complicated histories and have mixed with others multiple times, it's important to regularly update and confirm SNP panels using complete whole-genome datasets. Developing SNP panels using only a subset of populations can lead to ascertainment bias, another important consideration. Such bias could lead to underrepresentation of rare alleles that might be critical for local adaptation. SNP markers require constant sampling and revalidation to maintain these panels' diagnostic power over time.

Furthermore, high-throughput platforms like the iPLEX Mass ARRAY system have great benefits when it comes to cost-effectiveness and scalability, but they need a lot of money and expertise to obtain started. It is important to train staff in molecular techniques and data analysis so that the data they collect is correct and can be repeated. However, the benefits quick, accurate, and high-resolution information about genetic structure—far outweigh these problems, especially when you think about the risk of losing genetic integrity in endangered native populations (D. Henriques, Browne, et al., 2018).

17.7. In conclusion

SNP analysis is a big step forward in the way we study mitochondrial variation in *Apis mellifera*. Researchers have been able to identify SNP markers that are specific to certain strains or evolutionary lineages by comparing complete mitochondrial genomes from different bees. This affords them a level of detail that traditional PCR-RFLP methods (D. Henriques, Parejo, et al., 2019), can't. Using high-throughput genotyping platforms like the Agena BioScience iPLEX Mass ARRAY system has made this ability even better by letting us look at both mitochondrial and nuclear markers at the same time (Carina Ferreira Matias, 2021; D. Henriques, Browne, et al., 2018).

These diagnostic SNP markers are crucial for managing conservation because they provide precise molecular tools for finding non-native alleles. This makes it possible to find and control introgression in managed populations of *A. m. mellifera* (Buswell et al., 2023; Dogantzis & Zayed, 2019). SNP analysis will continue to be a big part of these efforts because conservation programs are using genetic data more and more to make breeding decisions and keep the genetic integrity of endangered honeybee subspecies.

Integrating SNP analysis into honey bee genetics not only helps us learn more about science, but it also helps protect a species that is important for the environment and the economy. Thanks to rapid advancements in technology and a stronger focus on careful conservation, high-throughput SNP genotyping will continue to be a key method for protecting the unique genetic history of *Apis mellifera*. Future research that combines detailed whole-genome data with targeted SNP panels will definitely improve the resolution even more. Such efforts will make sure that the most accurate and detailed genetic information is used to make decisions about conservation and management (Kim, Ah Rha, et al., 2019; Momeni et al., 2021).

Chapter 18

18. Integrating Nuclear and Mitochondrial Markers in a Single Genotyping Assay for Honey Bees

18.1. Introduction

Combining nuclear and mitochondrial markers into a single genotyping assay has become a key breakthrough that improves the accuracy and resolution of population genetic studies, especially in specics where conservation management is crucial. By using both nuclear and mitochondrial markers together, researchers can better understand the genetic makeup, mixing of species, and gene exchange patterns, which helps in conservation efforts. This combined method lets us see gene flow from the past and the present. It also lets us identify cytonuclear discordance and concordance that we can't see with just one type of marker. Tracking both maternal and biparental inheritance simultaneously in honeybees, which are crucial for pollination and farming, is crucial for monitoring conservation programs that aim to safeguard native subspecies from genetic homogenization through hybridization with introduced lineages (Bubnič et al., 2020).

18.2. Distinctive Information from Nuclear and Mitochondrial Markers

Nuclear and mitochondrial markers each provide unique data, which is a key benefit. Nuclear markers, like single nucleotide polymorphisms (SNPs), microsatellites, and even allozymes, show the genetic contributions from both parents. These markers provide us a strong way to measure the overall genetic diversity, population structure, and gene flow between different colonies or subspecies. Nuclear SNP panels have been shown to reliably separate population groups, measure hybridization events, and count introgression, which is crucial when checking the genetic health of conservation populations (Chahbar et al., 2013). On the other hand, mitochondrial markers like those in the tRNAleu-COX2 intergenic region or the COI region give a clear signal of matrilineal lineage. Their significance is because they are passed down from mother to child and don't usually recombine. Phylogeographic studies can utilize these markers to determine the divergence of subspecies over time (Chávez-Galarza et al., 2021).

18.3. Application in Hybrid Zones

Nuclear and mitochondrial markers work well together, which is especially advantageous when looking at hybrid zones where the genetic contributions from native and introduced populations are mixed. For example, several studies have shown that using nuclear SNP markers along with mitochondrial markers can help determine hybridization events between different types of honeybees. Researchers can see ongoing gene flow and admixture events from both parental contributions using nuclear markers. Mitochondrial markers, on the other hand, confirm maternal lineage and historical introgression from non-native queens (Bubnič et al., 2020). This two-marker strategy is especially important in areas where beekeeping has led to the hybridization of different subspecies, which could put the genetic purity of native populations at risk (D. Henriques, Parejo, et al., 2019).

18.4. Detecting Cytonuclear Discordance

Furthermore, tests that combine both markers make it easier to find cytonuclear incongruence, which is when the maternal (mitochondrial) and biparental (nuclear) genetic signals don't line up. Adaptive introgression, differences in how males and females move, and past hybridization events are some of the evolutionary processes that can lead to mismatches between mitochondrial and nuclear genetic signals. Researchers can get better estimates of the amounts of introgression and definitely find hybrid individuals when they look at both cytoplasmic (maternal) and nuclear markers at the same time (D. Henriques, Parejo, et al., 2019). Accurate estimates of introgression are critical for conservation programs because they help managers determine and protect populations that have high genetic integrity. They also help managers record of how well measures taken to limit gene flow from non-native lineages are working (Muñoz et al., 2014).

18.5. Methodological Advantages and High-Throughput Platforms

Using both nuclear and mitochondrial markers together not only makes it easier to determine things in hybrid zones, but it also has many methodological benefits. You can test DNA samples from queen exuviae or feces, for example, using minimally invasive or non-destructive methods. You can test both genomic compartments at the same time. This method sampling, which combines different types, reduces stress on important individuals, allows for repeated sampling over time, and makes it easier to do long-term studies that look at how gene flow and mate choice change in natural populations (Bubnič et al., 2020). High-throughput genotyping tools like SNP arrays and mass spectrometry can now create tests that measure hundreds to thousands of nuclear markers and some mitochondrial SNPs all at once. This cost-effective method not only saves time and money, but it also provides a consistent dataset that can be used for fine-scale population genetic analyses (Meixner et al., 2013).

18.6. Conservation Tracking of Native Subspecies

To effectively track the genetic effects of conservation practices, it is crucial to achieve higher resolution by combining both types of markers. Many European honeybee conservation programs, for instance, try to keep more cosmopolitan or commercially popular subspecies like *Apis m. ligustica* and *Apis m. carnica* from mixing with locally adapted subspecies like *Apis m. mellifera.* Researchers can use a combined genotyping test to follow both the maternal genetic line, which shows a bee's traditional background, and the nuclear genome, which might show if there has been mixing or recent breeding with other types that could harm local adaptations (Chávez-Galarza et al., 2015). This kind of detailed monitoring is essential because even small amounts of introgression can have big effects on traits that help organisms adapt, like disease resistance, foraging behavior, or thermal tolerance. These traits are all crucial for the long-term survival of native populations when the environment changes (Meixner et al., 2013).

18.7. Informing Targeted Management Decisions

Integrated genotyping assays exhibit superior resolution, facilitating informed targeted management decisions. Conservation managers can employ these tests to create genetic baselines, track the evolution of genetic diversity over time, and identify individuals or colonies with notable local genetic heritage. Studies combining nuclear SNP tests with mitochondrial DNA analysis have effectively differentiated between pure native colonies and those exhibiting evidence of mixing. This capability has enabled the establishment of breeding programs that exclusively choose bees with the appropriate genetic characteristics. This achievement is vital as maintaining genetic purity is essential for safeguarding biodiversity and ensuring that gene flow from commercially produced or introduced populations does not intermix with features already adapted to the region (Galindo-Cardona et al., 2013).

18.8. Detection of Hybridization Across Generations

The dual-marker method makes it easier to identify small hybridization events that might happen over several generations. Mitochondrial markers alone may show a pure maternal lineage, but when nuclear markers are added, they show gene flow between populations that may have happened through the paternal line. It is critical to be able to find things like these events on a small scale for species like honeybees, which have complicated mating systems and drones that can fly long distances. Researchers can obatin a full picture of the genetic makeup of a population by using both markers. This is because they can see all the genetic contributions that make up each individual (Eimanifar, Kimball, et al., 2018).

18.9. Phylogeographic and Evolutionary Insights

Combining nuclear and mitochondrial markers helps us learn more about phylogeographic patterns and the processes that lead to speciation and adaptation at the level of evolutionary studies. Because they change quickly and have clear patterns of maternal inheritance, mitochondrial markers have been used for a long time to figure out phylogenetic relationships. But mitochondrial DNA alone doesn't always show the whole demographic history, especially when there has been historical introgression or incomplete lineage sorting. Nuclear markers fill in these gaps by showing how genes move between generations and how populations are connected. Together, they make a powerful tool for figuring out how species have evolved over time and for finding past hybridization events that may have led to the current distribution and variety of subspecies (Meixner et al., 2013).

18.10. Resolving Cytonuclear Discordance in Practice

An integrative genotyping assay that includes both nuclear and mitochondrial markers also helps with research that tries to figure out cytonuclear discordance, which is when the two genomes don't match up. For instance, there have been reports of bees having a mitochondrial haplotype that doesn't match their nuclear genetic background. There may be cases of disagreement like these because of complicated evolutionary processes like sex-biased dispersal, selective breedings, or even queens moving around with the help of people. Researchers can determine these mismatches and their effects on population dynamics by looking at the mitochondrial and nuclear data from the same colonies. This in-depth knowledge is essential for making smart choices about conservation because it helps managers determine and remediate the genetic contamination sources in native populations (Muñoz et al., 2017).

18.11. Dual Confirmation in Conservation Contexts

In conservation situations where hybridization between native and introduced subspecies could threaten genetic purity, using both nuclear and mitochondrial markers together is a way to confirm results. For instance, in areas where conservation islands have been set up to protect native genetic stocks, mitochondrial markers show whether the maternal lineage is still pure, and nuclear markers show whether genes are moving from nearby populations that aren't protected. These integrative assays simplify the monitoring of cytonuclear congruence. This gives conservation biologists strong tools to judge how well isolation measures are working and to change management strategies based on what they see happening with introgression (Muñoz et al., 2014; Muñoz & De la Rúa, 2021).

18.12. Applications in Breeding Programs

The benefits of using both nuclear and mitochondrial markers go beyond conservation and into applied breeding programs as well. Breeders often have to improve traits like disease resistance, productivity, and temperament in managed honey bee populations. Genotyping tests that are quick and accurate can show both the maternal linet (hrough mitochondrial markers) and the overall genetic background (through nuclear markers). These tests are often used to figure out how genetically valuable a queen is. Studies have shown that combining both types of markers into one assay not only accelerates the genotyping process, but it also saves a lot of money by eliminating the need for separate assays. These kinds of integrated systems are especially useful in high-throughput settings, where hundreds or even thousands of individuals may be screened to help make selective breeding decisions that keep adaptive traits while lowering the risks of introgressive hybridization (Bubnič et al., 2020; D. Henriques, Browne, et al., 2018).

18.13. Advances in Multiplexed Assay Technology

From a technical perspective, the creation of multiplexed assays that look at both nuclear and mitochondrial markers at the same time is one of the best genetic tools we have right now. New technologies for genotyping, like microfluidic arrays, next-generation sequencing platforms, and mass spectrometry-based SNP detection, have made it possible to look at more than one genetic marker at a time in a single reaction. These technologies make it possible to create high-resolution datasets that are necessary for studying population genetics. One well-known example is the creation of a combined assay that genotyped more than 57 nuclear SNPs and 5 mitochondrial SNPs at the same time. This procedure gave very accurate and consistent estimates of genetic introgression and maternal lineage fidelity in honeybees (D. Henriques, Parejo, et al., 2019). Being able to look at so much data in one experiment makes genetic assessments much easier to understand. This process gives conservation strategies and breeder choices more accuracy than ever before (Meixner et al., 2013).

18.14. Advanced Statistical Integration of Multilocus Data

Using advanced statistical methods that combine data from multiple loci makes these tests even more useful for interpretation. For instance, scientists often apply Bayesian clustering algorithms on nuclear SNP datasets to categorize individuals into genetic clusters according to the frequency of their alleles. Adding mitochondrial data to these studies lets scientists find small signs of maternal introgression that might not show up in the nuclear genome alone. This combined analytical framework lets us not only accurately define population structure and hybrid zones but also estimate admixture proportions, which is an important measure for figuring out how much introgression there is from non-native lineages (Hereward et al., 2025; Muñoz et al., 2017).

18.15. Biological Insights and Cytonuclear Interactions

Integrated genotyping studies provide many biological insights that extend beyond merely analyzing numerical data. The interaction between nuclear and mitochondrial markers in honeybees reveals the relationship between behavioral and ecological traits. Mitochondrial inheritance has been linked to important physiological processes like energy metabolism and thermoregulation, which are necessary for survival in changing environments. When differences between nuclear and mitochondrial genomes are found, they can point to adaptive introgression processes in which beneficial alleles are kept in the mitochondrial genome even though genes are moving around in the nuclear genome. Researchers are becoming more and more interested in this kind of cytonuclear interaction. It shows how important it is to look at multiple markers at the same time to understand the evolutionary pressures on populations (Eimanifar, Brooks, et al., 2018; Kim, Ah Rha, et al., 2019).

18.16. Broader Taxonomic Applicability

The comprehensive perspective provided by combination tests is applicable not only to honeybees but also to several other animal groups facing analogous conservation challenges. Incorporating both nuclear and mitochondrial markers into investigations of managed and wild populations has proven beneficial for elucidating ambiguous genetic signals and differentiating between ancient polymorphisms and recent hybridization occurrences. Integrative techniques are particularly crucial for animals that can disperse readily or have been extensively relocated by humans, as the genetic signatures of historical and contemporary gene flow are intricately intertwined (Utzeri et al., 2021; Wragg et al., 2017).

18.17. State-of-the-Art Summary

In short, combining nuclear and mitochondrial markers in a single genotyping assay is the most advanced way to study population genetics. It not only shows how inheritance patterns are both nuclear and mitochondrial, but it also makes genetic tests more accurate and detailed. This integrated method makes it easier to idenetify cytonuclear incongruences, offers strong measures of introgression, and helps protect native subspecies by allowing for accurate tracking of hybridization events. Researchers and conservation managers now have the tools they need to protect genetic diversity in a world that is changing quickly. These tools include the technological advances that make these tests possible, as well as advanced analytical methods. These dual-marker assays are a must-have for protecting genetic purity and making sure that both wild and managed populations can survive (Bubnič et al., 2020; Momeni et al., 2021). This situation is especially true for honeybee conservation programs, where the risk of hybridization between native and introduced subspecies is always a concern.

18.18. Future Directions

Integrative assays are likely to help conservation genetics and population studies even more in the future. As the price of high-throughput sequencing keeps going down and bioinformatic tools get better, it will be easier to add a wider range of markers, such as complete mitochondrial genomes and large panels of nuclear SNPs, to streamlined workflows. This change will definitely help us better understand how gene flow works, improve our ability to track adaptive variation, and give us a clearer picture of how populations react to both natural and human-made selective pressures (Meixner et al., 2013; Rohde et al., 2025).

18.19. Multidisciplinary Collaboration

Researchers who study different aspects of population genetics, from phylogeography and evolutionary biology to applied conservation management, should work together to take a more complete approach to protecting biodiversity. When researchers combine integrated genotyping assays, they can effortlessly transfer knowledge from one system to another. This makes conservation strategies stronger for many groups of organisms that are facing similar threats (Chávez-Galarza et al., 2021; Parejo et al., 2020).

18.20. Practical Implementation Considerations

In practical applications, implementing these integrative assays in conservation programs requires careful consideration of several factors. First, we need to design sampling plans that ensure both nuclear and mitochondrial markers accurately reflect the range of genetic variation in the population under study. This could mean picking a group of samples from different colonies, geographic areas, or ecological niches that are representative of the whole. Second, the chosen markers should identify a lot of information. For example, nuclear SNP panels and mitochondrial regions were chosen because they can tell the difference between relevant lineages and show changes that are a result of adaptive processes. It is important to thoroughly compare these markers against known reference populations to set baselines and adjust later analyses (López-Uribe et al., 2017; Meixner et al., 2013).

18.21. Integration with Morphological Evidence

Finally, combining genetic data with ecological and morphological evidence can draw the conclusions of these kinds of studies even stronger. For instance, researchers have used geometric morphometrics in conjunction with molecular markers to distinguish subspecies that share significant similarities. This strategy provides an additional means of verifying the genetic data (Litvinoff et al., 2023; Oleksa & Tofilski, 2015). This interdisciplinary approach is necessary to fully understand how genetics, phenotype, and the environment interact, especially in species like the honeybee, which are important for both the environment and the economy.

18.22. Conclusion

In conclusion, using both nuclear and mitochondrial markers in a single genotyping assay is a revolutionary approach in population genetics that gives us the best information about inheritance patterns, hybridization dynamics, and introgression levels. This method is especially useful in conservation programs where keeping the genetic purity of native honeybee subspecies is essential for their survival. By using these tests to assess how well nuclear and mitochondrial DNA match and to acquire precise estimates of how much mixing has occurred, researchers and conservation managers can make informed decisions that help protect biodiversity and support responsible breeding practices (Bubnič et al., 2020; Muñoz & De la Rúa, 2021). The ongoing improvement and use of these integrated assays promise to change conservation genetics forever. They will guarantee that the genetic integrity and adaptability of important species can be preserved for future generations, even as environmental pressures and human activities grow stronger (Calfee et al., 2020; Hereward et al., 2025).

Table 13. Comparison of Nuclear vs. Mitochondrial Markers in Honey Bee Genotyping

Feature	Mitochondrial Markers	Nuclear Markers
Genetic Origin	Maternal inheritance only	Biparental inheritance
Primary Applications	Maternal lineage identification, subspecies determination, contamination detection	Genomic selection, trait analysis, subspecies authentication
Marker Types	tRNA^{leu}-COX2 fragment, COI barcoding, cyt *b* gene	Single Nucleotide Polymorphisms (SNPs), dCAPS markers
Sampling Method	Non-destructive wing clips, (honey eDNA)	Requires nuclear DNA (honey eDNA or bee tissue)
Analysis Complexity	Relatively straight forward sequencing	Requires high-fidelity polymerases to avoid replication errors
Key Advantages	- High lineage resolution - Non-lethal queen sampling - Species authentication	- Trait associatione.g., (varroa resistance) - Population genomic insights - Subspecies verification
Limitations	- Only reflects maternal history - Cannot assess nuclear traits	- Higher cost per sample -Population-specific baselines required - Requires $\geq$50 samples/colony for queen genotype

Integrated Assay Applications

Combined Approach: Genotyping-by-sequencing (GBS) assays now integrate both marker types using honey environmental DNA (eDNA), enabling:

Simultaneous subspecies authentication via mtDNA lineages and nuclear SNPs (Bovo et al., 2022; Utzeri et al., 2022).

Colony-level trait profiling without sacrificing bees (Bovo et al., 2022).

Honey origin verification through dual-genome analysis (Utzeri et al., 2022).

Methodology: The GBS approach analyzes 121 nuclear SNPs alongside mtDNA markers, achieving >53 million reads for high-confidence subspecies assignment. This dual-genome strategy overcomes limitations of single-marker systems by providing complementary data layers (Alsharhi et al., 2025; Bovo et al., 2022; Bubnič et al., 2020; Utzeri et al., 2022).

Chapter 19

19. Environmental DNA (eDNA) Analysis of Honey for Non-Invasive Mitochondrial DNA Research, Honey Authentication, and Honey Bee Hologenome Studies

Over the past ten years, environmental DNA (eDNA) methods have changed a lot. Now, researchers can sample biodiversity without harming it and use these methods to answer important questions in mitochondrial DNA (mtDNA) research (Bovo et al., 2018). Thanks to the development of high-throughput sequencing platforms and better extraction methods, it is now possible to obtain trace genetic material directly from environmental samples like honey. This procedure gives us new information about the genetic makeup of honeybee populations without having to sample bees directly (Bovo et al., 2020). These methods have gotten better over the past few years so that they can now collect not only bee mtDNA but also a wide range of genomic information from microorganisms, plants, and other organisms that live in the honeybee ecosystem. This process provides us a complete picture of the honeybee hologenome (Bovo et al., 2018).

19.1. Introduction

Honey is a complex food matrix that contains DNA fragments from the bees, the microbes that make them sick, and the plants they visit while looking for nectar (Bovo et al., 2018). The natural complexity of honey permits scientists a chance to acquire a full genetic picture of the ecosystem around the bees, which used to require time-consuming and invasive sampling methods (Soares et al., 2023). Scientists can look at bee mtDNA to see how genetically diverse a colony is, confirm the entomological origin of honey products, and determine subspecies-specific differences over large areas by taking eDNA from honey (Ribani et al., 2021). Researchers have also gotten around many of the problems that come with PCR-based methods, like amplification bias and limited marker representation, by using shotgun metagenomic sequencing methods along with strong bioinformatic pipelines (Bovo et al., 2020).

19.2. Advancements in eDNA Methodologies for mtDNA Research

The most important new thing in this field is taking bee mtDNA samples from honey without hurting or bothering the bee colonies (Bovo et al., 2018). Early research indicated that honey has detectable amounts of bee-derived mitochondrial fragments because bees leave behind genetic material when they collect nectar and make honey (Utzeri et al., 2018). Next-generation sequencing (NGS) platforms made this breakthrough possible. It has led to a shift from traditional, invasive sampling methods to a more sustainable, field-friendly method that doesn't harm bee populations (Galanis et al., 2022).

New methods for extracting DNA have greatly improved the amount and quality of eDNA found in honey, even when the DNA is very damaged or only present in small amounts (Soares et al., 2023). To obtain fragmented DNA out of a complicated, sugary biomatrix like honey, modified extraction methods that include better washing, centrifugation, and chemical treatment steps have been crucial (Bovo et al., 2018). These protocols are especially successful at keeping even the rare bee mtDNA fragments, which are important for later genomic studies that look at population structure and how species evolve (Bovo et al., 2020).

At the same time, the development of sequencing technologies, such as the use of Ion Torrent and Illumina platforms, has made it possible to obtain more and better coverage from honey-derived eDNA samples (Galanis et al., 2022). This progress not only makes it easier to determine bee mtDNA, but it also lets us obtain genomic sequences from other kingdoms that were caught in honey, like plants, bacteria, fungi, viruses, and parasites (Bovo et al., 2020). This development lets us build full metagenomic profiles that give us a complete picture of the bee hologenome. This data is very important for understanding how bees interact with each other and how diseases spread in and around the hive (Bovo et al., 2020).

19.3. The Holistic View of the Honey Bee Hologenome

One of the most important things that eDNA analyses of honey have taught us is how to obtain a complete picture of the honeybee hologenome (Bovo et al., 2018). In this sense, the word "hologenome" means the total amount of genetic material that comes from honeybees as well as the many organisms that live with them, help them, and make them sick (Soares et al., 2023). Studies of honey eDNA have indicated that bees add their mtDNA to honey along with DNA from the bacteria that live in their guts. This mixture gives us a look at the genetic diversity of both the host and the microbes (Ribani et al., 2022). This combined method lets scientists look at genetic diversity in mosaic patterns that show how bees forage, how healthy their hives are, and how they interact with the environment as a whole in a way that taking samples of bee tissues in small, separate areas cannot (Utzeri et al., 2019).

For example, having DNA sequences from plants is crucial for figuring out where honey comes from and what kind of plants it comes from. By comparing the plant DNA profiles found in honey to reference genomic databases, it is possible to confirm that the honey comes from a certain area and that the bees have eaten from certain flowers (Bovo et al., 2018). The presence of plant DNA, along with the identification of bee mtDNA, provides an independent line of evidence for the entomological origin of honey products. This aspect is crucial for verifying the quality and authenticity of honey products (Soares et al., 2023).

Furthermore, finding microbial DNA, which includes both favorable gut bacteria and undesirable ones, makes honey eDNA tests useful for more than just identifying species; they can also be used to verify the health of bee colonies (Ribani et al., 2021). The ability to identify multiple genomes in a single sample at the same time makes it easier to study complicated interactions between hosts and microbes that could lead to disease outbreaks or colony collapse (Galanis et al., 2022). In short, looking at the honeybee hologenome as a whole presents researchers a lot more information than studies that only look at host mtDNA. This lets them do more complex research that includes genetic, ecological, and environmental data (Bovo et al., 2018).

19.4. Honey Authentication and Subspecies-Specific mtDNA Haplotypes

The non-invasive sampling of honey for eDNA analysis has been very useful for proving that honey products are real by checking their insect origins and finding subspecies-specific mtDNA haplotypes in large areas (Bovo et al., 2022). Numerous studies have demonstrated the accuracy of identifying bee subspecies using the genetic information present in honey's eDNA. This information helps separate populations that have adapted to their local environment from those that have been affected by widespread migratory practices like queen trade (Bovo et al., 2022). This difference is important because people are worried about how human activities are making honey bee genetics more similar around the world (Ribani et al., 2021).

For example, shotgun metagenomic sequencing has been very advantageous in finding mtDNA haplotypes that are specific to certain subspecies. These haplotypes can then be used to make a biochemical fingerprint that can be compared to known genetic variants from other areas (Bovo et al., 2020). These detailed genetic fingerprints make it possible to prove that honey is real by checking that the genetic material found in the honey matches what is expected from local bee populations (Bovo et al., 2020). The fact that you can tell the subspecies of a bee without hurting it by sampling honey is crucial because it avoids the time-consuming and expensive process of sampling individual bees and presents a more accurate picture of the hive's overall genetic makeup (Bovo et al., 2022).

Furthermore, region-specific mtDNA haplotype information from honey eDNA has real-world uses for both managing beekeeping and certifying honey products (Ribani et al., 2022). Scientists can now make detailed maps of where honeybee populations are located and identify possible areas of admixture or genetic introgression that could happen because of migratory beekeeping practices by linking subspecies-specific markers with geographic and ecological data (Soares et al., 2023). These kinds of studies help protect and sustainably manage honey bee populations, and they also make people more sure that honey is real and comes from a reliable source, since these genetic markers can be used to examine product labels (Bovo et al., 2022).

Targeted genotyping-by-sequencing (GBS) tests that focus on important nuclear SNP markers and mitochondrial variants make it easier to tell the difference between different subspecies. This adds to the genomic dataset that honey eDNA can provide (Bovo et al., 2022). These GBS methods provide us two points of view. mtDNA markers tell us about the maternal lineage, while nuclear SNPs present us a different view of the colony's genetic mix and diversity (Bovo et al., 2022). This multilayered genotypic information makes honey eDNA an even better tool for authenticating honey. It helps protect local bee subspecies and guarantees that honey products on the market are of superior quality (Bovo et al., 2022).

19.5. Overcoming PCR Bias Through Shotgun Metagenomic Sequencing and Bioinformatics

People have long criticized traditional PCR-based methods for looking at bee mtDNA, like targeted amplification of certain barcoding regions, because they can cause amplification bias and make it harder to see genetic diversity (Bovo et al., 2018). Often, PCR bias made it challenging to see minor or degraded DNA fragments, which made the sample's true genetic diversity seem lower than it really was (Bovo et al., 2020). Shotgun metagenomic sequencing is a strong alternative that doesn't need primer-dependent amplification because it sequences all the DNA fragments in the sample (Soares et al., 2023).

This unbiased sequencing method lets us retrieve back DNA sequences that are both high and low in abundance. It also captures the full range of genetic diversity that is naturally present in honeybees and their microbiota (Bovo et al., 2020). Shotgun sequencing reduces the risk of preferential amplification by not using predetermined genetic markers. This makes it a better way to show allele frequencies and genetic variants (Utzeri et al., 2019). Adding complete bioinformatic pipelines makes these methods even more reliable by processing immense amounts of sequence data, using strict quality control measures, and correctly assigning taxonomic identities to the sequences that are found (Soares et al., 2023).

Advanced bioinformatic workflows perform carry out several essential tasks. They get rid of low-quality or contaminated sequences, break down complicated metagenomic datasets, and use alignment algorithms to match recovered sequences to large reference databases that cover a wide range of taxonomic groups (Bovo et al., 2018). These pipelines also make it easier to discover single nucleotide polymorphisms (SNPs) and indels in mtDNA. These are important for figuring out haplotypes that are specific to certain subspecies and genetic diversity within a species (Utzeri et al., 2019). By combining shotgun metagenomics with cutting-edge bioinformatics, we can get rid of the biases that come with PCR-based methods and greatly increase the range and detail of honey bee genetic studies (Soares et al., 2023).

These bioinformatic pipelines are even more reliable because they can correctly process DNA from different kingdoms that are present at the same time. This lets researchers look at bee mtDNA, microbial communities, and plant-derived DNA all in the same sample (Bovo et al., 2018). This all-encompassing analysis makes it possible to do multidimensional studies that look at species authentication, colony health, and ecological interactions all at once. The outcome reduces the potential for analytical artifacts caused by DNA degradation or contamination (Bovo et al., 2020). The genomic profiles that come out of this procedure are more accurate representations of the real biological environment inside and around the bee colony. This study shows how important non-invasive eDNA methods are for large-scale ecological and population studies (Soares et al., 2023).

19.6. Applications and Implications for Conservation and Honey Authentication

Using shotgun metagenomic sequencing on honey eDNA has big effects on conservation biology, managing beekeeping, and making sure food is safe to eat (Ribani et al., 2021). For conservation biologists, being able to monitor honey bee population genetics without hurting them is a crucial tool for keeping track of changes in genetic diversity, spotting admixture events, and finding new threats from invasive species or pathogens (Galanis et al., 2022). Many local honeybee subspecies are becoming more endangered due to human activities such as habitat loss, migratory beekeeping, and genetic homogenization, making these methods a useful and scalable way to protect and evaluate native genetic lineages (Bovo et al., 2022).

To stop honey from being mixed with other products and mislabeled, it is important to determine a clear genetic signature for honey made by certain bee populations (Soares et al., 2023). By finding unique mtDNA haplotypes that show the entomological origin of honey, producers and regulators can confirm that honey products are of superior quality and come from the places they say they do (Ribani et al., 2022). This genetic authentication process not only protects consumers but also provides local beekeepers with a financial boost by enhancing the appeal of region-specific, traditionally made honey to buyers (Bovo et al., 2022).

Additionally, shotgun metagenomic sequencing provides multi-kingdom genetic profiles that enable us to conduct integrative studies linking bee genetics with foraging ecology and environmental health (Bovo et al., 2018). For instance, finding plant, microbial, and bee DNA at the same time can show how bee foraging behavior changes with the seasons, how floral resources change, and how pathogen outbreaks happen from time to time. Such data all helps us understand how ecosystems work as a whole (Galanis et al., 2022). These kinds of analyses are very useful for creating adaptive management plans that make bee populations and the larger agroecosystems they support more resilient (Rishan, 2024).

In real life, eDNA sampling from honey is effortless to do, which makes it ideal for large-scale monitoring programs (Ruppert et al., 2022). Unlike methods that require capturing and processing bee specimens one at a time, eDNA extraction from honey lets you obtain genetic data from whole colonies with little disturbance (Bovo et al., 2018). This efficiency is especially useful in places where there is a lot of beekeeping or in remote areas where traditional sampling is difficult to do logistically. It allows for wide geographic coverage and long-term monitoring over time (Soares et al., 2023).

19.7. Future Directions and Research Opportunities

In the future, combining eDNA-based methods with new genomic technologies is likely to change honey bee research and conservation even more (Bovo et al., 2020). As sequencing accuracy, data processing speed, and the completeness of reference databases continue to advance, it will be easier to observe small genetic differences between honeybee subspecies and even the smallest changes in community composition within the hologenome (Bovo et al., 2018). Future studies may focus on improving bioinformatic algorithms that can sort through complicated metagenomic datasets. These advancements would enable real-time monitoring of colony health and prompt detection of invasive pests or pathogens through the use of eDNA signatures from honey (Utzeri et al., 2019).

Furthermore, as more complete reference libraries are made for honeybees and the other organisms that live with them, the accuracy of taxonomic assignments will keep getting better. This will provide researchers more confidence when they authenticate honey products and map genetic differences between regions (Soares et al., 2023). To obtain the most out of these non-invasive methods, geneticists, bioinformaticians, ecologists, and apicultural experts will need to work together across disciplines. Such collaboration will assure that all of the genetic data can be used to protect both bee biodiversity and the quality of honey products (Ribani et al., 2022).

Another promising area of research is looking at how the honeybee hologenome changes over time by taking samples of honey from different seasons and years (Galanis et al., 2022). Long-term studies like these could show how climate change, pesticide exposure, or new pathogens affect the genetic diversity and strength of local bee populations. This information is crucial for making decisions about conservation and policy (Ribani et al., 2021). Temporal genomic surveillance, along with traditional ecological assessments, could change how we approach how environmental factors affect the evolutionary paths of important species like the honeybee (Rishan et al., 2025).

19.8. Challenges and Considerations

Even with these big steps forward, there are still some methodological problems that need to be solved before honey-derived eDNA can be fully used in mtDNA research and honey authentication (Bovo et al., 2018). One of the most significant problems is that the quality and amount of DNA that can be recovered from honey can vary depending on things like how the honey is stored, how it is processed, and how much DNA has broken down (Soares et al., 2023). It will be crucial to set up standard procedures for collecting, storing, and extracting DNA from honey so that data from different studies and places can be compared (Bovo et al., 2020).

Furthermore, the honey matrix is challenging to work with because it has a mix of polysaccharides, proteins, and possible PCR inhibitors. This makes it difficult to extract DNA and use it for sequencing later (Utzeri et al., 2018). Researchers are looking into new ways to reduce these negative effects and obtain the highest-quality DNA possible for high-throughput sequencing. These include using additives or changing the buffer formulations (Soares et al., 2023). Also, bioinformatic pipelines need to be developed and improved all the time to accurately process and interpret the massive amounts of sequence data that shotgun metagenomics produces while keeping false positives and technical artifacts to a minimum (Bovo et al., 2020).

Another thing to address is the need for more complete and carefully chosen reference databases that have a larger collection of honey bee genomes, regional subspecies variants, and the genomes of the microbiota that live with them (Bovo et al., 2022). Without these strong databases, it may be harder to assign sequence reads to the right taxonomic group and figure out subspecies-specific haplotypes, which could lead to biases in later analyses (Soares et al., 2023). It is important to continue working on sequencing and annotating the genomes of local bee populations, as well as the microbes and plants that interact with bees. These efforts will help make eDNA-based honey authentication and population genetics studies more accurate (Bovo et al., 2018).

Furthermore, even though shotgun metagenomic sequencing is a big step forward in reducing PCR bias, there are still problems with how to interpret metagenomic data, especially when it comes to telling the difference between low-abundance real sequences and background noise (Utzeri et al., 2019). Careful validation experiments will be necessary to set reliable detection thresholds for different genetic markers and to compare the performance of these assays. These could include using mock communities and spike-in controls (Soares et al., 2023). Even though there are problems, the rapid pace of technological progress in sequencing and bioinformatics provides us hope that these non-invasive monitoring methods will keep getting better at being precise and accurate (Ruppert et al., 2022).

19.9. Conclusion

Combining environmental DNA methods with shotgun metagenomic sequencing is a big step forward in honey bee mtDNA research and honey authentication (Bovo et al., 2018). These methods have greatly expanded the range of genetic analyses by allowing researchers to collect bee genetic material directly from honey without harming the bees. Their work has given them a complete picture of the honeybee hologenome, which includes not only the bees' mtDNA but also a lot of other genomic information from microorganisms, plants, and other environmental factors (Bovo et al., 2018). This broad view is important for proving the authenticity of honey products, finding out where honey comes from in terms of insects, and finding mtDNA haplotypes that are specific to certain subspecies and vary across large areas (Bovo et al., 2022).

Furthermore, using shotgun metagenomic sequencing with advanced bioinformatic pipelines has gotten around the problems that come with traditional PCR amplification. This ensures the unbiased capture of the diverse genetic makeup of honey samples (Bovo et al., 2020). As the field grows, the continued use of new lab techniques, strong data processing algorithms, and large reference databases will make honey-derived eDNA even more useful for many uses in conservation genetics, apicultural management, and verifying the authenticity of food (Soares et al., 2023).

In summary, the new eDNA methods have not only expanded the scope of mtDNA research by enabling non-invasive honey sampling, but they have also equipped researchers with a potent tool for ecological and population genetic studies, enabling them to uncover minute genetic variations and intricate interactions among species on a scale previously unattainable (Ribani et al., 2021). The combination of advanced bioinformatics and shotgun metagenomic sequencing is starting a new era in honeybee research. This new era will help us learn important things about the genetic basis of bee health, biodiversity, and the complex web of ecological relationships that support our food systems and natural environments (Rishan, 2024).

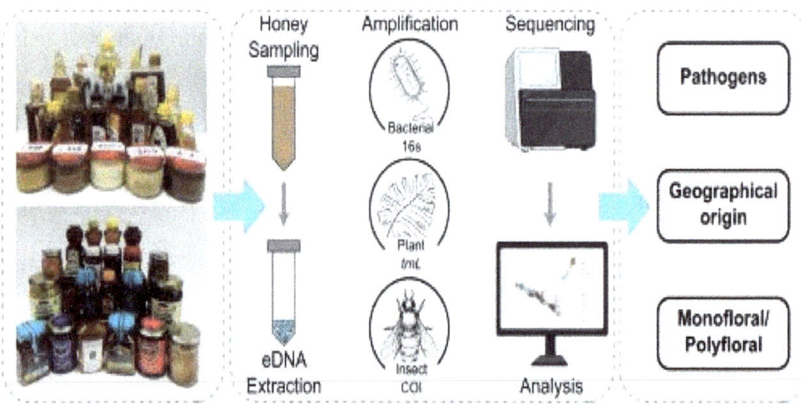

Fig. 27. Metabarcoding of eDNA for tracking the floral and geographical origins of bee honey (Pathiraja et al., 2023).

Chapter 20

20. A Comprehensive Bioinformatics Pipeline for Mitochondrial DNA Assembly, Annotation, and Population Genetic Analysis

The following is a full report on how a typical mitochondrial DNA (mtDNA) bioinformatics pipeline is set up and how the assembled mitogenomes are then analyzed and annotated. In modern mtDNA research, there are many important steps that must be followed to confirm that the final mtDNA sequences are correct and can be used for population and phylogenetic analyses. These steps include checking the quality of the raw sequence, trimming the reads, assembling the de novo contigs, annotating the genes, calling variants, and statistically evaluating genetic diversity.

20.1. Introduction

Mitochondrial genomes are small, circular DNA molecules that are found in large numbers in eukaryotic cells. They are commonly used as genetic markers in studies of evolution, forensics, and population genetics. There are many problems with putting together and analyzing mtDNA sequences from high-throughput sequencing data. For example, researchers must separate mtDNA reads from numerous nuclear sequences, including nuclear mitochondrial DNA segments, (or NUMTs), assemble repetitive regions, and accurately label functional features. Modern bioinformatics pipelines for mtDNA research have several steps that ensure that the sequence reconstruction is of high quality and the diversity metrics that follow are interpreted correctly. The pipelines include quality control of raw sequence data, trimming of sequences, de novo contig assembly, biological annotation to determine protein-coding regions, rRNA and tRNA genes, and noncoding regions, and finally variant calling and statistical analysis of genetic diversity, such as haplotype diversity, nucleotide diversity, and neutrality tests like Tajima's D. This kind of integrated approach not only makes it easier to handle the massive amounts of sequencing data that NGS instruments produce on a regular basis, but it also helps with strict downstream evolutionary analyses (Alqahtani, 2020).

20.2. Quality Control and Sequence Trimming

The first important step in any mtDNA bioinformatics pipeline is to assess the quality (QC) of the raw sequencing reads to assure that data artifacts like low base quality scores or adapter contamination don't affect the analyses that come after. Many pipelines start by carefully checking raw read data with software tools that demonstrate detailed quality metrics like per-base sequence quality, GC content distributions, sequence duplication rates, and the presence of adapters. There are a number of specialized QC tools, but many pipelines use widely available tools like Fast QC, which lets users look at quality profiles and recognize systematic problems across large datasets (Donato et al., 2020).

After the first QC report, sequence trimming is done to get rid of adapters and low-quality bases that can make it difficult to put short-read contigs together correctly. During this stage, standard software tools like Trimmomatic are used all the time. They scan the raw sequence reads remove any subsequences that don't meet a certain quality score or match a known adapter sequence. Many pipelines also use custom scripts to automate these steps and set specific thresholds that work best for mitochondrial data. These preprocessing steps are essential because they make it much easier for the computer to do the next steps of alignment and assembly, and they also lower the chances of making false variant calls later on in the process (Alqahtani, 2020; Andermann et al., 2020).

20.3. Contig Assembly and Genome Construction

Once the preprocessed reads pass quality checks, the next step is to put them together into contiguous sequences (contigs) that show the whole or almost whole mtDNA genome. Because mitogenomes are only about 14–20 kilobases long, de novo assemblers can put the genome back together, even when they only have a small amount of whole-genome sequencing data. The pipeline usually uses strong de Bruijn graph-based assemblers like SPAdes, which are excellent at putting together short sequencing reads that overlap to make contigs that show the mitochondrial genome (Donato et al., 2020). SPAdes is especially appealing because it lets you use different k-mer sizes during assembly. This method solves problems with low coverage and helps with resolving repetitive regions. Some workflows start with different assemblers like Velvet to create contigs before using SPAdes to enhance them, as seen in the SMART and SMART2 workflows for assembling the mitochondrial genome (Alqahtani, 2020).

Many pipelines use custom scripts that do multiple rounds of assembly and read filtering during the assembly phase. For example, after the first de novo assembly, contigs are filtered using BLAST searches against curated mitochondrial databases to get rid of any nuclear contaminants and to guarantee that only real mtDNA contigs are kept. After this, read filtering based on alignment is often done with tools like Bowtie to identify reads that align with high confidence to the assembled contigs and add them to the mitochondrial read pool. Some of the most advanced pipelines perform additional scaffolding steps by analyzing assembly graphs and applying maximum likelihood estimation methods to determine how to handle unclear repeat regions. This procedure makes sure that the final assembly is both contiguous and correct (Alqahtani, 2020; Formenti et al., 2021).

20.4. Annotation of Mitochondrial Genomes

Genome annotation is the next important step after getting a high-quality mitochondrial contig or scaffold from assembly. Annotation is the process of finding and describing the different parts of the mitochondrial genome and how they work. Protein-coding genes (PCGs), ribosomal RNAs (rRNAs), transfer RNAs (tRNAs), and noncoding control regions (known as the D-loop) are key components of the mitochondrial genomes in animals. Many researchers use annotation tools like MITOS and Mito Annotator for such tasks because they can automatically determine gene boundaries by comparing the assembled genome to reference datasets and by predicting new genes (J. Li et al., 2019). These pipelines usually use both ab initio prediction and homology-based search to correctly label gene features.

One important part of annotating the mtDNA genome is figuring out where the tRNA genes are. Tools like tRNA scan-SE are crucial here. They examine not only the nucleotide sequence of candidate tRNAs but also their secondary structure, which is necessary to confirm the biological relevance of predicted tRNA genes. Identifying the correct tRNA genes is crucial for understanding how genes work because these genes help produce proteins in mitochondria and may have unique features that are only present in mitochondrial genomes (Liu & Harbison, 2018; Winn et al., 2025).

The annotation process usually includes manually curating the data with sequence alignment tools like Geneious or MAFFT, in addition to finding coding and noncoding regions. Researchers use these tools to compare the predicted gene sequences to well-known reference mitochondrial genomes. The comparison lets them verify the annotations, change the gene boundaries if they need to, and verify that the overall gene order is correct. These kinds of comparisons often show that gene arrangements are the same across different groups of organisms. They also help identify rearrangements or duplications that might be important for evolution or function (Donato et al., 2020; J. Li et al., 2019).

20.5. Variant Calling and Quality Assessment

The next step in the mtDNA bioinformatics process is variant calling, which means finding differences between the assembled mtDNA genome and a standard reference sequence, (like the Revised Cambridge Reference Sequence for human mtDNA) or between different individuals. Identifying variants in mtDNA is difficult because of heteroplasmy, which means that different copies of the mitochondrial genome in one cell or organism can have different variations. To tackle these issues, pipelines usually align the high-quality trimmed reads back to the assembled or reference mtDNA genome using tools like Bowtie or BWA. Bowtie is particularly well-suited for the rapid alignment of short reads to a reference and is routinely used in combination with custom scripts for post-processing alignments (Alqahtani, 2020; Battle et al., 2022).

After the reads are lined up, popular tools and custom processes determine changes in the DNA sequence, like single nucleotide polymorphisms and short insertions/deletions, from the alignment data. The accuracy of finding variants is often improved by getting rid of PCR duplicates with tools like Picard and by applying quality filters that discard variants backed by few reads or unclear mappings. Quality checks at this stage also involve using statistical methods to determine how likely it is that a specific variant call is real, particularly in samples that have low coverage. Such careful filtering is important because even small errors can make it harder to analyze differences in genetic makeup and population genetics (Alqahtani, 2020; Battle et al., 2022).

20.6. Statistical Analyses of Genetic Diversity

After the mtDNA sequences are put together, labeled, and differences are identified, the process continues to statistical analyses that look at genetic diversity within and between populations. This usually includes figuring out different diversity measures (like haplotype diversityhow likely it is that two randomly chosen mtDNA sequences are different), nucleotide diversity (the average number of differences at each site between sequences), and neutrality tests like Tajima's D. These statistical parameters are central to understanding the demographic history and selective pressures acting on mitochondrial genomes. They help to elucidate whether a population has undergone recent expansion, experienced a bottleneck, or is under purifying or balancing selection.

For example, research on *A. mellifera* has used these methods to find out why there are many different haplotypes but not much difference in nucleotides, suggesting that the population may have quickly increased in size after a drop or experienced strong selection pressures. The computation of these metrics is often carried out using software tools such as DNASP, which enable researchers to incorporate standard population genetics models into their analysis pipelines. These analyses not only offer clarity about the evolutionary forces shaping mtDNA variation but also contribute to our understanding of the genotype–phenotype relationships across different geographic populations and evolutionary lineages (Eimanifar, Kimball, et al., 2018; Fan et al., 2021).

In particular, researchers use Tajima's D as a neutrality test to compare estimates of nucleotide diversity and the number of segregating sites. Negative values of Tajima's D are typically indicative of recent population expansion or the action of purifying selection, whereas positive values might suggest balancing selection or a structured population. By combining these statistical measures with the labeled mtDNA data, researchers can better understand how mitochondrial genes have evolved and draw conclusions about population changes like growth and reductions (Eimanifar, Kimball, et al., 2018).

20.7. Visualization via Haplotype Networks

Visualization is an important step in mtDNA research, as it offers a clear graphic way to show how different mitochondrial haplotypes are related. One frequently employed method for visualizing these relationships is the construction of haplotype networks using the median-joining algorithm. This algorithm works by linking haplotypes using the shortest mutational changes, which helps to show family relationships while also considering complex changes that might happen due to unclear mutations.

This type of visualization shows how haplotypes are connected and points out how common certain variants are, helping us understand how mitochondrial lineages are spread within and between different groups of colonies. Tools like Hapsolutely help researchers create and customize haplotype networks, allowing them to make clear and professional graphics that summarize complex evolutionary relationships. This method is particularly useful when regular phylogenetic trees can't fully show the complicated patterns of mitochondrial evolution, like when similar traits appear independently or when there are multiple possible relationships (Kong et al., 2023; Vences et al., 2024).

20.8. Integration of Custom Scripts and Software Tools

A hallmark of state-of-the-art mtDNA bioinformatics workflows is the integration of standard software tools with custom scripting to automate the processing of high-throughput sequencing data. Custom scripts are often created to handle tasks like repeatedly sampling trimmed reads, which helps assess the quality of the assembly; organizing output files from assemblers and annotators; and combining various quality control measures into one summary report. By using well-known software like Trimmomatic for trimming sequences, SPAdes for assembling them from scratch, and Bowtie for aligning reads, along with custom scripts, researchers can create adaptable and efficient systems that can tackle the different challenges posed by various sequencing methods and sample types (Alqahtani, 2020; White et al., 2025).

Such hybrid pipelines streamline the data processing workflow and ensure that the final mtDNA sequence and its associated variant calls are of high quality and reproducible across different studies. Furthermore, the modular nature of these pipelines allows for updates and modifications as new sequencing technologies emerge or as new bioinformatic tools become available.

20.9. Downstream Phylogenetic Analysis

After quality control, assembly, annotation, and variant calling, researchers use the assembled mtDNA sequences as the basis for phylogenetic and population genetic analyses. Researchers can make evolutionary trees and do comparative analyses that show how natural selection, adaptation, and historical demography work by accurately labeling coding and noncoding regions and providing detailed variant profiles. When looking at population growth or decline, haplotype diversity, nucleotide diversity, and Tajima's D are important statistical tools. Neutrality tests show what kinds of selective pressures are acting on mitochondrial genes (Eimanifar, Kimball, et al., 2018).

For accurate phylogenetic inference, it is also important to have strong mtDNA sequence alignment. To ensure that there are no misalignment artifacts, researchers often use multiple sequence alignment tools like MAFFT to make consensus alignments. Once we align the haplotypes, we can apply various phylogenetic methods to determine their relationships. We then subject the trees or haplotype networks derived from these analyses to additional statistical tests. Putting these methods together provides us a complete picture of how mitochondria have changed over time. It also helps us understand how genetics affects population structure and how species like *Apis mellifera* have changed over time (Eimanifar, Kimball, et al., 2018; Fan et al., 2021).

20.10. A Case Study: The Evolution of *Apis mellifera*

Studies of the population genetics of *A. mellifera* have found that measures of genetic diversity based on mtDNA are very useful. Researchers have found that there is a lot of haplotype diversity but not much nucleotide diversity. This pattern could mean that the population has gone through bottlenecks and then quickly expanded. Furthermore, tests for neutrality like Tajima's D are used to observe too many low-frequency polymorphisms that might be a sign of purifying selection or recent population growth. By adding these metrics to the mtDNA bioinformatics pipeline, researchers can see how the genetic evolution of *A. mellifera* has been affected by both natural and human-made factors. This approach presents them a full picture of the population dynamics at work (Eimanifar, Kimball, et al., 2018; Fan et al., 2021).

20.11. Conclusions

In short, mtDNA bioinformatics pipelines have many stages that depend on each other. They start with strict quality control of raw sequencing data and go all the way through read trimming, de novo contig assembly, and genome annotation to variant calling and statistical evaluation. Trimmomatic, SPAdes, and Bowtie are examples of standardized software tools that are essential for processing high-throughput sequencing data. Combining these tools with custom scripts results in efficient and reproducible workflows that generate high-quality mitochondrial genomes. Special tools like tRNAscan-SE label the assembled mitogenomes to accurately identify all the important coding and noncoding areas. Thereafter, the annotated sequences are put through in-depth statistical tests that figure out important genetic diversity measures like haplotype diversity, nucleotide diversity, and neutrality statistics like Tajima's D. These tests provide us with information about demographic history and selection dynamics. Finally, median-joining haplotype networks and other visualization methods show how mtDNA haplotypes are related to each other in a way that makes it easier to understand how organisms like *Apis mellifera* have evolved. This all-encompassing method not only guarantees the technical accuracy of mtDNA sequence reconstruction, but it also greatly enhances our knowledge of mitochondrial evolution and population history (Alqahtani, 2020; Donato et al., 2020; Eimanifar, Kimball, et al., 2018; Kong et al., 2023; J. Li et al., 2019; Vences et al., 2024).

These bioinformatics pipeline stages and annotation processes work together to create a strong framework for mtDNA research. They must ensure that future phylogenetic and population genetic analyses are based on high-quality, well-annotated sequence data. We need to use this integrated approach to understand how mitochondrial genomes change over time in different groups of organisms.

Chapter 21

21. Mitochondrial Phylogenetics of *Apis mellifera*: Methods, Insights, and Evolutionary Implications

Because it is passed down from maternal to offspring, has a relatively rapid substitution rate, and does not recombine, the mitochondrial genome (mtDNA) has long been an important molecular marker for tracing the evolutionary history of honey bees, especially *A. mellifera* (Carr, 2023; He et al., 2018). Researchers have combined data from all 13 mitochondrial protein-coding genes (PCGs) and ribosomal RNA (rRNA) genes, as well as complete mitogenome sequences, to create large and detailed datasets that help produce strong phylogenetic trees (Carr, 2023; Tihelka et al., 2020).

Researchers address potential issues of relying on a single gene, such as excessive changes or insufficient variation, by combining data from multiple mitochondrial genes. Datasets that include all 13 PCGs and rRNA genes present a much better picture of the genetic diversity across honeybee lineages (Eimanifar, Kimball, et al., 2018; D. S. M. Henriques, 2018). These kinds of concatenated methods improve the overall phylogenetic signal-to-noise ratio and make it easier to see how branches are ordered at deep nodes. This improvement is particularly evident when the sequences under examination originate from individuals that encompass the entire geographic range of *A. mellifera* (Carr, 2023; Y.-R. Li et al., 2024).

Both the use of concatenated datasets and the examination of whole mitogenomes offer advantages and disadvantages. Complete mitogenome analysis looks at both coding and non-coding regions, such as control regions and intergenic spacers, which may contain mutations that are useful for phylogenetic studies. However, these regions may need to be carefully aligned and filtered to avoid adding artifacts (Carr, 2023; D. S. M. Henriques, 2018). In contrast, the concatenation of 13 PCGs with rRNA genes omits areas that could cause problems, like hypervariable control regions. This makes the inferred tree topologies more reliable (Carr, 2023; S.-N. Song et al., 2016).

When building concatenated mitochondrial datasets, the first step is usually to carefully extract, align, and assess the quality of each gene sequence using the best bioinformatic tools available. For example, many researchers have used multiple sequence alignment programs like MAFFT to confirm that the alignments for both the protein-coding regions and the structural RNA genes are correct. This stage is often followed by manual curation to keep the open reading frames and secondary structure motifs in the rRNAs (Lu et al., 2021; Tan et al., 2011). Once we complete the individual gene alignments, we combine them into a single supermatrix. The result is then split up by gene or codon position, which lets different nucleotide substitution models be used on different parts of the dataset (He et al., 2018).

Model selection is an important part of phylogenetic reconstruction because it makes sure that the statistical models used in later tree-building methods accurately show how evolution works. Researchers often use software packages like Partition Finder to determine the best-fit models for each gene partition based on criteria like the Bayesian Information Criterion (BIC). This process makes the analyses more realistic and accurate (D. S. M. Henriques, 2018). Researchers then use both maximum likelihood (ML) and Bayesian inference (BI) methods to create phylogenetic trees. These two methods supply different levels of support for the resulting tree topologies (Carr, 2023; Tihelka et al., 2020).

Maximum likelihood methods search for the tree topology that maximizes the likelihood of observing the empirical sequence data under the specified substitution model. Modern ML programs like IQ-TREE, RAxML, and PHYML use advanced methods that can work with large combined datasets and include techniques to create many fake datasets by randomly picking data again and again, which helps to provide confidence levels for each part of the tree (J. D. Evans et al., 2013; Tihelka et al., 2020). Bootstrapping is the process of making a lot of fake replicate datasets by randomly sampling with replacements and reestimating the tree each time. Experts view high bootstrap percentages (typically above 70–80%) at significant nodes as evidence of the strength of those clades (Koeniger et al., 2011; Wang et al., 2018).

In contrast, Bayesian inference picks trees based on their posterior probability given the data and a prior distribution of trees. This method uses Markov Chain Monte Carlo (MCMC) techniques to look at the tree space all the time and figure out the posterior probability of each node in the phylogeny that was made (S.-N. Song et al., 2016; Tihelka et al., 2020). We can think of BI's posterior probabilities in the same way as bootstrap supports: values close to 1 mean that we have a lot of faith in the corresponding clades. These probabilities are very useful for comparing results from different methods (D. S. M. Henriques, 2018; Koeniger et al., 2011).

One of the most important benefits of using both ML and BI methods in phylogenetic studies is that they let researchers cross-validate the inferred tree topologies, which makes them more sure of the results. When both methods produce very similar trees and the nodes have strong support from high bootstrap values and high posterior probabilities, the resulting phylogeny is considered a reliable reflection of actual evolutionary relationships. This agreement of evidence is especially important when dealing with controversial topics like the significant differences between the African, European, and Middle Eastern lineages of *A. mellifera* (Carr, 2023; Dogantzis et al., 2021).

The phylogenetic trees derived from these comprehensive investigations have given us important information regarding the evolutionary history and biogeography of *A. mellifera*. Research utilizing concatenated mitochondrial datasets has consistently identified distinct clades corresponding to the principal geographic lineages. These clades exhibit profound divergences among populations from Africa, Europe, and the Middle East (Carr, 2023; Eimanifar, Kimball, et al., 2018). In these phylogenies, African lineages that occasionally appear basal or paraphyletic indicate an ancient radiation perhaps succeeded by colonization events in Europe and the Near East. Conversely, European and Middle Eastern lineages typically exhibit monophyletic clusters with robust statistical support. The evidence suggests that they have pursued distinct evolutionary trajectories and adapted to varying temperate and subtropical ecosystems (Carr, 2023; Ji, 2021).

For instance, some studies have found the basal position of *A. m. Mellifera* (which is usually thought to be from Northern Europe) compared to other subspecies. This has called into question older ideas that the species came from Africa or Asia (Carr, 2023; Tihelka et al., 2020). Other studies have shown a complicated pattern of diversification by looking closely at differences in mitochondrial sequences. These studies suggest that there were several ways for bees to move to Europe, including through the Strait of Gibraltar and Asia Minor (Carr, 2023). These results not only help us learn more about how different subspecies evolved, but they also have big effects on conservation biology and beekeeping because they show how genes have moved, diverged, and mixed over time to create the current diversity of honey bees (Buswell, 2022; D. S. M. Henriques, 2018).

There are several things that affect how reliable mitochondrial phylogenies are. The most important thing is the need for high-quality sequence data. Next-generation sequencing (NGS) technologies have changed the field by making it possible to quickly create full mitogenome sequences for many individuals and subspecies. The technology has made it possible to do large-scale meta-analyses that work around the problems of earlier studies that only used a few representative sequences per subspecies (Carr, 2023; D. Henriques, Browne, et al., 2018). By obtaining a wider range of samples, researchers can better understand intraspecific variation and identify potential issues such as mislabeling or sequencing errors that may complicate phylogenetic interpretation, particularly in regions where subspecies boundaries are unclear (Eimanifar, Kimball, et al., 2018; D. Henriques, Browne, et al., 2018).

Another important aspect of mtDNA phylogenetic analyses is the handling of different codon positions in protein-coding genes. For example, the third codon position changes quickly and requires more substitutions than the first and second codon positions. If such variation isn't taken into account, it could lead to substitution saturation and make it harder to see deeper evolutionary signals. To examine consistency in tree topology, many researchers create alternative datasets that exclude the third codon position. They often observe that adding third codon positions can improve resolution at shallow nodes, but leaving them out may be better for deeper nodes (Lu et al., 2021; Tan et al., 2011). Using tools like Partition Finder to apply these kinds of partitioning strategies makes sure that nucleotide substitution models are properly adjusted to the evolutionary dynamics of each partition. This makes the reconstructed trees more accurate and statistically strong (D. S. M. Henriques, 2018; Y.-R. Li et al., 2024).

Even though ribosomal RNA genes are very similar across species and provide a strong phylogenetic signal, they can be difficult to align because they have regions of different lengths and indels. Some studies choose to omit parts of rRNA that are very unclear or use strict filtering methods like Gblocks to get rid of poorly aligned blocks. This strategy keeps only the most phylogenetically useful parts of the sequences (He et al., 2018; Tan et al., 2011). Researchers have sometimes compared phylogenies made from only concatenated PCGs to those made from datasets that combine PCGs and rRNAs. Researchers discovered that the integrated approach can validate robust nodes and reveal subtle indications of evolutionary divergence that could otherwise go unnoticed (Eimanifar, Kimball, et al., 2018; D. Henriques, Browne, et al., 2018).

In phylogenetic studies, it is now common to use both ML and BI methods at the same time because they work well together and have different strengths and weaknesses. Maximum likelihood methods are especially useful for working with large datasets that have many partitioned models. They also offer reliable bootstrap support values that show how often a certain clade appears in replicate analyses. These bootstrap values are a direct measure of how reproducible the phylogenetic signal is (J. D. Evans et al., 2013; Tihelka et al., 2020). Bayesian inference uses the power of probabilistic sampling in MCMC frameworks to provide each node a posterior probability value that shows how confident you are in that clade. This method is especially advantageous when looking at different phylogenetic scenarios or figuring out how different prior assumptions affect tree topology (S.-N. Song et al., 2016; Tihelka et al., 2020).

When ML and BI methods yield the same results, such as identifying the same branching orders and strongly supporting the same clades, it makes the inferred evolutionary relationships much more likely to be correct. This method of cross-validation is critical in the study of honey bee phylogenetics because it has consistently supported the idea that there are deep differences between African, European, and Middle Eastern lineages. Furthermore, studies that include multiple individuals from different geographic areas for each subspecies tend to produce trees with very high statistical support. This makes them a useful way to understand the historical biogeography of *A. mellifera* (Carr, 2023; Eimanifar, Kimball, et al., 2018).

For instance, some studies have shown that African honey bee lineages are some of the oldest differences within *A. mellifera*. These lineages are often thought to have more genetic diversity and a paraphyletic arrangement in some analyses. These results support the idea that Africa was an early center of diversification for the species, which then spread to Europe and the Middle East during times of favorable weather (Carr, 2023; Kaltenpoth et al., 2012). European and Middle Eastern lineages, on the other hand, tend to form distinct, well-supported monophyletic groups. These patterns are not only related to the historical geographic separation of these groups but also to the ecological and behavioral changes that have set them apart over time (Carr, 2023; Ji, 2021).

The phylogenetic reconstruction of *A. mellifera* using mtDNA data not only helps us understand how populations have changed over time, but it also helps us understand demographic history in general, including events like population bottlenecks, expansions, and episodes of admixture that have left marks in the mitochondrial genome. For example, some European subspecies have a lower nucleotide diversity than African subspecies. The evidence suggests that their populations shrank in the past and then grew again recently. This aligns with glacial cycles and the events during which people returned to areas that had been uninhabited after the glaciers melted (D. S. M. Henriques, 2018; C. Ma et al., 2022). Discovering mitochondrial haplotype networks within each major clade demonstrates the functioning of microevolution and the structuring of populations at smaller geographic scales. This study is an important area of research for creating conservation plans and breeding programs for managed honey bee populations (Buswell, 2022; Eimanifar, Kimball, et al., 2018).

Adding mtDNA phylogenetic reconstruction to other types of evidence, like nuclear genomic data, morphological assessments, and ecological niche modeling, makes evolutionary studies in honeybees even more clear. Nuclear markers can show patterns of recombination, gene flow, and selection that are not visible in the mitochondrial genome alone. In contrast, mtDNA provides a clear representation of maternal lineage history. When mitochondrial and nuclear data agree on the same evolutionary hypotheses, the phylogenetic trees become more reliable and can be used to figure out how *A. mellifera* evolved across its entire range (D. S. M. Henriques, 2018; Koeniger et al., 2011).

Still, these kinds of analyses have to be cautious about a few problems and sources of error. One problem is that there could be alignment mistakes in areas with a lot of A-T content or repeated elements, like the control region of the mitogenome. Strict alignment methods and the exclusion of non-aligned segments from final analyses typically help researchers avoid these problems. This approach makes sure that the sequences used are both reliable and informative about phylogeny (D. S. M. Henriques, 2018). Incomplete lineage sorting is another possible source of error. Such errors can cause the mitochondrial genealogy and the species tree to not match up, especially when populations have recently diverged or undergone a lot of hybridization (Dogantzis et al., 2021; Tihelka et al., 2020).

Different protein-coding genes may evolve at different rates, which means they may not all contribute equally to the overall phylogenetic signal. This is another thing that needs to be carefully considered. Genes like nad 6 and atp 8 change more quickly than others, like cox1, which are more stable and are often used in studies that use DNA barcoding. Not only does dividing the dataset by gene and codon position make it possible to use better substitution models, but it also reduces artifacts that could change the tree topology (Lu et al., 2021; Tan et al., 2011). Furthermore, people often use methods to detect and address substitution saturation at third codon positions. This procedure makes sure that homoplastic mutations don't drown out deeper phylogenetic signals (He et al., 2018; Lu et al., 2021).

In addition to being methodologically sound, the information gained from mtDNA-based phylogenetic reconstructions can be used in conservation and beekeeping. Knowing how honeybee lineages have evolved and how they have changed over time helps us determine different genetic stocks that might be better suited to local conditions. This information is crucial for coming up with ways to keep genetic diversity, which is what makes honeybee populations strong enough to handle problems like new diseases, habitat loss, and climate change (Carr, 2023; D. S. M. Henriques, 2018). In managed populations, especially those used for commercial pollination, understanding the phylogeographic relationships between subspecies can help breeding programs combine traits like disease resistance, productivity, and tolerance to environmental stressors (Carr, 2023; Eimanifar, Kimball, et al., 2018).

Additionally, mtDNA analyses can help us understand the evolutionary history of honeybees and show us times in the past when subspecies mixed, which could have big effects on the genetic makeup of honeybee populations today. In some cases, finding mitochondrial haplotypes that don't match the expected subspecies designation has led to the discovery of hidden gene flow or recent mixing events. This study shows how dynamic honeybee population structure is (Eimanifar, Kimball, et al., 2018; D. S. M. Henriques, 2018). These results show how important it is to do big, integrative studies that look at molecular genetic data along with ecological and behavioral observations to provide a full picture of how honeybees have changed over time.

As sequencing technologies and analytical tools develop better, honey bee phylogenetics will become even clearer in the future. Researchers can now quickly sequence entire mitogenomes from dozens or even hundreds of individuals thanks to high-throughput sequencing platforms. This capability makes it possible to do population-level analyses that can demonstrate small-scale evolutionary processes (D. S. M. Henriques, 2018; Tihelka et al., 2020). With improved bioinformatics tools for organizing and testing data, these advancements allow researchers to build family trees that are both reliable and rich in detail, revealing both old separations and recent population changes (Carr, 2023; Lu et al., 2021).

Over the past ten years, combining mitochondrial phylogenetic studies has changed the way we approach the evolution of *A. mellifera*. Early research that only looked at a few mitochondrial markers gave us only a vague idea of how honeybee populations are genetically structured. But as datasets have grown in size and variety, including whole mitogenomes from a wide range of geographic locations, researchers have been able to see deep evolutionary splits and small differences between populations with more clarity than ever before (Carr, 2023; D. S. M. Henriques, 2018). These detailed phylogenies have led to new ideas about where *Apis mellifera* came from and how it spread. Some people have suggested a model in which Africa was the center of early diversification, with later events spreading honeybees to Europe and the Middle East (Kaltenpoth et al., 2012; Tihelka et al., 2020). This perspective is based on the fact that African lineages tend to be at the base of many trees.

The presence of strong groups that correspond to European and Middle Eastern lineages shows how being separated in the past and recent human movements have influenced where *A. mellifera subspecies* are found today. Many studies have shown that migratory beekeeping and the global trade in bee queens have caused unexpected patterns of genetic mixing, which complicates the phylogenetic story. Researchers can still figure out how these factors fit together and create a more accurate timeline of honey bee evolution by combining mitochondrial data with historical records and geographical information (Carr, 2023; Ji, 2021).

Conservation genetics also uses mtDNA-based phylogenetics in honeybees. As threats to pollinator populations grow around the world, it is important to protect the genetic diversity that provides them strength in the face of environmental changes. The detailed phylogenies that come from combining mitochondrial datasets not only show how the species has changed over time, but they also help us locate evolutionary significant units (ESUs) within the species. To protect the evolutionary potential of *A. mellifera* in the face of ongoing environmental challenges (Buswell, 2022; D. S. M. Henriques, 2018), these different genetic lineages, each with its set of adaptive traits, may be given priority.

In addition, the information gained from mitochondrial phylogenetic reconstruction goes beyond systematics and conservation; it also has important effects on how we understand how social insects adapt and become new species. Honeybees are a useful model for studying how complex social behaviors have changed over time. By looking at mtDNA data, we can see how social traits may have evolved along with genetic differences. For instance, different mitochondrial clades can be linked to differences in how colonies behave, how they reproduce, and how they respond to stress in their environment. This insight helps us understand the genetic basis of adaptation in these insects, which is important for both the environment and the economy (Carr, 2023; He et al., 2018).

The combination of mtDNA phylogenetic studies in *A. mellifera* also shows how important it is to use multiple methods in evolutionary biology. Mitochondrial markers present us a clear view of the maternal lineage, but when we combine them with nuclear genomic markers, morphological analyses, and ecological studies, we obtain a fuller picture of evolution. Integrated analyses have shown that the mitochondrial and nuclear genomes can tell slightly different stories about evolutionary history. Such variations can happen because of things like sex-biased dispersal, introgression, or selection on the mtDNA itself. These differences make researchers keen to learn more, which helps them understand how evolution works better. The evidence indicate that mtDNA should be used as one of several complementary tools instead of a stand-alone diagnostic marker (Eimanifar, Kimball, et al., 2018; D. S. M. Henriques, 2018).

With the advancements in computational phylogenetics, mitochondrial-based analyses have become even more reliable. New algorithms that better explore tree space and take into account complicated evolutionary models have made both ML and BI methods more powerful and easier to use. For example, Bayesian inference programs now commonly analyze data for tens of millions of generations, using strict checks like the average standard deviation of split frequencies and the potential scale reduction factor (PSRF) to ensure that the results are trustworthy and can be repeated (Lu et al., 2021; Tihelka et al., 2020). In the same way, ultra fast boots trapping methods have cut down on the time it takes to run ML analyses by a lot while still giving reliable support values for important nodes in the phylogeny (J. D. Evans et al., 2013; Wang et al., 2018).

These improvements to the methods have caused the phylogenetic results of different studies to agree more, which makes the inferred evolutionary relationships more reliable. When different research groups come to the same phylogenetic conclusions about the deep splits between African, European, and Middle Eastern lineages using different datasets and methods, the combined evidence strongly supports these evolutionary scenarios (Carr, 2023; Koeniger et al., 2011). This kind of agreement not only shows that mtDNA data can help answer difficult phylogenetic questions, but it also affords us a strong base for future research into the genetic, ecological, and behavioral differences that affect the evolution of honeybees.

In short, using mitochondrial DNA data to reconstruct phylogenies is still a key part of honey bee evolutionary studies. It affords us a level of detail that no other method can match when it comes to the deep divergence and biogeographic patterns that have shaped the evolution of *A. mellifera.* Researchers have built strong and statistically sound phylogenetic trees that show the evolutionary history of major lineages, including the deep splits between African, European, and Middle Eastern groups (Carr, 2023). They did this by using large datasets that came from either concatenated sequences of the 13 protein-coding genes along with rRNA genes or complete mitogenomes. Using both maximum likelihood and Bayesian inference methods, which each have their strengths and are backed up by strong statistical measures like bootstrap values and posterior probabilities, has made sure that these reconstructions show real evolutionary relationships instead of artifacts of the methods used (D. S. M. Henriques, 2018; Tihelka et al., 2020).

As we move forward, combining high-throughput sequencing technologies with advanced bioinformatic pipelines will help us attain better estimates of divergence times, show more subtle patterns of admixture, and shed light on the fine-scale population structure that makes honeybees so adaptable. As more mitogenome sequences are made from areas and subspecies that aren't well represented, the phylogenies will become more detailed. These discoveries will give us new information about how *A. mellifera* has spread and changed over time (Carr, 2023; Lu et al., 2021). In the end, these advances not only help us learn more about how honeybees have changed over time, but they also supply us with important information that can help us protect them and manage them in a way that is beneficial for the environment (D. S. M. Henriques, 2018; Ji, 2021).

In conclusion, the large amount of research that uses mtDNA data to build phylogenetic trees in honeybees shows how important it is to evolutionary biology. Researchers can create strong and statistically sound phylogenetic trees that show the deep evolutionary splits between African, European, and Middle Eastern lineages by using concatenated datasets that include 13 mitochondrial protein-coding genes and rRNA genes, as well as complete mitogenomes. Using both maximum likelihood and Bayesian inference methods at the same time makes sure that the reconstructed trees are not only mathematically sound but also biologically meaningful. They show the real history of evolution instead of artifacts of limited data or poor modeling (Carr, 2023; Tihelka et al., 2020). As methods improve, they will continue to improve the quality of data collection, alignment, and model selection. Mitochondrial phylogenetics will continue to be at the forefront of figuring out the complicated evolutionary story of *A. mellifera*, which is important for both basic evolutionary research and the practical management of one of the world's most important pollinators (Dogantzis et al., 2021; D. S. M. Henriques, 2018).

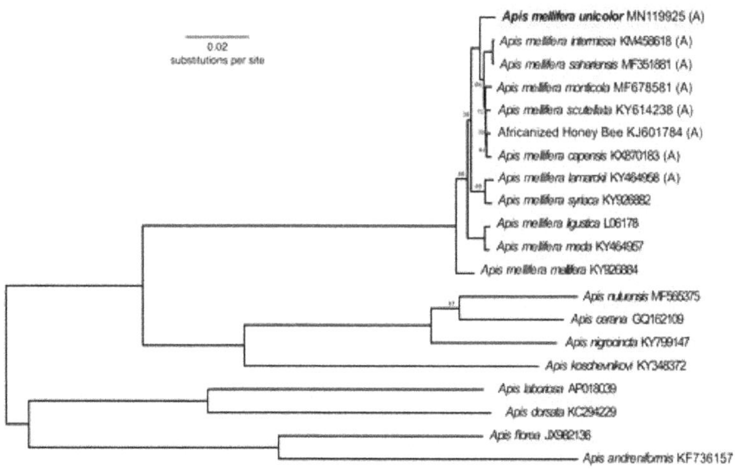

Fig. 28. Phylogenetic tree showing the relationship between *Apis mellifera unicolor* (GenBank: MN119925) and 19 other *Apis mellifera* honey bee mitochondrial genomes (GenBank accession numbers are listed after species names). A) indicates honey bees from the African A-lineage. The tree is mid point rooted. Node labels indicate bootstrap values, and unlabeled lineages are 100% (Boardman, 2019).

Chapter 22

22. Accelerated Evolution in Honey Bee Mitochondrial DNA: Insights from Comparative Genomic Analyses for Varroa Resistance and Adaptive Traits

Faster Evolution in Honey Bee Mitochondrial DNA: What We Can Learn from Comparing Genomes to Determine Varroa Resistance and Adaptive Traits

22.1. Introduction to Comparative Genomics in Honey Bees

Comparative genomic analyses have become an important tool for finding areas of rapid evolution in the mitochondrial DNA (mtDNA) of honeybees. This has led to the discovery of genetic differences that are strongly linked to traits that help honeybees adapt to their environment and resist disease. These studies use high-throughput sequencing, strict statistical analyses, and advanced bioinformatics methods to align and compare full mitogenome sequences from different honeybee strains. This process makes it easier to discover single-nucleotide polymorphism (SNP) markers that set resistant populations apart from susceptible ones.

22.2. Structure and Significance of the Honey Bee Mitogenome

The foundation of these studies is the complete sequencing of the honey bee mitogenome, which is around 16,400–16,450 base pairs long and includes key components like 13 protein-coding genes, 22 tRNAs, 2 rRNAs, and a non-coding A+T-rich control region. Early studies comparing the sequences showed not just differences in the DNA but also changes in the structure, like variations in the lengths of the spaces between genes and the A+T-rich region. The evidence suggests that these regions play an active role in adaptive evolution (Kim, Ah Rha, et al., 2019). These results show that even in a genome that is relatively small and stable, some areas can change more quickly than others. This shift can happen because of selective pressures like pathogen challenges and changes in the environment.

22.3. Advances in Sequencing and Bioinformatics for Mitogenomic Analysis

High-throughput sequencing technologies have made it possible to look at these genomic regions in much more detail. These platforms make it possible to sequence whole mitochondrial genomes in a thorough and cost-effective way. The process creates large datasets that include both rare and common genetic variations. Then, bioinformatics pipelines use algorithms based on the Burrows–Wheeler transform and de novo assembly methods to align short sequencing reads. These methods carefully rebuild the mitogenome, making it possible to observe small differences in nucleotides and changes in structure (Kim, Ah Rha, et al., 2019). Along with generating solid data, statistical analyses such as estimating SNP frequency and comparing sequence differences help identify areas with higher mutation rates than the normal rate of change in the mitogenome.

22.4. Discovery of SNP Markers Linked to Varroa Resistance

One of the most important things that came out of these studies was the discovery of specific mtDNA SNP markers that are linked to phenotypic traits like resistance to *Varroa destructor*. This parasite has a big effect on honeybee colonies worldwide, making it vital for the economy and the environment. We used next-generation techniques to sequence the complete mitogenomes of two strains of *A. m. ligustica*. One strain had high hygienic behavior (HHB) and was resistant to Varroa mites, while the other had low hygienic behavior (LHB). The comparative analysis of these sequences indicated that there are 23 SNPs spread out over 11 protein-coding genes. These SNPs have different patterns of nucleotide substitution that reliably tell the resistant and susceptible strains apart (Kim, Ah Rha, et al., 2019). Independent PCR amplification and sequencing of additional individuals from each strain confirmed the SNPs identified in these studies. This study indicates that they are useful as strong molecular markers for tracking Varroa resistance in breeding programs.

22.5. Functional and Evolutionary Implications of mtDNA Variability

Regions of the mtDNA that evolve quickly are often found in protein-coding genes, where even small changes in the sequence can have big effects on how mitochondria work and how cells make energy. For instance, protein-coding genes had the most sequence divergence at 3.7%, while non-coding regions like the A+T-rich control region had divergence values as high as 13.83% (Kim, Ah Rha, et al., 2019). These differences indicate that even though non-coding regions can change a lot, some protein-coding regions remain stable enough to distinguish between lineages and are also sensitive enough to detect beneficial mutations. The fact that these SNPs are spread out across many genes supports the idea that adaptive traits like Varroa resistance have a polygenic basis. This means that selective pressures may act on several mitochondrial functions that work together to make colonies stronger.

22.6. Maternal Inheritance and Marker-Assisted Selection (MAS)

The maternal inheritance pattern of mtDNA is another important part of these comparative genomic analyses. It presents a clear and simple signal of how a lineage has changed over time, without the recombination events that are common in nuclear DNA. This trait is especially useful for following how traits change in honeybees' haploid maternal lineages, where they are mostly expressed. When linked to traits that aid animal adaptation, such as cleanliness and disease resistance, mtDNA markers become crucial tools for monitoring and managing the genetic integrity of breeding programs. Marker-assisted selection (MAS) can use these confirmed mtDNA SNP markers to identify and select queens and maternal lines that have the desired genetic traits. This method helps spread resistant alleles in managed populations (Kim, Ah Rha, et al., 2019; Lefebre, De Smet, et al., 2024).

22.7. Statistical Modeling of SNP–Trait Associations

Statistical models are also crucial for putting SNP markers in context and figuring out what they mean for evolution. Researchers can determine how strongly certain SNPs are connected to traits like Varroa resistance by examining allele frequencies in various populations and using techniques such as one-way ANOVA, Kruskal–Wallis tests, and multiple comparison corrections. Studies that used modeling have shown that some mitochondrial markers, even though they are in very conserved areas, show allele frequency changes that are statistically linked to resistance phenotypes. For example, when looking at all the honeybee subspecies in Europe, researchers found that allele frequencies of genetic variants linked to drone brood resistance are linked to different *A. mellifera* lineages. This evidence supports the idea of using these markers in MAS strategies (Kim, Ah Rha, et al., 2019; Lefebre, Broeckx, et al., 2024; Lefebre, De Smet, et al., 2024).

22.8. Population-Level Insights and Historical Perspectives

Furthermore, comparative genomics has done more than just identify diagnostic SNP markers; it has also helped us understand how honeybee populations that are under a lot of parasitic pressure are evolving as a whole. Using samples from museums to look at the past has revealed that the arrival of *Varroa destructor* has led to a big decrease in mitochondrial haplotypic diversity, while nuclear diversity has remained fairly stable due to polyandrous mating behaviors (Mikheyev et al., 2015). These results indicate that the *mtDNA* may evolve faster, as shown by quick changes in allele frequency and less haplotype diversity. However, the overall structure of the genome can still support adaptive evolution.

22.9. Integration of Genomic and Statistical Frameworks

The methods used in these comparative genomic studies go far beyond just aligning sequences. They include a full workflow that involves collecting samples from different places, sequencing whole genomes at high speed, carefully checking the quality of the sequence data, and using advanced statistical tests to look for selection signatures. Next-generation sequencing has significantly transformed the field by enabling researchers to simultaneously examine thousands of SNPs across various individuals and populations. This method of using a lot of data not only makes it easier to observe small differences in allele frequency, but it also helps separate the effects of selection from those of genetic drift or gene flow. Researchers can make accurate predictions about whether bees are resistant or susceptible based on their mitochondrial genotype profiles (Grozinger & Zayed, 2020; Lefebre, Broeckx, et al., 2024) by using elastic net penalized regression and other advanced statistical methods.

22.10. Applications in Breeding and Conservation

Finding mtDNA SNP markers is an incredibly useful tool for breeding programs that want to make honeybee colonies more resilient. Breeders can use MAS by testing the mitochondrial genotypes of queen candidates and picking the ones that have the right SNP alleles. The procedure makes sure that the maternal lineage has the genes that make Varroa resistance and better colony health possible. This approach is especially advantageous for managed populations where keeping high colony survival rates is important for long-term beekeeping. Furthermore, because mtDNA is passed down from maternal to offspring, using these markers makes it easier to track resistance traits over time, which makes the breeding process easier and less complicated than nuclear genetic analyses (Kim, Ah Rha, et al., 2019; Lefebre, De Smet, et al., 2024).

22.11. Subspecies Differentiation and Structural Markers

Another benefit of using mtDNA to detect markers is that it can help tell different honeybee subspecies and populations apart. These groups often respond differently to Varroa infestations and other environmental stressors. Comparative studies have shown that mitochondrial sequences are different not only in the number of SNPs they have but also in the way their genes are ordered and how their non-coding regions are put together. These differences enhance the resolution, enabling the identification of adaptive signatures specific to a particular population. Changes in the A+T-rich area of mtDNA may have occurred due to recent or similar evolutionary changes, likely in response to challenges from the environment or diseases, and this area helps control how DNA is copied and read (Kim, Ah Rha, et al., 2019).

22.12. Complementarity of Nuclear and Mitochondrial Analyses

The majority of studies on adaptive markers have concentrated on nuclear DNA due to its larger size and potential for recombination. However, the mtDNA provides a different view. The maternal inheritance pattern makes the genetic background easier to understand, which makes it easier to link differences in phenotype to mtDNA variants. Researchers can obtain a better overall picture of the genetic structure behind resistance traits when they use both nuclear and mitochondrial markers together. This two-pronged approach makes predictive models in MAS programs more accurate and helps us learn more about the evolutionary forces that shape honey bees' adaptive traits (C. Chen, 2015; D. S. M. Henriques, 2018).

22.13. Environmental Pressures and Broader Adaptations

Selection for disease resistance may not be the only factor causing faster mtDNA evolution. Other ecological factors, such as adapting to climate and finding resources, may also play a role. For example, in some groups, certain patterns of selection in mtDNA have been linked to changes that help them live in high-altitude or temperate environments. These changes affect the health of the colony and its ability to fight off parasites. Researchers have been able to identify candidate regions in the mtDNA that likely help with better metabolic efficiency, stress tolerance, and, in the end, survival in adverse conditions by combining this kind of ecologically relevant data with genomic analyses. These results support the idea that the evolutionary changes in the mitogenome are closely linked to the larger adaptive processes that happen in honey bee populations (C. Chen, 2015; Grozinger & Zayed, 2020).

22.14. Future Directions and Conservation Implications

Finding mtDNA regions that evolve quickly has big effects on breeding and conservation programs. As honeybee populations are threatened by habitat loss, pesticide exposure, and parasitic stresses like Varroa mites, the ability to quickly use molecular tools to choose strong genetic lines is a long-term way to keep colonies healthy. By including tested SNP markers in regular screening processes, breeders can help bees resist *Varroa destructor* and enhance the overall health of the ecosystem, which will reduce the impact of various stressors on bee populations (Lefebre, Broeckx, et al., 2024).

22.15. Accuracy of Predictive Models and Geographic Adaptability

Furthermore, these studies use advanced statistical methods to ensure that the markers they uncover are statistically strong and can predict phenotypic performance in a wide range of genetic backgrounds. Elastic net penalized regression and other methods help cope with the large number of SNPs by finding a smaller set of markers that capture most of the genetic variation linked to the trait of interest. Not only are these smaller marker panels more affordable, but they also demonstrate high accuracy in classification. For example, some studies have indicated that they can correctly predict Varroa resistance phenotypes with an accuracy of about 76–88% (Kim, Ah Rha, et al., 2019; Lefebre, Broeckx, et al., 2024).

22.16. Longitudinal and Regional Monitoring

The method that looks at mtDNA can also help us monitor how populations change over time, along with using markers to select for disease resistance. Longitudinal studies that compare honeybee populations from the past and the present, for example, have shown that mitochondrial haplotype diversity changed a lot after *Varroa destructor* was introduced. These kinds of time-based studies not only help us understand the evolutionary history of resistance traits, but they also show us in real time how quickly selective pressures can change the genetic landscape (Mikheyev et al., 2015).

22.17. Conclusion

In summary, using comparative genomic studies, advanced sequencing, and thorough statistical analysis has revealed that parts of the mtDNA that change rapidly are closely related to traits that help organisms adapt, such as resistance to *Varroa destructor* and environmental adaptation. Finding SNP markers in these areas opens up a whole new area of honeybee breeding programs. It lets breeders choose strong strains that can handle parasites and tough environments. As our genomic tools and analytical methods develop better, they will help us protect honeybee populations and ensure that these important pollinators can survive in a world that is constantly changing (D. Henriques, Browne, et al., 2018; Kim, Ah Rha, et al., 2019; Lefebre, Broeckx, et al., 2024).

Chapter 23

23. Integrative Approaches in Mitochondrial DNA Analysis of *A. mellifera*: From Traditional Molecular Techniques to Genomic and Bioinformatic Innovations

In the past few decades, the study of mitochondrial DNA (mtDNA) diversity in *A. mellifera* has changed a lot because old molecular techniques have been used together with new genomic and bioinformatic methods. Recent changes in research methods have given scientists new ways to quickly and cheaply screen many samples. This process has enabled researchers to monitor genetic integrity across vast geographic regions. This report explains how combining different methods, such as SNP genotyping, PCR-RFLP, and HRM analysis, has made it much easier for us to understand the structure of honeybee populations and determine events where different populations mix.

23.1. Introduction

The western honeybee, *A. mellifera*, has complex population genetics shaped by its evolution, environmental adaptation, and human impact. In the past, traditional methods like PCR-RFLP and morphometric analyses were the first to show mtDNA variation by revealing maternal lineages and large-scale phylogenetic patterns (Ferreira, 2017). As time went on, though, problems with resolution and throughput led to the creation of new methods like SNP genotyping and High-Resolution Melting (HRM) analysis, which are much more sensitive and efficient (Alburaki et al., 2023). Modern bioinformatics pipelines also make it possible to combine nuclear markers with mtDNA data, giving a full picture of genetic mixing and population structure (Donthu et al., 2024).

23.2. Traditional Molecular Techniques

Traditional molecular methods have made it possible to study the differences in mtDNA within *A. mellifera*. We widely use PCR-RFLP (Polymerase Chain

Reaction-Restriction Fragment Length Polymorphism) to amplify important mitochondrial regions, such as the tRNAleu-COII intergenic region. This phase is followed by digestion with specific restriction enzymes (like DraI) to show haplotype differences based on fragment length polymorphisms (Alburaki et al., 2023). These tests have been very cost-effective, especially in places where resources are limited, and they have helped to classify major evolutionary lineages for the first time (Alsharhi et al., 2025). We used traditional sequencing methods to directly Sanger-sequence amplified mtDNA segments, in addition to PCR-RFLP. This method, even though it had a lower throughput, gave us high-confidence sequence data that was crucial for phylogenetic reconstructions (Ferreira, 2017).

23.3. Advancements in High-Throughput SNP Genotyping

Along with traditional methods, SNP genotyping using arrays or targeted sequencing has changed the way we identify small genetic differences. Researchers can test hundreds or thousands of single nucleotide polymorphisms at the same time with SNP panels. Such an approach gives them a more detailed genetic fingerprint that can tell the difference between closely related haplotypes (Chávez-Galarza et al., 2015). SNP genotyping platforms are now used in many studies to identify useful markers that set haplotypes apart or demonstrate intrapopulation admixture with amazing sensitivity (Donthu et al., 2024). High-throughput screening has proven to be very useful when monitoring genetic integrity over large areas, particularly in areas where the native genetic diversity is threatened by introgression from non-native subspecies (D. S. M. Henriques, 2018).

23.4. High-Resolution Melting (HRM) Analysis

High-Resolution Melting (HRM) analysis is a powerful method that finds tiny sequence changes by looking at how short PCR amplicons melt. It works well with SNP genotyping and PCR-RFLP. There are many benefits to HRM analysis: it is quick and cost-effective by nature and does not require sequencing of the target regions. This means that large sample sets can be screened in a single assay (Kahya, 2020). HRM is excellent for tracking small changes in mtDNA between populations because it can pick up on single nucleotide differences. This aspact feature makes it easier to tell apart closely related maternal lineages and observe heteroduplex formations in mixed samples (Kahya, 2020). These traits make HRM analysis especially advantageous when dealing with the problem of introgression in mixed colonies where there may be some low-level heterogeneity (Porrini et al., 2021).

23.5. Integration of Modern Genomic Data

In the past ten years, genomic data has become more widely available than ever. Whole-genome resequencing and next-generation sequencing (NGS) methods have made it possible to obtain complete datasets that include both nuclear and mitochondrial genomes. These genomic platforms make it possible to create large SNP datasets, which makes it easier to accurately study nuclear–mtDNA interactions, admixture events, and phylogeographic patterns (D. S. M. Henriques, 2018). When looked at together with mtDNA sequences, detailed nuclear SNP panels have shown complicated patterns of co-inheritance that weren't possible to detect with standard single-marker methods (Chávez-Galarza et al., 2015). Whole mitogenome sequencing enhances this analysis by examining variations throughout the entire mitochondrial genome. Such analysis provides us information about how lineages have changed over time and historical demographic events that single-gene studies might miss (D. S. M. Henriques, 2018).

23.6. Robust Bioinformatics Pipelines

Both new experiments and the rise of strong bioinformatics pipelines have caused changes in honeybee genetics. These pipelines combine multilocus datasets from both old and new methods, making it easier to systematically study population structure, admixture, and evolutionary dynamics (Alsharhi et al., 2025). Researchers can attain a complete picture of how populations differ by combining nuclear SNP data with mtDNA sequences using advanced computer models. This process helps them figure out how to resolve differences between biparental and maternal inheritance patterns (Donthu et al., 2024). Furthermore, these bioinformatic methods include filtering SNP datasets for quality, clustering analyses like those done by STRUCTURE or

DAPC algorithms, and making phylogenetic networks that show how complicated evolutionary relationships are (Chávez-Galarza et al., 2015).

Bioinformatics also helps us see and analyze the levels of admixture between populations, which helps us determine gene flow and introgression events. Researchers can observe small differences in genetic variation using tools like principal component analysis (PCA) and model-based clustering. They then look at these differences in light of geographic barriers, historical refugia, and human-mediated dispersal events (D. S. M. Henriques, 2018). When two different species breed together, it is especially important to combine nuclear and mitochondrial markers. This is because mtDNA alone may not provide a complete picture, as it is inherited only from one parent (Düttmann et al., 2022).

23.7. Applications in Monitoring Genetic Integrity

The above-mentioned fast and inexpensive screening methods enable the monitoring of large numbers of honeybee colonies. The use of PCR-RFLP, HRM analysis, and high-throughput SNP genotyping together has made it possible to do large-scale surveys of mtDNA diversity in many places, such as the US, Europe, North Africa, and the Arabian Peninsula (Alburaki et al., 2023; Alsharhi et al., 2025). For instance, studies in the US have shown that managed honeybee populations have low mtDNA haplotype diversity, which raises concerns about genetic bottlenecks and how they can be affected by environmental stressors (Alburaki et al., 2023). At the same time, these methods can identify introgression from non-native subspecies, which could put locally adapted gene pools at risk (Düttmann et al., 2022).

In places where Africanization has happened, scientists have used molecular methods to tell the difference between African and European mtDNA lineages. In the past, we used PCR-RFLP and sequencing of mtDNA markers like the COI-COII intergenic region to sort these lineages. Now, we can identify mixed populations even better with SNP-based methods (Düttmann et al., 2022). For example, in Nicaragua, studies that combine morphometric data (like measurements of forewing length) with mtDNA analyses have given us a more detailed picture of how populations change and how much African maternal lineage has mixed in (Düttmann et al., 2022).

23.8. Precision in Genotyping and Haplotype Discrimination

Using SNP genotyping and HRM analysis has also greatly improved the accuracy with which mtDNA haplotypes are distinguished. SNP tests concentrate on nucleotide positions that provide abundant information. This methodology lets them observe even small differences in a population (Chávez-Galarza et al., 2015). This level of accuracy is essential when telling very similar subspecies apart and sharing haplotypes because they have recently diverged or genes have flowed back and forth (D. S. M. Henriques, 2018). HRM analysis adds to this strategy by giving a quick way to observe single nucleotide differences without having to do more sequencing runs, which saves both time and money (Kahya, 2020).

In addition, these new genotyping methods can be scaled up and used in high-throughput formats, which makes it possible to do large population surveys that were not possible with only traditional methods. For example, multiplex SNP panels have been made to quickly put bees into different subspecies or to figure out how mixed up hybrid zones are (D. S. M. Henriques, 2018). These high-resolution genotyping tests furnish important information for both conservation management and breeding programs. They ensure the preservation of the genetic integrity of native honeybee populations (Chávez-Galarza et al., 2015).

23.9. Integration of Nuclear and Mitochondrial Data

Combining data from the nuclear and mitochondrial genomes is a significant improvement in current population genetics research on *A. mellifera*. Nuclear markers, like those from genome-wide SNP arrays, add to the maternally inherited mtDNA by showing how both parents pass on genes and giving us a better idea of how the genome is mixed overall (D. S. M. Henriques, 2018). Using integrated bioinformatic pipelines to look at these datasets at the same time allows for a multidimensional look at population structure. Differences between nuclear and mitochondrial markers can point out things like sex-biased gene flow or historical introgression events (Donthu et al., 2024; D. Henriques, Browne, et al., 2018).

The process of integration often includes strict data filtering, statistical modeling to estimate admixture, and phylogenetic reconstructions that take into account the different evolutionary histories of the nuclear and mitochondrial genomes at the same time (Chávez-Galarza et al., 2015; D. Henriques, Browne, et al., 2018). This dual-marker method is especially useful for solving cases where mtDNA alone inaccurately reflects population history because it doesn't recombine. It provides a more complete picture by including both maternal and biparental inheritance (Alsharhi et al., 2025).

Bioinformatics pipelines support these combined analyses with a multitude of advanced software tools. For instance, people often use tools like principal component analysis (PCA), Bayesian clustering methods (like STRUCTURE or ADMIXTURE), and discriminant analysis of principal components (DAPC) to figure out complicated genetic patterns (Chávez-Galarza et al., 2015; D. Henriques, Browne, et al., 2018). These computer-based methods have been very useful for figuring out the fine-scale structures of populations, finding small signs of admixture, and following the paths of gene flow over time (D. S. M. Henriques, 2018).

23.10. Implications for Conservation and Breeding

These methodological advances have important real-world effects that go beyond academic interest; they are essential for the effective conservation and long-term management of honeybee populations. Genetic diversity is critical for keeping colonies healthy and strong, especially when new diseases, pesticides, and rapid changes in the environment happen (Alburaki et al., 2023). Modern SNP genotyping, through high-throughput screening and HRM analysis, can identify potentially harmful genetic bottlenecks and monitor the intermingling of commercial or non-native strains (Düttmann et al., 2022). This, in turn, helps selective breeding campaigns that want to keep the unique genetic heritage of native subspecies while not losing traits that help them adapt to their environment (D. S. M. Henriques, 2018).

In the real world, beekeepers and conservation managers can use these integrated genetic tools to regularly examine colonies for their mtDNA haplotype diversity and nuclear genetic makeup. This lets them identify colonies that still have a lot of native genetic integrity (D. S. M. Henriques, 2018). Regular genetic monitoring helps keep a balance between increasing colony productivity and protecting genetic diversity in places where hybridization is common, like where commercial queens are brought in or migratory beekeeping is practiced (Alsharhi et al., 2025).

23.11. Case Studies and Regional Applications

Many case studies show how combining old and new methods can make things work better. In Europe, researchers have used PCR-RFLP and SNP genotyping to discover where native mtDNA haplotypes are located and how much mixing occurs between native and introduced subspecies (Chávez-Galarza et al., 2015; D. Henriques, Browne, et al., 2018). For example, studies in the Iberian Peninsula have found that mtDNA lineages are spread out in complex patterns related to past ice age refuges and later recolonization events (Chávez-Galarza et al., 2015). Bioinformatic studies of nuclear SNP data in these studies have helped scientists learn more about secondary contact zones and uneven gene flow, which has led to better conservation plans (D. S. M. Henriques, 2018).

Similar methods have been used in places like the USA and the Arabian Peninsula to monitor the spread of non-native haplotypes caused by beekeeping and bee movement (Alburaki et al., 2023; Alsharhi et al., 2025). These studies use scalable SNP genotyping and HRM analysis platforms that can handle large sample sizes without losing accuracy. This technique lets researchers assess the genetic integrity of populations in real time over large areas (Donthu et al., 2024; D. Henriques, Browne, et al., 2018).

Furthermore, research on Africanized honey bee populations has shown how important it is to use both genetic markers and phenotypic assessments like wing morphometrics, which provide different types of evidence for identifying different lineages (Düttmann et al., 2022; Porrini et al., 2021). Using bioinformatics to combine these datasets not only confirms the traditional markers but also reveals hidden genetic variation. This process helps us learn more about how hybrid swarms work and what happens in the long term when genes are mixed (Porrini et al., 2021).

Improvements in whole-genome resequencing technologies have made it possible to sample all of the mitochondrial and nuclear markers in Africanized populations. This has revealed complex patterns of gene flow that go against what was previously thought to be true based only on mtDNA polymorphisms. These studies show how useful it is to use an integrative approach that combines traditional PCR-based tests with in-depth genomic analyses to make strong, high-resolution datasets (D. S. M. Henriques, 2018).

23.12. Future Perspectives and Ongoing Innovations

In the future, combining traditional molecular methods with cutting-edge genomic and bioinformatic methods will help us better understand the structure of *A. mellifera* populations. New technologies, like third-generation sequencing platforms, will make it possible to read even longer sequences and figure out complicated genomic areas, like those with repetitive sequences or structural variants (D. S. M. Henriques, 2018). At the same time, improvements in machine learning and artificial intelligence are expected to improve bioinformatic pipelines by automating the finding of small admixture signals and predicting gene flow patterns from big datasets (Chávez-Galarza et al., 2015).

In addition, as these technologies become easier to acquire and cheaper, it is expected that monitoring both nuclear and mitochondrial genomes at the population level will become common practice. By combining real-time sequencing data with cloud-based processing pipelines, we will be able to obtain almost instant feedback on the genetic health of honeybee colonies. Such information will help researchers make better decisions about how to manage bees and protect them (D. S. M. Henriques, 2018).

This kind of technological convergence will also make it easier to study adaptive genetic variation, which is when fine-scale resolution of SNP markers across the genome can identify loci that are being selected for in response to environmental stressors. This can then help breeding programs that want to make colonies more resilient to new problems like pathogens, climate change, and habitat fragmentation (Donthu et al., 2024).

Research that mixes old genetic markers with new genomic methods is beginning to explore the complicated areas where different species mix, influenced by both nature and human activities. A thorough study of these patterns through robust bioinformatics will continue to enhance our understanding of how *A. mellifera* evolved and will assist us in developing more targeted conservation plans (D. S. M. Henriques, 2018).

23.13. Conclusions

In short, the use of both old and new molecular techniques, like PCR-RFLP, SNP genotyping, and HRM analysis, has entirely changed how we can study mtDNA diversity in *A. mellifera*. These new methods have facilitated the rapid and cost-effective screening of large sample sets, enabling the close monitoring of genetic integrity across vast regions (Alburaki et al., 2023; Alsharhi et al., 2025). Advanced bioinformatics pipelines that combine nuclear and mitochondrial markers also present us a new way to look at population structure, admixture, and evolutionary history (Donthu et al., 2024).

This combined method not only helps us learn more about how maternal lineages change over time, but it also helps us determine introgression events and hybridization patterns in a wide range of ecological settings. These kinds of insights are crucial for making decisions about conservation and breeding, since they help protect the genetic resources of native honeybee populations in the face of new and difficult environmental problems (Düttmann et al., 2022; D. Henriques, Browne, et al., 2018).

The development of genomic technologies and advanced bioinformatics tools shows that population genetics research is going through a big change. These advances make it possible to observe subtle patterns of genetic variation and admixture with an unprecedented level of detail. Such discoveries will help us better manage and protect one of the world's most important pollinators for the economy and the environment. It is expected that new technologies will continue to combine sequencing and computational methods, making it possible to obtain even more accurate and useful information about *Apis mellifera* populations (Chávez-Galarza et al., 2015; D. S. M. Henriques, 2018).

Thereby, the past, present, and future of honeybee genetics are all connected by a series of methodological advances that have always helped us learn more about how diverse populations are. The combination of traditional molecular techniques and modern genomic approaches, along with cutting-edge bioinformatics analyses, creates a powerful and integrated framework that not only improves our basic understanding of *mtDNA* diversity in *A. mellifera* but also serves as an essential tool for protecting and managing honey bee populations around the world in a way that is positive for the environment (Ferreira, 2017; D. Henriques, Browne, et al., 2018).

In conclusion, this integrated research framework is necessary for solving the difficult problems that modern beekeeping faces. Researchers can now more effectively understand the complicated genetics of *A. mellifera* by using a mix of traditional DNA tests and modern genomic technologies along with powerful computer analyses. The results will help future generations understand and protect the evolutionary legacy of this important pollinator (Alburaki et al., 2023; D. Henriques, Browne, et al., 2018).

Chapter 24

24. Standardization of Methodologies in Honey Bee Mitochondrial DNA (mtDNA) Research

24.1. The Importance of Standardization in mtDNA Research

Standardizing methods is an important and ongoing topic in mitochondrial DNA (mtDNA) research on honeybees. Improvements over the years have ensured the repeatability and comparability of results across labs and geographical areas (J. D. Evans et al., 2013). This strict standardization covers many steps of the work, from the first collection of bee samples to the DNA extraction protocols, the carefully optimized PCR conditions, and the full bioinformatics analyses needed to make sense of the data. This process makes sure that researchers all over the world can work together and compare their results with confidence (Meixner et al., 2013).

24.2. Sample Collection Protocols

The standardization of sample collection protocols is a crucial first step. There are now clear rules in place to protect the integrity of samples. These rules say that biological material must be collected in controlled conditions and using methods that reduce contamination while keeping the tissues or whole specimens in their biological state (J. D. Evans et al., 2013). Since mtDNA is passed down from maternal to offspring (J. D. Evans et al., 2013), samples must come from defined sources, like worker bees from the brood area or carefully chosen individuals that reliably represent a single maternal lineage. These protocols ensure that subsequent analyses reveal authentic genetic signals, unaffected by drift, contamination from unrelated individuals, or damage from improper storage (Meixner et al., 2013).

24.3. Standardized DNA Extraction Techniques

Standardized DNA extraction methods ensure consistency across all samples. Getting high-quality DNA out of the samples is just as important as collecting them. Over the past few decades, researchers have worked on and improved DNA extraction methods suitable for different kinds of bee tissues. To extract mtDNA from individual bee legs, for example, researchers have developed sophisticated techniques utilizing Chelex® resin. The procedure ensures the mitochondrial genomes remain intact during cell breakdown (J. D. Evans et al., 2013). Other extraction methods, such as those based on CTAB, have also been used to address problems caused by inhibitors in bee tissues or that compromise the integrity of the cell wall. We have thoroughly tested these methods to ensure they yield high-quality DNA suitable for subsequent steps (J. D. Evans et al., 2013). Standardized extraction methods are crucial because they affect how sensitive PCR amplification is and how accurate genotype calls and population genetic inferences are (J. D. Evans et al., 2013).

24.4. PCR Optimization and Standardization

Another important part of standardizing methods in mtDNA research is the conditions for the Polymerase Chain Reaction (PCR). To obtain the most specific results and the fewest mistakes, optimized PCR protocols have clear rules about the types of reagents, the order of the primers, and the thermocycling conditions (J. D. Evans et al., 2013). For instance, a widely used protocol starts with a five-minute denaturation step at 94°C, followed by 35 cycles of denaturation, specific annealing at exact temperatures (for example, 48°C for certain mtDNA regions), and extension steps. These steps guarantee the constant amplification of intergenic regions with variable sequences (J. D. Evans et al., 2013). These details are important because even small changes in temperature or reagent concentrations can make a big difference in how well amplification works, which in turn affects the haplotype frequencies that are seen (Meixner et al., 2013). Furthermore, the creation of protocols for multiplex PCR assays makes it possible to amplify more than one marker at a time. This capability makes it easier to compare and analyze complex genetic data sets (J. D. Evans et al., 2013).

24.5. Bioinformatics Pipelines and Data Interpretation

As PCR conditions develop better, so do bioinformatics pipelines, which are necessary for turning raw sequence data into useful molecular diagnoses and comparative analyses. Standardized bioinformatic workflows include steps like quality trimming, demultiplexing, primer removal, and sequence alignment with well-known programs like CLUSTAL W or MUSCLE. Next, we perform phylogenetic analysis using methods such as Neighbor-Joining and Maximum Likelihood (J. D. Evans et al., 2013). Working with mtDNA sequences requires careful description and comparison across studies, which makes these methods especially useful (J. D. Evans et al., 2013). Furthermore, using standardized data formats (usually FASTA) and publicly available sequence databases like GenBank and SILVA makes analyses even more repeatable and provides researchers a way to compare their results to known benchmarks (Darby et al., 2020; Deiner et al., 2017).

24.6. Practical Applications and Collaborative Research

The focus on standardized protocols isn't just for show; it offers real-world benefits that go beyond the classroom and help with collaborative research, conservation, and the creation of diagnostic tools. When different labs use the same methods for collecting samples, extracting DNA, amplifying it, and analyzing it with bioinformatics, it is possible to combine data from different places and times to look at evolutionary patterns on a global scale (Chua et al., 2023). This shared framework makes it easier to do meta-analyses and big surveys, which are important for figuring out how honey bee populations have changed over time and how things like breeding practices, environmental pressures, and human actions affect them (Meixner et al., 2013). It also supports the creation of diagnostic tools that can confirm genetic purity, which is essential for both breeding programs and conservation strategies that aim to protect native *A. mellifera* populations (Meixner et al., 2013).

24.7. Enhancing Reproducibility and Quality Control

These standardized methods have helped international collaborations a lot by letting researchers build large genetic databases that combine sequence data from different parts of the world. These kinds of databases help us learn more about how different honeybee populations are related through evolution. They also present us important starting points for measuring genetic variation and finding introgression events or hybridization (J. D. Evans et al., 2013). Using these standardized methods makes sure that genetic markers, like the COI-COII intergenic region, which is commonly used to identify maternal lineages, are consistently amplified and analyzed. This ensures that observed differences reflect genuine biological diversity, not merely the outcome of the employed methods (Meixner et al., 2013).

Setting up detailed quality control steps at every stage of the process makes it even easier to reproduce and compare studies. For instance, researchers consistently use spectrophotometric methods, such as measuring absorbance at 260 and 280 nm, to verify the purity and concentration of DNA. This procedure makes sure that only high-quality samples move on to the next analytical steps (J. D. Evans et al., 2013). Furthermore, using housekeeping genes or spiked control templates as internal and external controls in PCR tests stops false positives and negatives, making genetic tests more reliable (J. D. Evans et al., 2013).

24.8. Evolution of Methodologies and Technological Innovation

Technical problems and new discoveries have led to changes in the creation and improvement of these standardized methods over time. For instance, early mtDNA studies in honeybees had to address problems like allele dropout and PCR artifacts. These problems got better over time as primer design and reaction conditions improved (J. D. Evans et al., 2013). Next-generation sequencing platforms like Illumina MiSeq and 454 pyrosequencing have also changed the field by making sequencing faster and cheaper. At the same time, the use of strong bioinformatics pipelines has made it possible to turn raw data into useful information faster and more accurately than ever before (Von & Hardulak, 2020). These small changes show how flexible methodological standardization is in mtDNA research and how new ideas, along with following established rules, can move the field forward (Liu Bsc, 2021).

24.9. Diagnostic Tools and Conservation Strategies

One critical use of standardized mtDNA research is to make diagnostic tools that can tell how genetically pure native *A. mellifera* populations are. As researchers become increasingly concerned about how mixing with other bee types is damaging local genetic diversity, tests using mtDNA have become crucial for protecting and breeding bees (Meixner et al., 2013). Standardized methods assure that tests reliably identify maternal lineages, notice changes in haplotype frequencies, and track the genetic health of colonies over time (Soares et al., 2023). This ability is important not only for scientific research but also for practical uses in beekeeping and conservation management. For instance, it assists in selecting queens and devising breeding strategies that maintain populations tailored to their local environment (Chua et al., 2023).

24.10. Phylogenetic and Evolutionary Applications

Standardized mtDNA methods are also crucial for studies of evolution and phylogenetics. These protocols make it possible to accurately compare sequence data from different studies and species. This procedure helps scientists put together evolutionary histories and look at population structure and gene flow. Researchers often use statistical models and methods like Neighbor-Joining, Maximum Likelihood, and Bayesian inference to figure out how honey bee lineages are related to each other evolutionarily (J. D. Evans et al., 2013). Comparing data on an equal basis has facilitated our understanding of how honeybees have evolved and adapted to various environments (J. D. Evans et al., 2013).

24.11. Error Mitigation and Data Quality Control

In bioinformatics analyses, standardization also applies to how sequence artifacts and data quality are handled. The workflow that ensures the accuracy of mtDNA sequences, for instance, includes crucial steps like alignment, chimera checking, and strict quality filtering. By following these careful steps, researchers can avoid mistakes in sequencing and misreading data, ensuring that the final family trees are strong and the results are reliable (Deiner et al., 2017; Soares et al., 2023). This degree of precision in bioinformatics analysis is essential to the acceptance of molecular data as the gold standard for species identification and evolutionary research, as well as to the general reproducibility of genetic studies.

24.12. Training, Collaboration, and Capacity Building

The international community benefits from the standardization of approaches in addition to technical reproducibility and a culture of cooperation and data exchange. Combining datasets and conducting meta-analyses spanning hundreds of samples and several continents is made simpler when labs follow standard procedures and exchange comprehensive methodological data (Darby et al., 2020; J. D. Evans et al., 2013). This collaborative attitude accelerates scientific research and enhances our overall understanding of the biology and evolution of honeybees by pooling knowledge and resources. More significantly, it lays the groundwork for coordinated conservation efforts to protect native bee populations, which are essential for maintaining ecological equilibrium and agricultural output (Meixner et al., 2013).

24.13. Future Directions and Ongoing Innovation

Additionally, the quality of research outputs has changed dramatically as a result of the move toward standardized methodologies. Clear guidelines for methods help create better tools to assess genetic purity and enhance the effectiveness of PCR and sequencing experiments (J. D. Evans et al., 2013; Soares et al., 2023). For example, researchers are able to develop diagnostic assays with high sensitivity and specificity that can consistently differentiate between native and non-native genetic lineages by carefully defining the parameters for sample handling, reagent quality, thermocycling profiles, and bioinformatic thresholds (Meixner et al., 2013). Thus, conservation policies and breeding programs inform efforts to preserve or reestablish the genetic integrity of native *A. mellifera* populations (Meixner et al., 2013).

24.14. Impact and Educational Value

The consistent rise in data quantity and quality over the past few decades is indicative of the influence of these methodological advancements. Detailed studies that use complete genome data and high-quality mtDNA markers have taken the place of older research, which often had problems with inconsistent results and limited sequence information. The progress made through standardization is shown by the move from simple PCR-RFLP tests, which used one restriction enzyme like DraI to distinguish between mtDNA lineages, to advanced next-generation sequencing (NGS) methods (J. D. Evans et al., 2013). These advancements provide reliable, large-scale data that are crucial for understanding how honeybees evolve and for monitoring changes in their populations as human activities and environmental pressures change (Liu Bsc, 2021).

It's also important to remember that the standardized procedures described in the literature act as training materials for new researchers in the field. In-depth procedures and detailed instructions, occasionally supported by online databases and video tutorials, guarantee that novices in honey bee molecular research can quickly become accustomed to best practices and incorporate their work with international studies (J. D. Evans et al., 2013). Both applied and scholarly aspects of bee conservation, where quick and precise genetic monitoring can guide management decisions right away, depend on this capacity building (J. D. Evans et al., 2013).

24.15. Conclusion

In conclusion, the standardization of methods in mtDNA research is one of the most important developments in honey bee molecular genetics in recent decades. To guarantee that results are repeatable and comparable across various studies and labs, comprehensive protocols for sample collection, DNA extraction, PCR amplification, and bioinformatics analyses have been created and improved (J. D. Evans et al., 2013). These standardized techniques speed up our understanding of honeybee evolution and help develop successful conservation strategies by facilitating international collaboration by offering a common framework for data generation and interpretation (Chua et al., 2023; Darby et al., 2020).

Additionally, clear guidelines are important for creating tests that verify the genetic purity of local *A. mellifera* populations and for enhancing the quality of scientific research (Meixner et al., 2013). As a result, breeding programs that focus on preserving locally adapted genetic variants are essential for maintaining the resilience of native bee populations against anthropogenic and environmental stressors (Soares et al., 2023). Together, these efforts have produced a body of work that is reliable, repeatable, and extremely pertinent to honeybee genetics research at both the fundamental and applied levels.

In the end, a strong foundation for honey bee molecular research has been created by consistently following standard procedures, starting from when a bee sample is taken to the final steps of analyzing the genetic data. In addition to strengthening our knowledge of honeybee evolution, the scientific community has made sure that the data produced are trustworthy resources for the preservation of this vital pollinator species by closely following established procedures and constantly improving them in light of new technological developments (J. D. Evans et al., 2013; Meixner et al., 2013). Because of these standardized methods, honey bee research appears to have a bright future and will continue to be essential to international efforts to monitor, conserve, and manage honey bee populations in a sustainable manner for future generations (Darby et al., 2020).

Chapter 25

25. Harnessing Mitochondrial Markers and Honey-Derived eDNA for Honey Authentication and Bee Conservation: A Molecular Approach to Food Traceability and Biodiversity Management

25.1. Abstract

Combining mitochondrial markers with environmental DNA (eDNA) taken from honey samples is a game-changing method for verifying food and managing conservation. This method takes advantage of the natural benefits of mitochondrial DNA (mtDNA), such as its high copy number, maternal inheritance, and fast evolutionary rate, to accurately trace the entomological origin of honey and identify specific honeybee subspecies. Recent studies that used this method have taken mtDNA research into new areas. For example, they have been able to verify the biological source of honey products and track the geographic distribution of native and exotic bee lineages. This study has big effects on beekeeping, protecting biodiversity, and making sure food is safe to eat. This report explains the technical basis, methodology, and implications of using mitochondrial markers with honey-derived eDNA. It uses examples from several studies to show how this new approach supports strong authentication protocols and helps with conservation management.

25.2. Introduction

In both the food industry and bee conservation, the authenticity and origin of honey have become critical issues. Traditional methods of characterizing honey, such as physicochemical profiling or melissopalynological techniques, have proven insufficient when confronted with advanced methods of adulterating the product or mislabeling it. As a result, molecular techniques have become a popular option because they can quickly, accurately, and without harming the honey determine where it came from. Mitochondrial markers are especially suitable for this job because they are very stable, common, and able to tell species and subspecies apart (Kek et al., 2017). When combined with environmental DNA (eDNA) extraction, which uses the complex genetic material found in honey, the resulting method lets scientists acquire DNA signatures from not just the bees that make honey but also from the plants, pathogens, and microbiota that live with them (Bovo et al., 2018). This chapter discusses the application of mitochondrial markers in conjunction with honey eDNA. This article concentrates on the technical aspects that enable distinguishing between different species, identifying subspecies, and gaining a comprehensive understanding for food authentication and monitoring bee lineages in various regions.

25.3. Mitochondrial Markers and Their Relevance

Mitochondrial markers are mostly used because of the unique properties of mtDNA. First, they have many copies in cells, which makes it more likely that PCR will work, even when the matrix (honey) has broken DNA pieces in it (Mohamadzade Namin et al., 2022). Second, mitochondrial sequences transmit from maternal to offspring and undergo rapid changes over time. This process lets scientists tell closely related species and subspecies apart (Kek et al., 2017). For example, scientists often use the cytochrome oxidase subunit I (COI) gene and the 16S rRNA gene to trace the family tree of honeybees and to distinguish between species like *A. mellifera*, *Apis cerana*, and *Apis dorsata* (Kek et al., 2017; Soares et al., 2019). Furthermore, mitochondrial markers show differences between lineages that have been linked to where they came from, which helps us understand where honey bee populations live (Soares et al., 2023; Taurisano et al., 2024).

25.4. Extraction of Environmental DNA from Honey

Honey is a complicated biological matrix that contains eDNA from many different sources, such as bee tissues, glandular secretions, and bits of plants that the bees have collected. The extraction protocols usually have to accommodate the problems caused by the high sugar content and the presence of PCR inhibitors (Mohamadzade Namin et al., 2023). New DNA extraction methods have made it possible to acquire high-quality mtDNA from honey, which is often present in tiny amounts and in a very fragmented state (Mohamadzade Namin et al., 2022; Moškrič et al., 2021). Using silica-column-based kits that are sold in stores and have been optimized for these types of samples greatly increases the amount and purity of the eDNA that is recovered.

This optimized eDNA extraction is critical for the next steps in PCR amplification that look for mitochondrial markers (Soares et al., 2023).

25.5. Methodological Approaches

25.5.1. PCR Amplification and Target Selection

The method of using mitochondrial markers with honey eDNA involves creating PCR primers that focus on short mitochondrial pieces that are most effective for amplifying damaged samples. Researchers usually choose parts of the COI gene, the 16S rRNA gene, or areas between genes like tRNA^leu-cox2 because these segments are the right size (between 100 and 500 base pairs) for amplifying honey samples that have been old or heated (Chávez-Galarza et al., 2021; Soares et al., 2019). Some studies have created species-specific primers that make amplicons that are only found in certain honeybee lineages. For example, there are primers that can tell the A, M, and C lineages apart in European honey bee populations (Soares et al., 2019; Taurisano et al., 2024).

25.5.2. Restriction Fragment Length Polymorphism (RFLP) and High-Resolution Melting

After PCR amplification, different analytical methods improve the ability to distinguish species apart. RFLP analysis uses restriction enzymes to cut the PCR products, creating unique banding patterns for each species or subspecies that can be observed with gel electrophoresis. For example, the enzyme AseI effectively distinguishes different *Apis species* when applied to PCR products made from mitochondrial 16S rRNA fragments (G. Zhang et al., 2024). High-resolution melting (HRM) analysis, however, helps to identify small changes accurately in the DNA sequence by monitoring the temperature at which the PCR product melts. Such analysis is a quick, closed-tube option that reduces the risk of contamination (Kek et al., 2017; Mohamadzade Namin et al., 2022).

25.5.3. Next-Generation Sequencing Approaches

Recent improvements in next-generation sequencing (NGS) and metabarcoding have made it possible to analyze honey eDNA in more than one way at the same time. Deep shotgun metagenomic sequencing makes it possible to detect a wide range of mitochondrial markers in one run. This procedure includes not only the main bee species but also other organisms that live with them, like parasitese (e.g., *Varroa destructor*) (Bovo et al., 2018; Utzeri et al., 2019). This all-encompassing method makes things more sensitive and clear, especially when looking at more than one mitochondrial region at the same time (Moškrič et al., 2021; Soares et al., 2023).

25.5.4. Tracing Entomological Origin and Detecting Specific Subspecies

Using mitochondrial markers to determine the insect origin of honey is a big step forward in this field. Researchers can connect the genetic signatures to the bee species that made the honey by looking at the mtDNA sequences found in honey-derived eDNA. For instance, researchers have successfully used mitochondrial markers to distinguish the difference between *Apis cerana* and *Apis mellifera*. This distinction is important for markets where the two types have completely unique commercial values (Mohamadzade Namin et al., 2022; Y. Z. Zhang et al., 2019). Researchers have also been able to identify different mtDNA haplotypes that match native or exotic honeybee lineages by looking at areas that have polymorphisms that are specific to certain subspecies, like the tRNA^leu-cox2 intergenic region (Chávez-Galarza et al., 2021; Soares et al., 2019). This ability is critical in situations where conservation laws require the identification of native subspecies, like *A. m. ligustica* in protected areas (Taurisano et al., 2024).

25.5.5. Food Authentication Implications

Using mitochondrial markers and eDNA from honey samples together has big effects on checking the authenticity of food. First, the clear identification of the insect origin provides a genetic "fingerprint" that can be used to examine product labels and keep consumers safe from false labeling (Mohamadzade Namin et al., 2023; Soares et al., 2023). Being able to identify even small amounts of adulteration (like mixing honey from different species at a rate of 1%) makes sure that the products sold in stores are really from the source they say they are (Mohamadzade Namin et al., 2022; Y. Z. Zhang et al., 2019). In addition, these molecular methods are quick and cheap, which makes them useful for regular quality control in the honey business (Moškrič et al., 2021; G. Zhang et al., 2024).

Furthermore, using honey as an eDNA source instead of taking samples from live bee colonies is less invasive, which accelerates the authentication process and reduces stress on bee populations. This benefit is especially important because bee populations are declining around the world, and conservation priorities mean that natural populations should be disturbed as little as possible (Bovo et al., 2018; Utzeri et al., 2019).

25.5.6. Geographic Tracking of Bee Lineages and Conservation Management

In addition to verifying the food, mitochondrial marker analysis of honey eDNA is a very useful way to track bee populations across different areas. Researchers have long used differences in mtDNA haplotypes to map out where different bee lineages live (Soares et al., 2019; Tihelka et al., 2020). Researchers can figure out how bees move around and what types of bees are in different areas by mapping these haplotypes. For example, the C1 mitotype in *A. m. ligustica* is more common in certain geographic areas where native populations are known to live. This technique makes it easier to protect native bee genotypes (Soares et al., 2023; Taurisano et al., 2024).

This geographic tracking is crucial for conservation management in many ways. First, it helps to identify genetic introgression from non-native or hybridized populations, which could hurt the health of native bee stocks (Soares et al., 2023; Taurisano et al., 2024). Second, mtDNA biomarkers can help regulators monitor changes in bee populations due to environmental stress or human activities like queen trading and colony migration (Mohamadzade Namin et al., 2023; Taurisano et al., 2024). Third, mapping the distribution of bee lineages helps create breeding programs and conservation policies that are specific to certain areas. These policies aim to keep genetic diversity, which is essential for the resilience and sustainability of bee populations and the ecosystems that depend on them (Soares et al., 2023; G. Zhang et al., 2024).

25.5.7. Novel Applications and Future Directions

Researchers have recently started to look into new uses for mtDNA research that go beyond just identifying species. Combining mitochondrial markers with eDNA from honey samples opens up many new possibilities. Researchers are utilizing this method to monitor the spread of pathogens, such as monitoring *Varroa destructor* haplotypes in honey, which can aid in integrated pest management (Utzeri et al., 2019; Y. Z. Zhang et al., 2019). Discovering parasite DNA in conjunction with bee mtDNA provides a comprehensive view of the health of the bees and the hive, an insight previously unattainable through conventional methods.

Researchers are expanding molecular authentication techniques by combining mitochondrial markers with nuclear genetic markers in multi-marker analyses. For example, researchers have used Exon-Primed Intron Crossing (EPIC) markers of nuclear genes along with mtDNA markers to acquire a more complete picture of the bees' genetics (Moškrič et al., 2021). This two-part method not only confirms the mother's lineage, but it also provides us information about the contributions of both parents. This type of analysis is especially advantageous for finding hybridization events. These kinds of thorough genetic tests are essential for keeping breeding programs in equilibrium and making sure that conservation efforts protect not only the unique traits of native bee subspecies but also the overall genetic diversity (Mohamadzade Namin et al., 2022; Moškrič et al., 2023).

Another new use is in police work to find out about honey fraud. With better molecular tools, we can now figure out tricky cases of honey being wrongly labeled or misrepresented by comparing the mtDNA haplotype profile from suspect honey samples to existing databases of bee population genetics (Soares et al., 2019; Y. Z. Zhang et al., 2019). This forensic use not only helps stop economic fraud, but it also boosts consumer trust and helps regulatory agencies keep food quality standards high.

The use of mitochondrial marker analysis is growing even more because of the rapid growth of high-throughput sequencing technologies like DNA metabarcoding and next-generation sequencing. These technologies let us do broad-spectrum analyses of the eDNA in honey, which include both qualitative and quantitative assessments of the types of species present. These methods make it easier to observe changes in bee populations early on, which could act as an early warning system for changes in ecosystems and environmental stressors (Bovo et al., 2018; Soares et al., 2023).

25.5.8. Case Studies and Comparative Analyses

Numerous studies have demonstrated the effectiveness of the combined mitochondrial-eDNA approach in real-world situations. A thorough study used mitochondrial markers to examine honey samples and determine the origins of the bees. The results were very similar to what is already known about the geographic distribution of the different bee subspecies. For example, honey from Europe showed different genetic patterns that were linked to the C-lineage, while honey from places where *Apis cerana* lives had its own unique mtDNA markers (Soares et al., 2019; Y. Z. Zhang et al., 2019). These kinds of comparisons show that the PCR-RFLP and HRM methods are excellent at telling species and subspecies apart with a lot of sensitivity.

Another study used next-generation sequencing to do a metagenomic analysis of honey eDNA. The researchers found the most common honeybee species, small plant parts the bees foraged on, and the microbes that lived there. By connecting the plant's origin with the insect's fingerprint, this complete genetic profile provided crucial information about the honey-making environment. Therefore, the integration of mtDNA markers and deep sequencing evolved into a comprehensive approach for environmental verification and monitoring (Bovo et al., 2018; Utzeri et al., 2019).

Comparative studies of traditional authentication methods, like spectroscopic methods, and the mitochondrial-eDNA approach have shown that the mitochondrial-eDNA method is much better at distinguishing between different types. mtDNA typing has a very high resolution, which lets it determine small amounts of adulteration, like mixing honey from different bee species at levels as low as 1%. This level of precision is much better than what standard compositional analyses can find (Mohamadzade Namin et al., 2022; Y. Z. Zhang et al., 2019).

25.5.9. Implications for Industry and Regulatory Frameworks

Using mitochondrial markers with honey eDNA has big effects on the honey industry around the world. Food producers and regulatory agencies would benefit from a quick, cheap, and reliable way to determine the authenticity of products that can be added to standard quality control procedures. Being able to verify that honey comes from the right bee species and subspecies helps fight fraud, keep high-quality products from being mixed with lower-quality ones, and keep geographical indications honest. In places like Europe and Asia, where honey from native bee subspecies is worth more on the market, it's critical to verify that the labels are correct (Mohamadzade Namin et al., 2023; Soares et al., 2023).

Furthermore, honey-based eDNA sampling doesn't hurt bee colonies, which is beneficial for ethical research and sustainable beekeeping. This is especially useful given the worries about bee populations dropping and the environment getting worse. Regulatory frameworks can certify honey products and even use these molecular methods as evidence in legal disputes over mislabeling. These measures would protect both consumers and honest producers (Moškrič et al., 2021; Taurisano et al., 2024).

25.5.10. Conservation Management and Biodiversity Preservation

There are many ways in which mitochondrial marker analysis can help with conservation management. First, it allows you to monitor the distribution of bee populations and their changes over time. This aspact is important for finding places where native bee subspecies are in danger of being hybridized or invaded by non-native species (Soares et al., 2023; Taurisano et al., 2024). Second, being able to quickly verify the genetic health of bee populations makes it possible to create targeted management plans that protect genetic diversity. For example, conservation programs that are specific to a region can help local subspecies like *A. m. ligustica* while keeping exotic lineages from spreading (Soares et al., 2019; Taurisano et al., 2024).

Mitochondrial markers can also be utilized for ongoing molecular monitoring, assisting adaptive management in addressing environmental stressors such as climate change, habitat fragmentation, and pesticide exposure. The genetic information from honey eDNA provides us a starting point for figuring out how healthy bee populations are and how conservation efforts might affect them. Such data are very useful for conservation groups working at home and around the world to protect pollinator diversity and ensure that agricultural ecosystems can last (Soares et al., 2023; Utzeri et al., 2019).

25.5.11. Challenges and Limitations

There are still some problems with combining mitochondrial markers with honey eDNA, even though it offers many benefits. The amount and quality of DNA from honey can vary based on storage, processing, and natural inhibitors (Mohamadzade Namin et al., 2023; Moškrič et al., 2021). To attain results that can be repeated and trusted, it is important to ensure that DNA extraction protocols are strong and that PCR conditions are as satisfactory as they can be. Furthermore, the lack of complete reference databases for honey bee mtDNA haplotypes can make it difficult to accurately assign sequences to certain subspecies, especially in areas with a lot of genetic diversity or where new hybridization events have happened (Chávez-Galarza et al., 2021; Soares et al., 2019).

Furthermore, while techniques like PCR-RFLP and HRM analysis provide quick and easy ways to analyze data, they need to be carefully standardized across labs to assure that results are consistent and can be compared (Mohamadzade Namin et al., 2022; G. Zhang et al., 2024). Next-generation sequencing techniques could help solve these problems by allowing multi-marker analyses and deeper sequence coverage. However, these methods are also pricier and need specialized bioinformatics knowledge (Bovo et al., 2018; Soares et al., 2023).

25.5.12. Future Perspectives

There is a lot of promise for mitochondrial marker research in honey authentication in the future, as studies are still going on to improve and expand on current methods. One expected improvement is the creation of multiplex PCR systems that can amplify several mitochondrial regions at once. These improvements will make the test more accurate. Some recent studies have looked into how combining nuclear markers with mtDNA might make it easier to tell subspecies apart and provide us more information about the biparental genetic structure of bee populations (Mohamadzade Namin et al., 2022; Moškrič et al., 2021).

Furthermore, standardizing protocols and making global reference databases for honeybee mtDNA haplotypes will be crucial for these methods to be widely used in both academic research and business. Collaboration among regulatory agencies, research institutions, and industry stakeholders could potentially establish international standards for the use of genetic markers in determining the authenticity of honey. These types of regulations would safeguard consumers and producers and contribute to the preservation of native bee subspecies by closely monitoring genetic diversity and lineage integrity (Soares et al., 2023; Taurisano et al., 2024).

Furthermore, as next-generation sequencing technologies develop better, they will probably become easier to use and less expensive for honey eDNA analysis. This will make it possible to study bee populations in more detail and at higher resolutions in a wider range of geographic areas. The results will give us new information about the environmental and evolutionary factors that affect the diversity of honeybees. Bringing together high-throughput sequencing and bioinformatic tools that can analyze complex DNA data from honey samples will expand research in this field, helping scientists discover tiny genetic changes that might indicate new dangers to bee populations (Soares et al., 2023; Utzeri et al., 2019).

25.5.13. Conclusion

In conclusion, combining mitochondrial markers with eDNA from honey samples allows for detailed tracking of the entomological origin of honey. This makes it possible to identify species and subspecies through strong genetic analyses. This new method uses the high copy number, maternal inheritance, and rapid evolutionary divergence of mtDNA to accurately identify honey products and also provide useful information for tracking locations and managing conservation efforts. These methods protect against false labeling and adulteration by confirming the geographic and entomological origins of honey. They also increase consumer trust and support regional conservation efforts to protect native bee lineages. There are still problems to solve, especially when it comes to improving DNA extraction methods and creating standardized reference databases. However, the progress made in PCR-based techniques, HRM analysis, and next-generation sequencing is making it possible to do more accurate, cheaper, and thorough tests of honey authenticity. As the field grows, future research will probably identify more uses for these molecular tools in non-traditional settings. Such developments will make genetic analyses even more important for sustainable beekeeping and protecting biodiversity.

The index includes terms such as honey authentication, mitochondrial DNA, environmental DNA, honey bee subspecies, food traceability, conservation management, PCR-RFLP, high-resolution melting analysis, and next-generation sequencing.

The text integrates references into its citations. The text utilizes citations to substantiate the information. Studies showing that these methods are possible (Bovo et al., 2018; Kek et al., 2017; Mohamadzade Namin et al., 2022) strongly support the current state of research. This integrated method not only confirms the insect origin of a honey sample, but it also provides important information about where different honeybee lineages live, which is necessary for making effective plans for protecting them (Soares et al., 2023; Taurisano et al., 2024).

In conclusion, combining mitochondrial markers with eDNA from honey samples makes a strong, new, and versatile tool that can be used for both food authentication and conservation management. As methods improve, this approach is expected to become more and more important for making sure that honey products are safe and easy to trace, protecting the genetic identity of native bee populations, and ultimately helping to achieve the larger goal of sustainable beekeeping and protecting biodiversity (Utzeri et al., 2019; G. Zhang et al., 2024).

Chapter 26

26. Next-Generation Genomic Tools for Honey Bee Conservation: Integrating Nuclear–Mitochondrial Genotyping, eDNA Innovations, and Computational Advances to Safeguard Pollinator Resilience

This topic discusses how new techniques, such as combined nuclear–mitochondrial genotyping tests, improved ways to collect environmental DNA (eDNA), and better computer systems, are likely to advance the next generation of mitochondrial DNA (mtDNA) research in honeybees. These innovations will make it possible to better track changes in populations, observe introgression events more easily, and make better decisions about how to manage conservation (Adams et al., 2019).

26.1. Introduction

Honeybees *(Apis mellifera)* are an important part of both agriculture and natural ecosystems. Keeping the genetic integrity of honeybees is important to protect biodiversity and ensure that pollination services remain available for a long time (Barbosa et al., 2021). Molecular biology has made it much easier to track genetic variation within and between populations, especially by improving the analysis of mitochondrial and nuclear DNA markers (Adams et al., 2019). By combining nuclear and mitochondrial genotyping, we can learn more about both maternal genetic lineages and genetic variation that comes from both parents. This methodology revolves around the problems with single-marker approaches (Dong et al., 2021). Recent advances in eDNA extraction and next-generation sequencing have made it possible to do large-scale surveys without having to touch the organisms directly (Andrews et al., 2021). The detailed framework shown here shows that these new methods are about to change honeybee genetics by providing better ways to track population structure, identify introgression events, and plan conservation management strategies (Adams et al., 2019).

26.2. Combined Nuclear–Mitochondrial Genotyping Assays

Combined nuclear–mitochondrial genotyping assays are an important advancement because they look at two different parts of the genome together to provide a full understanding of honey bee genetic diversity. This two-part approach aims to lessen the problems that mtDNA markers have by using high-resolution nuclear SNP data (Adams et al., 2019). These problems can stem from things like mitochondrial introgression or incomplete lineage sorting. Researchers can determine maternal lineages and the more complicated biparental signatures that make up the population structure (Dong et al., 2021) at the same time by sequencing both nuclear and mitochondrial markers. These integrated tests excel in identifying minor introgression events that a single genomic compartment may overlook. The result presents a more complete picture of gene flow between managed and wild honeybee populations (Bovo et al., 2022). Furthermore, using these markers together makes it easier to tell the difference between species and subspecies, which is important for protecting local ecotypes and managing hybrid zones (Bovo et al., 2022). Using these combined tests not only helps us understand the family tree of species better, but it also makes it simpler to locate traits that help bees adapt to their local environment and resist diseases (Adams et al., 2019). This kind of methodological synergy ultimately helps create a strong monitoring system for changes in honeybee populations and encourages conservation practices that are based on evidence (Dong et al., 2021).

26.3. Improved Environmental DNA (eDNA) Extraction Techniques

Recent improvements in eDNA extraction methods have made it much easier to acquire beneficial DNA from complicated and damaged environmental samples. This is especially important for studies that use bee products and hive-related samples (Mohamadzade Namin et al., 2023). These kinds of improvements make it possible to obtain nuclear and mitochondrial DNA fragments from a wide range of materials, including honey, pollen, and hive debris. This capability makes it possible to use large-scale, non-invasive sampling methods (Bovo et al., 2020). Better protocols include optimizing the conditions for cell lysis, using high-efficiency binding matrices, and improving the purification steps that together raise the yield and quality of the DNA that is recovered (Adams et al., 2019). These methods also use new ways to remove inhibitors, which are important for lowering PCR biases and lessening the effects of contaminants that are naturally found in honeybee samples (Buswell et al., 2023). These protocols improve the spatial and temporal resolution of honeybee genetic monitoring (Andrews et al., 2021) by making sure that even small amounts of DNA can be reliably extracted from environmental samples. This means that conservationists and beekeepers can now do more organized surveys of honeybee populations over large areas without having to use invasive or time-consuming collection methods (Gueuning, 2019).

26.4. More Sophisticated Computational Pipelines

The rapid increase in sequencing output means that we need to create equally advanced computational pipelines that can accurately process and understand large, complicated genomic datasets (Andrews et al., 2021). Modern bioinformatics tools now include algorithms for fixing mistakes, reconstructing haplotypes, and estimating allele frequencies. Such technology makes it possible to quickly tell the difference between real genetic signals and noise (C. Chen et al., 2022). These pipelines also use advanced statistical models that combine nuclear and mitochondrial data in a way that makes it easier to determine introgression events and piece together the histories of populations (Dong et al., 2021). One important thing about these tools is that they can work with mixed samples from colony-level pooling or eDNA-derived datasets, which is better than what traditional individual-based sequencing methods can do (Buswell et al., 2023). Using machine learning techniques in these pipelines makes it even easier to tell apart population clusters on a small scale and attain better estimates of admixture proportions, even in populations with a lot of genetic diversity (Gueuning, 2019). As computers evolve better, these pipelines should become even easier to use, more scalable, and more reliable, giving conservation managers real-time feedback on their decisions (Law et al., 2024).

26.5. Enhanced Monitoring of Population Dynamics

The use of improved eDNA extraction and advanced computational pipelines with combined nuclear–mitochondrial genotyping assays directly helps keep track of honey bee population changes over time and space (Adams et al.,

2019). High-resolution markers from nuclear DNA show genetic drift, migration, and local adaptation clearly. Mitochondrial markers, on the other hand, show how maternal lineages have changed over time (Bovo et al., 2020). Monitoring at such a high resolution is crucial for promptly identifying changes in population parameters, which can stem from environmental shifts, disease outbreaks, or human activities such as queen trade and migratory beekeeping (Buswell, 2022). When combined with improved eDNA methods, these genetic tests enable scientists to monitor populations without causing harm to them. This process makes it possible to regularly and thoroughly assess the structure of honeybee populations (Adams et al., 2019). By constantly taking samples from the environment and using computer programs to acquire genetic metrics, conservationists can acquire data that is almost real-time and helps them plan ahead (Gueuning, 2019).

26.6. Enhanced Detection of Introgression Events

Gene flow from one genetic lineage to another is called introgression. It can have a big impact on the ability of honeybee populations to adapt and their genetic integrity (Dong et al., 2021). Because mitochondrial markers are passed down from maternal to offspring, they may not fully capture the bidirectional nature of gene flow (Bovo et al., 2022), which is why traditional mtDNA-based studies often miss these kinds of events. However, combining genotyping assays with nuclear markers resolves this problem by showing the complicated patterns of admixture that happen during introgression events (Bovo et al., 2022). Advanced computer analyses support the distinction between incomplete lineage sorting and true introgression, providing us with statistically robust conclusions about the exchange of genetic material between managed and wild subspecies (Buswell, 2022). It is crucial to correctly interpret these signals because introgression can weaken local adaptations and even cause the loss of unique genetic traits that are essential for conservation efforts in the area (Adams et al., 2019). Researchers can determine low-frequency introgression events earlier and more accurately by combining nuclear and mitochondrial datasets. This procedure allows for early intervention and conservation actions (Bovo et al., 2022).

26.7. Informed Decision-Making in Conservation Management

The enhanced capacity to monitor population fluctuations and identify introgression events serves as a robust foundation for making informed decisions in conservation management (C. Chen et al., 2022). Conservation managers can objectively assess the genetic health of honeybee populations when they have access to detailed genetic data. They can then achieve a balance between agricultural productivity and the need to protect native genetic resources (Donthu et al., 2024). Integrated genotyping tests reveal both the current genetic diversity and past trends, helping to determine if populations are decreasing due to inbreeding or mixing with non-native bees (Mohamadzade Namin et al., 2023). These kinds of detailed datasets provide us a lot of information to work with when creating targeted breeding programs and rules that protect the genetic uniqueness of local subspecies (Utzeri et al., 2021). Furthermore, these data can help set conservation priorities, like finding areas with a lot of genetic diversity that need habitat protection or separating populations that are at risk of introgression (Utzeri et al., 2021). By combining genomic data with ecological and environmental data, decision-makers can choose the best actions to take to help honeybee populations survive and thrive in the long term (Gueuning, 2019).

26.8. Integration with Practical Apiculture and Biodiversity Conservation

Combining these different methodological advances not only helps us learn more about how *A. mellifera* has changed over time, but it also directly helps with practical uses in beekeeping and protecting biodiversity (Adams et al., 2019). In commercial beekeeping, the ability to quickly and non-invasively assess the genetic makeup of colonies makes it possible to identify unwanted hybridization events that could hurt the quality of the honey and the strength of the colony (Bovo et al., 2020). Beekeepers can obtain detailed information about their bees' ancestry that helps them choose which bees to breed to strengthen local adaptations and lower their susceptibility to pathogens (Buswell, 2022). Conservation biologists can also use these tools to make large-scale genetic maps that show how genes move across landscapes. This helps them set baselines for tracking the effects of environmental change and human activity (Gueuning, 2019). The dual focus on applied apiculture and conservation management shows a change in how genetic data are used in practice, connecting basic research with operational strategies (Andrews et al., 2021).

26.9. Integration with Practical Apiculture and Biodiversity Conservation

As combined genotyping methods, *eDNA* extraction protocols, and computer tools continue to improve, they will aid us in closely monitoring and safeguarding honeybee populations (Ferrari, Ricardo, Dias, de Souza Araujo, et al., 2024). One big problem is still making sure that protocols are the same in all labs and field studies so that data from different sources can be compared directly (García-Olivares et al., 2021). To solve this problem, academic researchers, government agencies, and industry stakeholders will all need to work together. Each of these groups is important for moving the science of honeybee conservation forward (Gueuning, 2019). Researchers are looking into how far they can push ultra-deep sequencing and single-cell genomic analyses to acquire even more detail, especially in places where DNA degradation is a big problem (Bovo et al., 2020). Ongoing improvements in cloud-based computing platforms will make advanced analytical tools more accessible to everyone. Such advances will allow researchers in resource-limited areas to add useful data to global honeybee monitoring networks (Andrews et al., 2021). As these integrated methods develop better, we can expect the creation of automated workflows that combine DNA extraction, sequencing, and data interpretation in almost real time. This change has the potential to fundamentally alter the management and protection of honeybee populations (Dong et al., 2021).

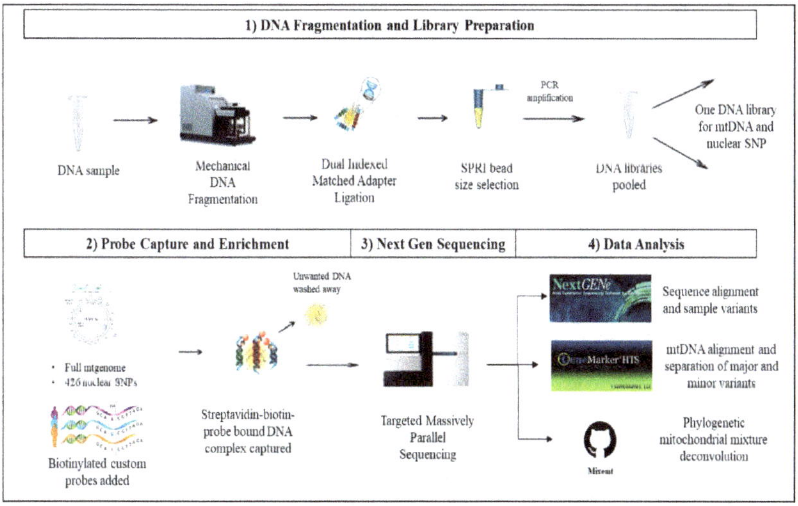

Fig. 29. Steps of Environmental DNA*eDNA)* Extraction Techniques

26.10. Case Studies and Applied Research

Several recent case studies show how these new methods could change the way we protect honeybees and keep bees (Buswell et al., 2023). Using colony pool-sequencing techniques and nuclear-mitochondrial genotyping together gave a full genetic picture of honey bee populations. The results revealed previously unseen levels of introgression in managed colonies (Buswell et al., 2023). Another study used a rapid method to extract DNA from honey samples to identify where the honey came from and track the spread of mitochondrial haplotypes in different regions, linking genetic information to environmental and business factors (Bovo et al., 2022). These examples show how advanced molecular techniques can be used in the real world. They show that the most advanced genetic tools can help with both scientific discovery and practical management (Olate-Olave et al., 2021). As more information is gathered through these methods, meta-analyses and comparative studies will help us learn even more about how honey bee populations evolve and how to protect their genetic legacies (Gueuning, 2019).

26.11. Implications for Global Pollination Services

The genetic health of honeybee populations has a direct effect on global pollination services. Evolutionary resilience is a key factor in how well colonies do when they are stressed by things like disease, climate change, and habitat fragmentation (Adams et al., 2019). These new technologies make it possible to do high-resolution genetic monitoring, which can identify small changes in allele frequencies that could be early signs of a loss of adaptability (Andrews et al., 2021). These kinds of insights are important not only for keeping beekeeping practices strong but also for making sure that wild bee populations can still pollinate native plants and crops (Utzeri et al., 2021). Researchers can separate the evolutionary history of honeybees from recent changes made by humans by combining data from nuclear and mitochondrial markers. This process presents us a deeper understanding of how natural adaptation and managed breeding work together (Adams et al., 2019). In the end, these improvements are part of a bigger plan that will ensure the future of pollination services, which are essential for global food security and the health of ecosystems (C. Chen et al., 2022).

26.12. Synergistic Effects on Conservation Policy

Outside of the lab, the new methods used to collect data have important policy implications (Donthu et al., 2024). When decision-makers have access to accurate, high-resolution genetic data, they can make conservation policies that protect different groups of people, enforce breeding rules, and control the movement of beehives to keep native genetic diversity (Bovo et al., 2022). In particular, collaborations between agencies can use standardized genomic data to figure out what conservation efforts should be focused on in both the region and the world. Such efforts will help stop the spread of invasive genotypes and ensure that local honeybee populations stay strong (Utzeri et al., 2021). Furthermore, using genomic information along with ecological risk assessments helps create management plans that prevent hybridization and changes in gene flow patterns before they occur (Buswell, 2022). This kind of proactive, data-driven approach is a big change in conservation policy, and it is strongly supported by the combination of advanced genomic and computational methods (Gueuning, 2019).

26.13. Integration with Breeding Programs

These new technologies have many practical uses, especially in breeding programs (Ferrari, Ricardo, Dias, Araújo, et al., 2024). By using combined nuclear–mitochondrial genotyping tests during regular breeding checks, beekeepers can choose traits that improve the health, productivity, and resilience of bee colonies while minimizing the risk of unwanted genetic mixing (Bovo et al., 2022). Detailed genomic profiling makes sure that selective breeding programs don't accidentally reduce the ability of native honey bee populations to adapt (Adams et al., 2019). At the same time, better methods for extracting eDNA make it possible to sample many colonies across large areas without hurting them. This process presents breeding programs with large datasets that show the genetic diversity that exists in the field (Mohamadzade Namin et al., 2023). Admixture estimation and haplotype phasing are two detailed computer analyses that help breeding programs by finding specific parts of the genome linked to beneficial traits like disease resistance or the ability to adapt to climate changes (Dong et al., 2021). The combination of advanced genotyping, eDNA sampling, and computational analytics sets the stage for a new era in precision beekeeping, where genetic information is the basis for every management decision (Ferrari, Ricardo, Dias, de Souza Araujo, et al., 2024).

26.14. Towards a Comprehensive Genomic Framework

Using a mix of nuclear and mitochondrial genotyping tests, improved eDNA extraction methods, and better computer analysis creates a complete system for studying mtDNA in honeybees (Andrews et al., 2021). This framework encompasses a wide range of topics in honeybee genetics, from fundamental evolutionary processes to practical issues in beekeeping. This approach makes sure that monitoring, conservation, and breeding efforts are based on a strong and varied genetic understanding (Bovo et al., 2022). Researchers can make predictive models that guess how population dynamics will change in the future and look at the possible effects of environmental stressors by combining high-quality genomic data with strict analytical methods (Buswell et al., 2023). The resulting integration not only helps us learn more about how *A. mellifera* evolved, but it also connects academic research with real-world uses in the field (Gueuning, 2019). With this framework in place, conservation management can move from reacting to problems to taking steps to protect honeybee populations for the long term (Adams et al., 2019).

26.15. Challenges and Future Directions

Despite the immense potential of these advanced methods, there remain certain issues that require resolution (García-Olivares et al., 2021). For instance, standardizing eDNA extraction protocols is critical for guaranteeing the comparability of genetic data across studies and geographic regions. However, establishing such standards is still an area that needs more work (Andrews et al., 2021). Furthermore, the full integration of nuclear and mitochondrial datasets needs bioinformatic workflows to keep getting better, especially when it comes to correctly telling the difference between real genetic signals and artifacts like NUMTs (Dong et al., 2021). Furthermore, understanding complicated patterns of admixture and introgression requires advanced statistical models that are both quick to run on computers and biologically accurate (C. Chen et al., 2022). To solve these problems, molecular ecologists, computational biologists, and conservation practitioners will need to work together across disciplines. They will also need to continue investing in research infrastructure (Gueuning, 2019). Future research should strive to develop standardized, open-access computational pipelines that can adapt to various bee-related environmental samples. These initiatives would help make the global effort to monitor honeybee genetic diversity more organized (Adams et al., 2019).

26.16. Conclusion

In conclusion, the combination of nuclear-mitochondrial genotyping assays, better eDNA extraction methods, and more advanced computational pipelines will bring about a major change in mtDNA research in honeybees (Adams et al., 2019). These advances will provide researchers and practitioners the tools they need to monitor population dynamics in high resolution, observe introgression events with greater accuracy, and make better decisions in both conservation management and commercial beekeeping (Bovo et al., 2022). These new ideas have led to the creation of a comprehensive framework that not only helps us understand how *A. mellifera* evolves but also turns that knowledge into useful ways to keep the bees' genetic integrity and support sustainable beekeeping practices (Utzeri et al., 2021). In the end, the coming together of molecular, bioinformatic, and ecological methods marks the start of a new era in honeybee research. In this new era, integrated genetic insights will guide adaptive management and conservation policies that protect these important pollinators for the future (Gueuning, 2019).

To fully take advantage of these new technologies, we will need to continue doing interdisciplinary research and create standardized protocols (Ferrari, Ricardo, Dias, Araújo, et al., 2024). Academic research will not be the only application for the tools and methods discussed here as they become more accessible and improve over time. They will also be used to help with practical conservation, improve breeding programs, and protect the biodiversity that supports both natural ecosystems and agricultural productivity (Andrews et al., 2021). In this new environment, the combination of advanced genotyping methods, strong eDNA assays, and cutting-edge computational pipelines promises to create a strong and flexible monitoring system that will protect the genetic legacy of *A. mellifera* for future generations (Dong et al., 2021).

This combined method not only helps us learn more about how honeybees have changed over time, but it also helps us use this knowledge in real life in the fields of apiculture and biodiversity conservation (Gueuning, 2019). Using these methods consistently will help researchers observe even small changes in population structure and genetic makeup. The findings will help them respond more quickly and effectively to new problems like habitat destruction, disease outbreaks, and stress caused by climate change (Buswell, 2022). By doing this, they will set the stage for a new generation of management strategies that are based on data and can change to meet the needs of honeybee populations around the world (Law et al., 2024).

To sum up, the future of mtDNA research in honeybees will be shaped by a combination of new methods that together create a complete framework for genetic monitoring (Adams et al., 2019). The field is in a favorable position to provide precise, useful information about honey bee population dynamics, introgression events, and overall genetic health (Bovo et al., 2020). This progress is made possible by the combined use of nuclear and mitochondrial markers, improved eDNA extraction methods, and cutting-edge computational analyses. These changes will lead to better conservation policies, better beekeeping practices, and, in the end, a sustainable future for the ecological services that honeybees provide (Buswell et al., 2023).

The way forward includes ongoing validation and standardization, linking new ideas from the lab with real-world applications that solve problems, and building collaborative networks that make it easier to share data and improve methods (Andrews et al., 2021). This spirit of working together is important to ensure that the progress made in genetic monitoring leads to conservation actions and breeding strategies that protect the genetic diversity and functional resilience of honeybee populations (Gueuning, 2019).

These different methods work together to show us new ways to understand how evolution, gene flow, and climate change affect honeybees (Dong et al., 2021). They offer a plan for the next phase of genetic research that expands our scientific knowledge and leads to real improvements in how we care for and protect these important pollinators. These measures will help honeybees continue to thrive in a world that is changing quickly (Ferrari, Ricardo, Dias, de Souza Araujo, et al., 2024).

Chapter 27

27. The Role of Mitochondrial DNA Markers in Honey Bee Phylogenetics and Biogeography

27.1. Introduction to Mitochondrial DNA in Honey Bee Phylogenetics

Mitochondrial DNA (mtDNA) markers are the most important part of honey bee phylogenetic studies because they are passed down from mother to child, don't mix with other DNA, have high mutation rates, and are relatively simple to amplify. All of these things make them a clear record of maternal lineages over long periods of time (Alburaki et al., 2023). These traits make mtDNA a great tool for studying the structure of populations, the divergence of subspecies, and historical biogeographic patterns in *A. mellifera*. This is because the haplotypes created only show the maternal genetic contribution and not the confusing effects of recombination that are common in nuclear markers (Cleary, 2022).

27.2. The COI-COII Intergenic Region as a Primary mtDNA Marker

The intergenic region between cytochrome c oxidase subunits I and II, also known as the COI-COII or tRNAleu-COX2 intergenic region, is a critical mtDNA marker in honey bee phylogenetics. It has a lot of structural variation because it has repeat sequences of different lengths and indels (D. S. M. Henriques, 2018). The COI-COII region has polymorphic elements called P and Q units that change in both number and sequence. These polymorphisms have been used to show the main evolutionary lineages, such as the African (A), Western European (M), and Eastern European (C) lineages (D. S. M. Henriques, 2018). Because this area is so structurally complex, PCR-based restriction fragment length polymorphism assays (PCR-RFLP) can quickly and cheaply screen it. Typically, these tests employ the DraI enzyme to reveal distinctive length polymorphisms and variations in restriction sites associated with evolved maternal lineages (Alburaki et al., 2023).

27.3. The ND2 Gene: A Complementary Coding Region Marker

The ND2 gene, which codes for one of the subunits of the NADH dehydrogenase complex, adds to the phylogenetic resolution of the COI-COII intergenic region because it has a mix of moderate conservation and nucleotide substitutions that act as informative polymorphisms among closely related haplotypes (Cleary, 2022). The coding sequence of ND2 is more stable than non-coding regions, which change more quickly and may show homoplasy. This stability makes it easier to understand detailed evolutionary relationships when combined with the more changing intergenic regions (D. S. M. Henriques, 2018). Subsequently, using both the COI-COII region and the ND2 gene together provides the best phylogenetic resolution by balancing the hypervariability of non-coding sequences with the structured variability of coding regions (Chávez-Galarza et al., 2021).

27.4. Methodological Approaches to mtDNA Marker Analysis

Researchers often use established methods to separate mitochondrial DNA from individual worker bees in a way that increases the amount and quality of the DNA. For example, they might use Chelex extraction methods. Then, they use region-specific primers to amplify the targeted regions using PCR (Coulibaly et al., 2019). In RFLP assays, the amplified products are either digested with restriction enzymes or sequenced directly. Both methods allow for precise haplotype assignment and make it easy to quickly screen large sample sets, which is important for studies at the population level (Düttmann et al., 2022).

27.5. Application of COI-COII in Population Structure and Biogeographic Studies

The COI-COII intergenic region is even more useful because it can show both the big evolutionary splits and the small differences in the structure of regional populations. Researchers have used this marker to identify over a hundred different haplotypes around the world. Specific geographic areas and known historical introductions link many of these haplotypes. This tool has allowed researchers to piece together colonization routes and guess how populations grew or shrank in the past (D. Henriques, Chávez-Galarza, et al., 2019). For instance, in places like the United States, COI-COII markers have helped to identify the main maternal lineage that comes from the North Mediterranean C lineage. They have also found small contributions from the M and African A lineages, which suggests that there are complicated patterns of international bee importation and mixing (Alburaki et al., 2023).

27.6. ND2 Gene Contributions to Subspecies Delineation

Along with the COI-COII region, ND2 is a useful genetic marker because its coding sequence evolution is affected by slightly different selective pressures. This implies that it generates phylogenetic signals that are comparable to those from the intergenic spacer (D. S. M. Henriques, 2018). Comparative studies have shown that ND2 phylogenies often match the broad groupings seen in complete mitochondrial genome analyses. The result supports lineage assignments and shows how strong mitochondrial markers are at defining subspecies (Techer, Clémencet, Simiand, Preeaduth, et al., 2017).

27.7. Integration of COI-COII and ND2 Markers in Phylogenetic Studies

Researchers have been able to not only catalog haplotype diversity but also figure out the exact phylogenetic relationships between honeybee subspecies by combining data from the COI-COII and ND2 markers. For instance, analyses that look at both markers have found distinct maternal sublineages within the larger African lineage. These sublineages are defined by specific patterns of insertions, deletions, and point mutations within the COI-COII region. This classification is supported by similar differences in ND2 sequences (Angelique & Diversité, 2017; Chávez-Galarza et al., 2015). These results have important effects on how we consider the evolutionary history of honeybees, especially in areas where there are many different maternal lineages because of both natural spread and human introduction (Coulibaly et al., 2019).

27.8. Geographic and Environmental Correlations of mtDNA Haplotypes

Research has found that the patterns of haplotypes in the COI-COII region are closely related to environmental and geographic factors, showing how useful mtDNA markers can be. For example, mitochondrial sequence data have revealed a new African subgroup that seems to be unique to the islands in the South West Indian Ocean. This discovery has giant implications for our understanding of regional biogeography and the historical patterns of honeybee dispersal (Techer, Clémencet, Simiand, Preeaduth, et al., 2017). The fact that mtDNA haplotypes match up with known climatic zones or biomes supports the idea that events in the past, like glacial refugia and the expansions that followed, have had a big impact on the genetic makeup of honey bee populations today (Angelique & Diversité, 2017).

27.9. The Role of mtDNA in Detecting Admixture and Introgression

Furthermore, mtDNA markers provide us important information about a mother's lineage that helps us determine historical admixture events and introgression. This is especially true in managed bee populations, where queens are often moved and replaced during commercial beekeeping. Admixture studies that use COI-COII markers have found hybrid mitotypes in some areas. This study shows how human-mediated transfers can affect genetic diversity and local adaptation (Düttmann et al., 2022; D. Henriques, Chávez-Galarza, et al., 2019). These insights are crucial for coming up with conservation plans that will help protect the genetic integrity of native honeybee subspecies. They are also important for breeding programs that want to keep colonies healthy and strong in the face of environmental stresses and new pathogens (Tozkar, 2020).

27.10. Integration with Nuclear Markers and Mitogenome Data

By combining mtDNA data with other genetic markers like nuclear microsatellites and single nucleotide polymorphisms (SNPs), we can obtain a better picture of how honeybee populations change over time by looking at both maternal and biparental inheritance patterns. Mitochondrial markers reveal the maternal background of a population, but when combined with nuclear markers, they can help researchers spot differences that may come from things like male gene flow or incomplete lineage sorting (Veisi et al., 2020). More and more studies these days are using integrative methods to obtain better at classifying subspecies and to improve our understanding of honey bee phylogenetics in a world that is changing quickly (Donthu et al., 2024).

27.11. Mitogenomics as a High-Resolution Extension of mtDNA Studies

Furthermore, new methods for sequencing the whole mitochondrial genome (mitogenome) are starting to work with traditional single-marker methods, giving us a clearer picture of the evolutionary history of honeybees. Comprehensive analyses of the mitogenome have confirmed the broad maternal lineages initially revealed by the COI-COII region. They have also clarified phylogenetic relationships that were previously unclear because short mtDNA segments are very variable and can lead to homoplasy (D. S. M. Henriques, 2018). These whole-genome methods confirm the phylogenetic frameworks made with single markers and provide us a new look at how quickly and in what ways molecular evolution happens in honey bee mitochondria (D. S. M. Henriques, 2018).

27.12. Multilocus Approaches and Congruent Phylogenetic Signals

Researchers have found that while the COI-COII analysis is still very beneficial for identifying major lineages, the ND2 gene offers additional information that strengthens ideas about recent changes and evolutionary speeds among subspecies (D. S. M. Henriques, 2018; Techer, Clémencet, Simiand, Preeaduth, et al., 2017). They achieved this by carefully comparing the evolutionary trees from the COI-COII intergenic region and the ND2 gene with those from full mitogenome sequences. Using both markers in a multilocus framework has become a standard procedure in the field, making sure that phylogenetic inferences are both strong and repeatable (Oleksa et al., 2021).

27.13. Historical Biogeography and Climatic Influences on Lineage Dispersal

Using mtDNA markers, historical biogeography studies have also been able to show how climate change has affected the spread and diversity of honeybee populations. For example, the COI-COII region has been very useful in mapping out the paths that people took to colonize Europe and Africa after the last Ice Age. It has shown how geographic barriers and changes in climate have affected the spread and isolation of different maternal lineages (Chávez-Galarza et al., 2021; Coulibaly et al., 2019). These studies indicate that different lineages have different geographic patterns. Some lineages stay in certain areas, while others have spread out a lot, partly because of human activities like bringing in commercial queen stocks (Alburaki et al., 2023).

27.14. Implications for Conservation, Breeding, and Management

In places where there is a lot of mixing, like parts of Anatolia and the Mediterranean basin, mtDNA markers have helped scientists detect small introgression events that would have gone unnoticed otherwise. The molecular evidence from the COI-COII region not only proves that hybrid individuals exist, but it also helps show which way genes flow between subspecies. The result provides us more information about the evolutionary effects of long-term hybridization (Tozkar, 2020). These kinds of results are crucial for figuring out how evolution works to make honeybees so adaptable and strong in their environment (Düttmann et al., 2022). These mitochondrial markers can be used for more than just academic research; they can also help with honeybee conservation and management strategies around the world. By quickly genotyping large numbers of colonies using mtDNA markers such as the COI-COII region, beekeepers and conservationists can monitor genetic diversity and identify non-native lineages that could potentially threaten the adaptation of local populations (D. Henriques, Chávez-Galarza, et al., 2019). Furthermore, the use of molecular tools like HBeeID, which combines mtDNA data with genome-wide SNP data, shows how modern high-throughput methods can work with traditional mtDNA markers to help identify subspecies more accurately and improve breeding strategies (Düttmann et al., 2022).

27.15. Conclusion

In short, mitochondrial DNA markers, especially the COI-COII intergenic region and the ND2 gene, have changed the way we approach how honeybees evolved. They are a reliable and high-resolution way to trace maternal lineages, figure out subspecies boundaries, and figure out how *A. mellifera* has changed over time and mixed with other species (Chávez-Galarza et al., 2021; D. Henriques, Browne, et al., 2018). We will be able to better understand honey bee phylogenetics as we continue to combine these traditional markers with whole mitochondrial genome approaches and nuclear genetic data. Such information will help us protect and manage this important pollinator for the long term (Coulibaly et al., 2019).

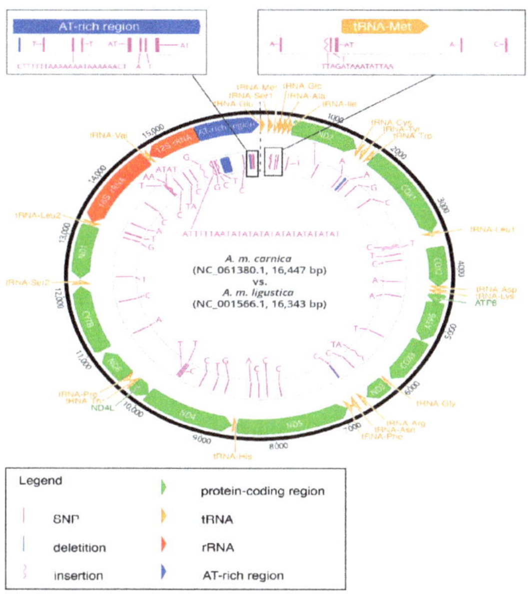

Fig. 30 presents a circular representation of the complete mitochondrial genome of *A. m. carnica* from Slovenia (SICarnica), along with a comparison with the mitochondrial genome of *A. m. ligustica* (REF Ligustica).

The outer circle presents the newly obtained mitochondrial genome of SI Carnica, complete with all corresponding annotations. The direction of the arrows indicates either light- or heavy-strand coding. The transparent inner circle presents the mitochondrial genome of REFLigustica. The detected differences in comparison to the mitochondrial genome of A. m. carnica are outlined. The two most variable regions (the AT-rich region and the region around tRNA-Met) are enlarged to enable a detailed view of the comparisons (Moškrič et al., 2022).

The mitochondrial genome of *Apis mellifera* (Western honey bee) comprises 37 genes and a non-coding control region, with a total length of 16,248–16,590 bp and high A+T content (84–85%). Below is a comprehensive table listing all mitochondrial markers, their functions, and their applications in research and industry.

Table 14. Mitochondrial Markers of *Apis mellifera*

Marker Name	Type	Function	Application Purpose
Protein-coding genes13)			
ATP6	Protein-coding	ATP synthase F0 subunit 6; energy production	Phylogenetic studies, subspecies differentiation (Nakagawa et al., 2018)
ATP8	Protein-coding	ATP synthase F0 subunit 8; energy production	Evolutionary lineage tracing (Nakagawa et al., 2018)
COI	Protein-coding	Cytochrome c oxidase subunit I; electron transport	Genetic diversity assessment, subspecies identification (Alsharhi et al., 2025)
COII	Protein-coding	Cytochrome c oxidase subunit II; electron transport	Lineage assignment (e.g., A/M/C lineages) (Utzeri et al., 2021)
COIII	Protein-coding	Cytochrome c oxidase subunit III; electron transport	Population structure analysis (Nakagawa et al., 2018)
CytB	Protein-coding	Cytochrome b; electron transport	Phylogeography, stress response studies (Alsharhi et al., 2025)
ND1–ND6, ND4L	Protein-coding	NADH dehydrogenase subunits; electron transfer	Evolutionary divergence tracking (Alsharhi et al., 2025; Nakagawa et al., 2018)
Transfer RNA genes22)			
tRNA-*Ala*, *Arg*, etc.	Transfer RNA	Amino acid delivery for mitochondrial protein synthesis	Structural variation studies (e.g., clover leaf arm absence in *Gln*, *Ser1*, *Thr*) (Nakagawa et al., 2018)
Ribosomal RNA genes2)			

16S rRNA	Ribosomal RNA	Large ribosomal subunit; protein synthesis	PCR-RFLP lineage differentiation (A/M/C lineages) (Attardo et al., 2019)
12S Rrna	Ribosomal RNA	Small ribosomal subunit; protein synthesis	Subspecies clustering (Attardo et al., 2019)
Non-coding regions			
Control Region	A+T-rich	Replication/transcription initiation	Subspecies-specific length variation analysis (Nakagawa et al., 2018)
COI-COII intergenic	Non-coding	Evolutionary signal region	DraI restriction test for lineage/haplotype identification (A/M/C) (Alsharhi et al., 2025)

Key Applications

Honey Authentication:

COI-COII and 16S rRNA markers distinguish mitochondrial lineages (A, M, and C) to verify honey's entomological origin (Alsharhi et al., 2025; Utzeri et al., 2022).

C-lineage mitotypes (C1 for *A. m. ligustica*; C2 for *A. m. carnica*) authenticate geographic origin in commercial honey (Alsharhi et al., 2025; Utzeri et al., 2022).

Subspecies Identification:

Protein-coding genes *(COI, CytB)* and tRNAs enable precise subspecies classification. For example, *A. m. jemenitica* is found in Africa and Arabia (Alsharhi et al., 2025).

Conservation Genetics:

Control region and rRNA variation monitor genetic integrity of populations facing import restrictions (e.g., New Zealand hybrids) (Nakagawa et al., 2018; Utzeri et al., 2021).

Phylogenetic Studies:

Gene start/stop codons *(ATT, ATG, TAA)* and strand distribution (9 heavy-strand vs. 4 light-strand PCGs) reveal evolutionary relationships (Alsharhi et al., 2025; Nakagawa et al., 2018).

Functional Notes

Start/Stop Codons: Protein-coding genes use *ATT* (most common), *ATG*, *ATA*, or *ATC* starts, with universal *TAA* stops (Nakagawa et al., 2018).

Strand Bias: The heavy strand encodes 9 PCGs + 14 tRNAs; the light strand encodes 4 PCGs + 8 tRNAs + 2 rRNAs (Nakagawa et al., 2018).

Conservation: Gene order is identical across subspecies, with minor length/composition variations (Nakagawa et al., 2018).

This genomic framework supports critical applications in food authenticity, conservation, and evolutionary biology, leveraging subspecies-specific markers for precision analysis (Nakagawa et al., 2018; Utzeri et al., 2021).

Chapter 28

28. Case Studies of *Apis mellifera* mtDNA:

The conservation of honeybees relies on the diversity, evolution, and applications of mitochondrial DNA. This chapter brings together and critically looks at several case studies from different geographic areas and subspecies. It shows how mtDNA analyses have been used to uncover the evolutionary history, genetic diversity, and conservation challenges of the western honeybee. The case studies look at honeybee populations in Africa, Europe, the Middle East, and other areas. They show both new methods and ways to help protect these bees.

28.1. Introduction

Researchers have used mitochondrial DNA (mtDNA) a lot to study the genetic diversity and evolution of *A. mellifera*. This chapter shows how mtDNA analyses can help us understand phylogeographic patterns, determine hybridization and introgression events, and make better decisions about how to protect honey bee populations. The case studies show both global and regional trends in honeybee evolution by combining studies that use PCR-based analyses, sequencing the whole mitogenome, and other molecular markers. These case studies have provided us important information about how to define lineages, how genes have moved through time, and how human activities affect genetic diversity.

28.2. Case Studies in *Apis mellifera* mtDNA Research

The next sections present a summary and analysis of important case studies that have helped us learn more about the diversity of *A. mellifera* mitochondria.

28.2.1 African and Middle Eastern Honey Bee Populations

Many studies have examined how African and Middle Eastern subspecies have evolved and how they differ. Abou-Shaara et al. (2021) studied 21 types of *A. mellifera* from Africa, Europe, and Asia by analyzing their mtDNA and using PopART TCS networks. Their research indicated that there were geographically structured branches. For example, African subspecies like *A. m. sahariensis* were genetically isolated, while Middle Eastern specimens from Saudi Arabia showed clear patterns of hybridization with both local and European subspecies (Abou-Shaara, Al-Ghamdi, et al., 2021). This case study shows how important it is to use mtDNA to tell subspecies apart and to learn how historical dispersal events and secondary contact zones have changed genetic patterns over time. Furthermore, mtDNA studies in the Middle East have found unexpected lineage sorting events. For instance, based solely on mitochondrial haplotype data (Abou-Shaara, Al-Ghamdi, et al., 2021), researchers moved some morphologically classified *A. m. lamarckii* to different genetic lineages (Abou-Shaara, Al-Ghamdi, et al., 2021).

Avval et al. (2024) used a different method by studying the Iranian honeybee, *A. m. meda*, and analyzing the mtDNA from a specific part called the cytochrome oxidase I (COI) region. The Iranian subspecies groups are separate from other regional groups, and there is also evidence of introgression and high polymorphism within the COI region (Sarhaddi Avval et al., 2024). These results back up the idea that the Middle East is a hotbed of new honeybee species. Mitochondrial DNA can distinguish the difference between hidden lineages and provides us with information about how bees adapt to their environment (Sarhaddi Avval et al., 2024).

28.2.2. Iberian Peninsula and Southern Europe

Researchers have done a lot of case studies on honeybees from the Iberian Peninsula. They have used mtDNA haplotype analyses to look at how glacial refugia and human-mediated dispersal affect mtDNA diversity. A study by Arias et al. in 2006 looked at populations from southern France to northern Morocco using allozyme, morphometric, and mtDNA haplotype analyses. Their results showed a clear geographic separation between honeybee subspecies. For example, *A. m. iberiensis* in the southwest had mostly African (A lineage) haplotypes, while populations in the northeast (close to *A. m.*

mellifera) had different patterns with European markers (Arias et al., 2006). Another thing that stood out was the presence of sharp genetic breakpoints instead of gradual clines. This means that gene flow between some populations is limited. Further analysis of these results indicated that there are multiple ecotypes within subspecies like *A. m. carnica* and *A. m. macedonica.* This evidence shows how natural barriers and human activities interact to shape local mtDNA variation (Arias et al., 2006).

These studies show how important mtDNA analyses have been for both recreating the history of biogeography and supporting conservation efforts that are still going on. Finding unique maternal lineages and spotting introgression events are key parts of making plans to protect the native honey bee diversity in southern Europe.

28.2.3. South African Honey Bee Subspecies

Research on South African honeybee subspecies has also produced a lot of very useful case studies. Eimanifar et al. (2018) studied the complete mitochondrial genomes of *A. m. capensis* and *A. m. scutellata* populations found in South Africa. Their research indicated that the mtDNA markers were not always useful for identifying the two subspecies because their haplotype distributions overlapped (Eimanifar, Kimball, et al., 2018). The study found that there were many different haplotypes but not many different nucleotides. This pattern suggests that there was a bottleneck in the past followed by a quick expansion (Eimanifar, Kimball, et al., 2018).

Phylogenetic analyses of all 13 protein-coding genes and two rRNAs that were put together showed that RSA honeybees made a monophyletic clade and that the subspecies mixed with each other (Eimanifar, Kimball, et al., 2018). This case study is crucial because it highlights a significant problem with relying solely on mtDNA to identify different subspecies, particularly in situations where extensive hybridization occurs due to natural contact zones and human-mediated colony exchanges.

Furthermore, the case studies from South Africa show that mtDNA alone is not enough to tell subspecies apart. This led to suggestions to combine nuclear markers with mtDNA data to obtain a full picture of the genetic makeup (Eimanifar, Kimball, et al., 2018).

28.2.4. European Introgression and Conservation of Native Honey Bees

For a long time, conservationists have been worried about introgression and hybridization in populations of European honeybees. Ellis et al. (2018) wrote about introgression in native populations of *A. m. mellifera*, also known as the dark European honey bee, which is native to northwestern Europe. Their study used mtDNA analysis, focusing on the COI-COII intergenic spacer and nuclear markers, to show that commercial practices have caused a lot of gene flow from imported subspecies like *A. m. ligustica* and *A. m. carnica* (Ellis et al., 2018). In Germany, Switzerland, and some parts of the UK, mtDNA markers have been crucial in finding maternally inherited introgression patterns. These results have big effects on how conservation programs should be set up to protect the native genome of *A. m. mellifera* (Ellis et al., 2018).

The European case demonstrates the practical application of mtDNA in conservation management. mtDNA studies help beekeepers and conservationists identify and protect gene pools that are at risk of being replaced by non-native lineages by showing how much hybridization has happened and how native haplotypes have stayed the same even after a lot of introgression. In regions such as Cornwall (UK) and parts of central Europe, closely monitoring mtDNA lineages remains a crucial tool in combating genetic erosion in native populations.

28.2.5. Environmental DNA (eDNA) Approaches in Honey Bee mtDNA Studies

Researchers are starting to look into how to use environmental DNA (eDNA) from honey samples to study the diversity of *A. mellifera* mtDNA at the colony level. Bovo et al. (2022) demonstrated that eDNA from honey, which contains tiny bits of DNA from many bees, can be used to identify genetic information like mtDNA haplotypes and even nuclear SNPs (Bovo et al., 2022). This new method lets researchers obtain samples from a large part of a colony's genetic diversity without having to take samples from individual bees. Their research not only shed light on how mtDNA lineages are spread out across places like Italy, but it also made it possible to prove that honey comes from insects (Bovo et al., 2022). Researchers can now track maternal lineages and even estimate colony-level admixture by combining targeted PCR amplification and next-generation sequencing technologies. This method provides a non-destructive approach to monitor diversity and identify potential introgression events without the need for direct observation.

28.2.6. Mitochondrial SNP Marker Development for Conservation Programs

Another important use of mtDNA research is to create mitochondrial SNP (mtSNP) markers that can be used in conservation programs to track the ancestry of evolutionary lineages. Henriques et al. (2019) talked about finding five very informative mtSNP markers from important mitochondrial genes like COX1 and ND4. These markers have been used with nuclear SNP assays (M1 and M4 panels) to create a complete molecular tool for measuring the extent of C-lineage introgression in native populations of *A. m. mellifera* (D. Henriques, Parejo, et al., 2019). The fixation index values (F_ST) for these mtSNPs are close to one, which means that the lineages are almost entirely different from each other. This property makes them valuable markers for monitoring conservation (D. Henriques, Parejo, et al., 2019). These tools have been advantageous in places where high levels of introgression put unique genetic adaptations at risk. They give conservation managers useful information that helps them decide which native mtDNA lineages to protect or restore first.

28.2.7. Nuclear-Mitochondrial Interactions and the Influence of Numts

Researchers have mostly looked at the maternally inherited genome when studying mtDNA, but they have also found that nuclear copies of mitochondrial sequences (numts) make things more difficult. Behura, (2007) showed that NUMTs can be a big problem because they can mess up PCR amplification and phylogenetic reconstructions if they aren't properly identified or omitted (Behura, 2007). Studies that used honeybee mtDNA have shown how important it is to carefully design assays and use bioinformatics filtering to verify that only real mitochondrial sequences are looked at. Many conservation case studies didn't focus on numts, but their presence shows how important it is to have integrated methodological frameworks that combine strong molecular protocols with strict bioinformatic pipelines to confirm mtDNA data.

28.2.8. Additional Insights from Genetic Variation in Turkey and Beyond

Regional studies from Turkey have helped us learn more about how mtDNA diversity shows how populations are spread out over space. Kandemir et al. (2006) looked at differences in mitochondrial DNA in honey bee populations from Turkey. They used a combination of PCR-RFLP methods and sequencing parts of mitochondrial genes such as ND2 and the COI-COII intergenic spacer (Kandemir et al., 2006). Their findings, which involved identifying different genetic types and creating a family tree using neighbor-joining methods, clearly indicated that there are African, European, and other genetic lines in the area (Kandemir et al., 2006). This research adds to studies done in Europe and the Middle East by showing how geographic barriers and ecological conditions affect mitochondrial diversity and lineage differentiation (Kandemir et al., 2006).

28.2.9. Strain Identification and Trait Association through mtDNA SNPs

Along with studies of population structure and phylogeography, mtDNA has been advantageous in connecting mitochondrial genotype to certain phenotypic traits. Kim et al. (2019) used high-throughput sequencing technologies to make single-nucleotide polymorphism (SNP) markers for the entire mitogenome. Their study found SNPs that were unique to certain strains and linked to traits like being resistant to *Varroa destructor* and being clean in honeybee strains (Kim, Ah Rha, et al., 2019). Breeding programs, aiming to make colonies healthier and more resistant to pathogens, can use these markers to diagnose problems and make predictions. This case study clearly shows how mtDNA research can be used in real life by demonstrating how small differences in DNA can be linked to traits that are important for the economy and the environment.

28.2.10. European-Wide Genetic Diversity and Introgression Patterns

Muñoz and De la Rúa (2021) looked at populations of honeybees from central and southeastern Europe and did a more in-depth study of genetic diversity in these bees. Their research used microsatellite markers and mtDNA analyses to demonstrate that native subspecies such as *A. m. carnica, A. m. ligustica, A. m.*

macedonica, and A. m. cecropia have high genetic diversity and complicated mixing patterns (Muñoz & De la Rúa, 2021). The mtDNA data showed that there were different haplotype distributions that reflected historical colonization routes and local adaptations. It also indicated that there was a lot of mixing caused by migratory beekeeping and queen importation. This case study shows how useful mtDNA analysis can be over a large area and also warns that conservation efforts need to take into account how much human activities affect the genetic integrity of native populations (Muñoz & De la Rúa, 2021).

28.3. Discussion

The case studies in this chapter show how mtDNA analysis can be used in many different ways for research and conservation of *A. mellifera*. The fact that mtDNA has been used as a molecular marker in so many studies, from early RFLP-based studies in Turkey (Kandemir et al., 2006), to whole-mitogenome analyses in South African populations (Eimanifar, Kimball, et al., 2018), shows that it is the best way to figure out phylogeographic patterns and tell subspecies apart through maternal inheritance. One thing that all of these case studies have in common is the need to combine mtDNA data with other genetic markers, like nuclear SNPs and microsatellites, as well as morphometric and allozyme data. Researchers can get around the problems with mtDNA alone, like how sensitive it is to introgression, how complicated numts ara (Behura, 2007), or how hard it is to sort lineages when there are multiple subspecies in the same area.

Each case study supports the idea that environmental factors in different regions, such as the climate barriers in southern Europe and the unique ecological conditions in Africa and the Middle East, are crucial in determining the genetic diversity seen in mtDNA. For instance, the various maternal lineages found in the Iberian Peninsula and the South African hybrid zones clearly show that both natural and human-made factors influence the structure of mtDNA (Eimanifar, Kimball, et al., 2018). These results have big effects on how we protect the environment. To protect the genetic integrity of native lineages, it is important to closely watch mtDNA haplotypes and carefully manage beekeeping practices that could lead to unwanted introgression.

The use of environmental DNA from honey (Bovo et al., 2022) and the creation of highly informative mtSNP markers (D. Henriques, Parejo, et al., 2019) are two new techniques that make mtDNA analyses even more useful. These methods not only let you watch whole colonies without touching them, but they also let you do quick genetic tests that can help you make decisions about conservation actions in real time.

28.4. Conclusions

The case studies in this chapter show how mitochondrial DNA research can be used in many different ways to help save honeybees. mtDNA has been essential for understanding the evolutionary history of honeybees and guiding conservation efforts. For example, it has helped us figure out how African and Middle Eastern subspecies are genetically isolated and hybridized (Abou-Shaara, Al-Ghamdi, et al., 2021; Muñoz & De la Rúa, 2021) and how complex patterns of introgression affect European native populations (Ellis et al., 2018; Muñoz & De la Rúa, 2021). Studies in South Africa (Eimanifar, Kimball, et al., 2018) have also helped us understand how it is to use mtDNA alone when subspecies look different but are genetically mixed. The use of cutting-edge tools like whole-mitogenome sequencing, high-throughput SNP genotyping platforms, and environmental DNA analyses shows that the field of mtDNA research in *A. mellifera* is not static but continues to change to meet the needs of conservation, management, and breeding.

In general, these case studies show how important mtDNA analysis is for finding lineage-specific markers, keeping track of introgression and admixture events, and backing up molecular diagnostics that are necessary for effective conservation programs for endangered native honey bee populations. Future research that combines mtDNA with large nuclear genomic datasets and high-resolution bioinformatics methods will definitely help us learn more about the diversity of honeybees and protect this important pollinator for future generations.

To sum up, the case studies in Chapter 6 make the following main points:

- Studies in Africa, the Middle East, Europe, and South Africa show that mtDNA is a strong marker for showing both how individuals moved around in the past and how genetic exchange is still happening (Abou-Shaara, Al-Ghamdi, et al., 2021; Muñoz & De la Rúa, 2021).

- Integrated approaches that use both mtDNA and nuclear markers get around the problems with the mitochondrial genome and make phylogenetic reconstructions and conservation assessments stronger (Ellis et al., 2018; D. Henriques, Parejo, et al., 2019).
- New methods, like using environmental DNA from honey samples and making targeted SNP assays, have added to the tools we can use to non-invasively track population genetics at both the colony and regional levels (Bovo et al., 2022; D. Henriques, Parejo, et al., 2019).
- Finally, the fact that molecular and bioinformatics methods are constantly changing means that mtDNA research will continue to be important for managing honey bee biodiversity, especially since human activities are still changing the genetic landscape of *A. mellifera* around the world (Eimanifar, Kimball, et al., 2018; Sarhaddi Avval et al., 2024).

These case studies provide a comprehensive overview of the latest advancements in mtDNA research on honeybees. They show how natural evolutionary processes and human actions affect each other in complicated ways. This is the basis for the global conservation and sustainable management of *A. mellifera.*

Chapter 6 brings together the different geographical and methodological insights from these studies. It shows how far mtDNA research has come and calls for more multidisciplinary approaches to protect the genetic legacy of our most important pollinator species (Arias & Sheppard, 2006; J. D. Evans et al., 2013).

Chapter 29

29. Conservation Genetics

This chapter looks at how mitochondrial DNA (mtDNA) markers can help us keep track of genetic health, trace maternal lines, and support sustainable beekeeping, while also exploring how conservation genetics helps protect the genetic variety and future adaptability of *A. mellifera*. Creating plans to protect local honeybee subspecies from harm caused by human activities and mixing with non-native bees relies on studying their population structure, evolutionary background, and genetic mixing, all of which fall under conservation genetics (Alburaki et al., 2023).

Since *A. mellifera* is an important pollinator that has many different subspecies adapted to specific environments around the world, the importance of conservation genetics is very clear. The genetic differences found in their mitochondrial and nuclear DNA are important because they affect traits like disease resistance, productivity, and how well they can adapt to climate changes (Alburaki et al., 2023). Keeping track of mtDNA diversity is essential because it helps us understand how honeybee populations have evolved through their mothers and how human actions have affected them over time (Alburaki et al., 2023).

Mitochondrial DNA markers, like the COI-COII intergenic region and tRNAleu-cox2 region, are useful in conservation genetics because they are passed down from mothers (maternal), change quickly, and don't mix much, allowing scientists to identify different maternal lineages and understand the evolutionary connections between subspecies, which helps track events that could threaten the genetic health of local populations. These markers allow scientists to identify various maternal lineages and establish the evolutionary connections between subspecies, thereby facilitating the tracking of introgression events that could threaten the genetic integrity of native populations (Sarhaddi Avval et al., 2024). Detailed studies have shown that populations in places like the United States have low mtDNA haplotype diversity, mostly made up of a few common haplotypes, which indicates they may be at risk from new diseases and changes in the environment (Alburaki et al., 2023).

The use of molecular techniques based on mtDNA sequencing has greatly improved conservation genetics in honeybees by allowing for quick and precise reconstructions of their evolutionary history. Scientists have been able to identify different evolutionary groups, such as the North Mediterranean (C) lineage, West Mediterranean (M) lineage, and African (A) lineage, by examining differences in mitochondrial regions; each of these groups reveals important information about past migrations, population declines, and how genes have spread between continents (Bertrand et al., 2015). Molecular insights are absolutely essential for developing conservation plans that can help preserve or restore the natural genetic reservoirs of *Apis mellifera*, especially in areas where human activity has upset the natural gene pool (Bertrand et al., 2015).

Molecular studies of honeybee populations have indicated that conservation genetics is not only concerned with cataloging genetic diversity. Instead, it also is significant in clarifying the evolutionary mechanisms producing and preserving diversity over time. Studies done in the USA and Europe have shown that looking at mtDNA differences can help us understand how honeybees spread and grew in new areas after they arrived (Chávez-Galarza et al., 2021). Additionally, finding new haplotypes and sub-lineages in bee populations from different areas shows that these bees are still evolving and helps us understand how natural selection, genetic drift, and human activities interact (Chávez-Galarza et al., 2021). Finding and controlling genetic introgression—which can arise from queen trade and colony movement by commercial beekeepers—is one of the main difficulties in using conservation genetics to honey bee populations. Introgression from imported subspecies, such as those derived from the C or M lineages, into native populations disturbs locally adapted gene complexes and can erode special genetic traits essential for survival under local conditions (Ellis et al., 2018). Research on both managed and wild bee colonies has shown that even places with strong conservation efforts can still be affected by foreign genes, as seen through hybrid mitochondrial markers and nuclear DNA signs of mixing (Ellis et al., 2018).

Traditional mtDNA tests have improved thanks to new methods like high-throughput sequencing and single nucleotide polymorphism (SNP) analyses, allowing conservation geneticists to create complete genetic profiles that include both nuclear and mitochondrial data. For example, scientists have used nuclear SNP tests along with mitochondrial SNP markers from genes like

COX1, ND4, and LSU rRNA to track the family history and see how much mixing has happened in endangered subspecies like *A. m. mellifera* (D. Henriques, Parejo, et al., 2019). These combined methods provide a clearer view of the genetic health of populations, which helps in planning better and more effective conservation efforts (D. Henriques, Parejo, et al., 2019).

Furthermore, extending the use of conservation genetics in honeybees involves building and monitoring mating stations and conservation centers. These projects aim to preserve locally adapted gene pools by establishing isolated reserves that minimize genetic drift and unwanted introgression. Protective actions have been taken in many areas of Europe to preserve the native dark European honeybee, *Apis mellifera*, whose genetic integrity has been further validated with mtDNA markers, including the DraI COI-COII test (Rortais et al., 2011). Despite the existence of surrounding hybrid zones, these conservation centers have successfully maintained pure mother lineages (Pinto et al., 2014).

Beyond the conventional methods, conservation genetics now mostly monitors genetic diversity in an affordable and effective way using non-invasive sampling techniques and in silico genotyping. Recent studies indicate that this switch to high-throughput, non-invasive techniques allows regular, broad monitoring of queen lineages and colony genetic structure without upsetting the bees (Madella et al., 2021). These techniques provide real-time genetic surveillance, thus enabling quick reactions to new hazards, including the invasion of non-native haplotypes or sudden reduction in genetic diversity (Madella et al., 2021).

Current conservation plans locate a useful backdrop in historical patterns of genetic diversity. For example, research on the mtDNA diversity of *A. mellifera* in the Canary Islands and Iran has revealed different evolutionary paths shaped by initial population sizes, sharp reductions in population, and the unique geography of these areas (Miguel et al., 2015). These studies not only highlight the rich genetic legacy of honeybees but also act as warning stories about the effect of human-mediated introductions, which have repeatedly resulted in the erosion of native genetic diversity by hybridization (Muñoz et al., 2014).

Conservation genetics has helped to show how African mtDNA haplotypes have mixed into mostly European populations in Africa and those that have come from there to the Americas. For example, research in East-Central Europe has shown that African mitotypes occasionally appear, mainly due to the import of non-native queens and the intentional breeding of hybrids like Buckfast bees (Oleksa et al., 2021). Detailed phylogenetic studies further clarify the gene flow dynamics between African, European, and hybrid populations, highlighting the difficulty of controlling admixture in areas with great genetic diversity and active beekeeping (Oleksa et al., 2021).

High-resolution molecular studies, including those using whole-genome SNP data in conjunction with mtDNA sequencing, have strengthened our knowledge of the genetic composition of honey bee populations even more. These methods have been valuable in revealing subtle patterns of genetic differences both within and between conservation areas, thereby guiding the design of breeding programs intended to preserve local adaptations while reducing the risks of inbreeding and loss of heterozygosity (Parejo et al., 2016). Studies carried out in Switzerland, for example, have revealed that some degree of nuclear genomic introgression from imported lineages persists even within conservation areas, underlining the need for ongoing genetic monitoring and selective breeding to maximize the efficacy of conservation activities (Parejo et al., 2016).

Apart from broad genomic approaches, conservation genetics has gained advantages from more conventional approaches, including geometric morphometrics and microsatellite markers. Particularly in areas where migratory beekeeping has changed natural genetic distributions, these techniques have given complementary insights into honeybee population structure (Kelomey et al., 2018). Combining morphometric analyses of wing venation patterns with mtDNA studies provides a dual method for distinguishing between subspecies and identifying hybridization events, thereby supporting important information for conservation management plans (Kelomey et al., 2018).

Resolving taxonomic uncertainty among *A. mellifera subspecies* has especially benefited from the combination of several molecular markers. Molecular markers, which include both mitochondrial and nuclear data, are now commonly used to identify and define subspecies because traditional methods based on physical traits often fail to detect small genetic differences (Ostroverkhova & Konusova, 2022). These combined techniques help to settle disputes about the categorization of ecotypes and produce more exact conservation units, which form the foundation for focused conservation and breeding programs (Sarhaddi Avval et al., 2024).

One prominent instance of effective conservation genetics is the continuous endeavor to maintain the genetic integrity of *A. m. mellifera* in Europe. Although significant hybridization with non-native subspecies has occurred, the establishment of conservation centers employing rigorous breeding practices and genetic monitoring has helped to preserve populations with high frequencies of native mtDNA haplotypes (Pinto et al., 2014). Long-term genetic studies have demonstrated that populations in these protected areas exhibit significantly less introgression compared to those in unprotected areas, thereby supporting the validity of conservation policies based on molecular genetic monitoring (Pinto et al., 2014).

Programs for conservation also have to address the inherent difficulties presented by the haplodiploid mating system of the honeybee, which renders the species especially susceptible to inbreeding depression. Directly related to colony health and resilience against pathogens, parasites, and environmental stresses is genetic diversity at both colony and population levels (Porrini et al., 2022). Efforts to keep or increase genetic variety through planned mating techniques and well-placed conservation areas have successfully helped managed honeybee populations lessen the negative impacts of genetic drift and inbreeding (Porrini et al., 2022).

Conservation genetics has been crucial outside of Europe and North America in the management of honeybee populations in areas with different evolutionary backgrounds, including the Hawaiian Islands. Unique mtDNA haplotypes absent in mainland populations have been found in some isolated island populations, reflecting the long-term geographical isolation and special environmental adaptation that have occurred (Szalanski et al., 2016). The identification of such island-specific haplotypes, including new variants, illustrates the importance of using customized conservation plans that acknowledge and protect these special genetic fingerprints (Szalanski et al., 2016).

Another important focus of conservation genetics is understanding the temporal variations in genetic diversity brought about by human activities. Because of their constant introduction of non-native queens, selective breeding techniques, and habitat changes, managed honeybee populations have shown rapid evolutionary changes according to longitudinal studies (Tanasković et al., 2021). These changes in genetic diversity caused by human actions require management plans that include regular genetic checks to ensure that conservation efforts remain effective despite ongoing human impacts.

Legislative support has enabled the use of conservation genetics to preserve the native subspecies *A. m. ligustica* in the Emilia-Romagna region of Italy. Here, detailed studies of mtDNA haplotypes—especially in the tRNAleu-cox2 region—have shown that there is not much variety in haplotypes, mostly showing a unique C1 haplotype that indicates the presence of native ligustica populations (Taurisano et al., 2024). While some mixing with non-native, commercially popular subspecies has been seen, the strong presence of native haplotypes highlights the positive impact of strict conservation laws and breeding rules (Taurisano et al., 2024). In such areas, comprehensive conservation plans depend on ongoing genetic monitoring, controlled queen importation policies, and help for local beekeepers to preserve these distinct mtDNA lineages (Taurisano et al., 2024).

Broadly speaking, conservation genetics for *A. mellifera* has evolved into a multidisciplinary project combining molecular genetics, ecological research, and socioeconomic issues. High-resolution genomic tools have strengthened our knowledge of honeybee diversity and enabled legislators to apply data-driven conservation policies. Large-scale mitochondrial DNA studies, for example, have exposed hitherto unreported African subgroups unique to island ecosystems in the Southwest Indian Ocean, which calls for recommendations for the development of region-specific conservation policies preserving these endemic lineages (Techer, Clémencet, Simiand, Preeaduth, et al., 2017).

Furthermore, the reconstruction of thorough phylogenetic trees grounded on whole mitochondrial genome data has shed light on the evolutionary roots and divergence of more than thirty *A. m. subspecies*. These evolutionary trees show that different groups (like M, C, O, A, Y, and the new S lineage) still exist and can be reliable indicators for conservation efforts, even with the complicated mixing of genes seen in many areas (Tihelka et al., 2020). These results drive home the importance of conservation plans that not only aim to preserve existing genetic diversity but also to keep the evolutionary processes producing fresh genetic variants over time under focus.

Ultimately, by studying mitochondrial DNA diversity, conservation genetics of *A. mellifera* provides valuable information about the genetic makeup, history, and ability to adapt of honey bee populations around the world. By using various molecular markers and advanced genomic methods, scientists and conservationists can develop and implement effective plans that protect local genetic diversity, reduce the harmful effects of mixing with non-native populations, and support sustainable beekeeping practices. In the face of rapid environmental changes and continuous anthropogenic pressures, these kinds of strategies are crucial to guarantee that honeybees keep offering necessary pollination services, support agricultural output, and help to stabilize ecosystems (Alburaki et al., 2023; Ellis et al., 2018; Muñoz et al., 2014).

Future research in conservation genetics should aim to better combine mtDNA data with nuclear genomic information, develop standard methods for non-invasive and on-site genetic monitoring, and explore how new bioinformatics tools can predict how honey bees will adapt to environmental challenges. The integration of these contemporary technologies with conventional conservation practices will critically depend on refining breeding programs, creating strong genetic refuges, and guaranteeing the preservation of the evolutionary legacy of *A. mellifera* for future generations (Parejo et al., 2016; Pinto et al., 2014).

Finally, conservation genetics helps us understand and protect the genetic makeup of *A. mellifera* by focusing on mtDNA variation. It not only clarifies the dynamics of honeybee populations both historically and currently but also offers specific, scientifically based solutions for mitigating the dangers of hybridization and genetic erosion. In conjunction with international efforts to legislate and support genetic conservation, the ongoing development of genetic technologies promises to secure the genetic resilience and ecological functionality of honey bees worldwide, ensuring that these essential pollinators continue to thrive in an ever-changing environment (Requier et al., 2019; Rortais et al., 2011).

Conservation genetics is a crucial area focused on protecting and restoring the unique genetic traits of honey bees, *A. mellifera*, by tackling issues like mixing of different bee breeds, loss of genetic variety, and the effects of commercial bee management. Whether by means of isolated conservation centers, high-throughput genomic monitoring, or targeted breeding programs, the techniques covered in this chapter represent a proactive approach to managing the genetic resources of one of the most significant pollers in the world (Madella et al., 2021; Oleksa et al., 2021).

Looking ahead, it is essential for conservation policies to remain adaptable in response to new genetic discoveries and environmental challenges, thereby ensuring that the conservation genetics framework evolves in tandem with our understanding of honey bee biology. By means of coordinated worldwide efforts combining scientific research with pragmatic beekeeping and legislative action, we can guarantee that the genetic diversity of *A. mellifera* is maintained and allowed to flourish, ensuring the future of sustainable apiculture and the ecosystems dependent on these amazing insects (Taurisano et al., 2024; Tihelka et al., 2020).

This chapter has provided a thorough review of the ideas, tools, and applications of conservation genetics in honeybees, emphasizing the crucial role of mtDNA diversity analysis in revealing the intricate evolutionary tapestry of *A. mellifera*. Researchers have built a solid plan that helps guide current management practices and future efforts to protect these important pollinators by combining historical knowledge, advanced molecular techniques, and practical conservation strategies. Success of conservation genetics in honey bee management will ultimately rely on ongoing cooperation among geneticists, ecologists, beekeepers, and legislators. By working together, we can protect the genetic legacy of *A. mellifera* from the numerous challenges posed by globalization, climate change, and habitat loss, thereby preserving genetic resilience and ecological vitality for future generations (Muñoz et al., 2014; Rortais et al., 2011).

Chapter 30

30. Future Directions in Conservation Genetics

Presents a detailed plan for research that aims to increase our knowledge of the diversity of *A. mellifera* mitochondrial DNA (mtDNA) and how it can be used in real life to protect honeybees. This chapter talks about four new areas of research: functional mitogenomics, which connects mtDNA variants to metabolic traits like Varroa resistance; studies of rare paternal mtDNA transmission in hybrid populations; using ancient DNA methods to recreate pre-industrial genetic diversity; and combining nuclear genomic data with traditional mtDNA analyses to settle long-standing taxonomic disagreements. These research directions will help bridge the gap between molecular variation and phenotypic outcomes. This chapter will give conservationists new tools to improve breeding programs and rebuild native populations.

30.1. Functional mitogenomics: linking mtDNA variants to metabolic traits

The mitochondrion is important for making energy in cells because it helps with the oxidative phosphorylation (OXPHOS) pathway, and its genome has important instructions for making the proteins that are needed for this process. Functional mitogenomics tries to find out how natural differences in mtDNA affect how well honeybee colonies make energy and adjust to their surroundings. Researchers who looked at different types of *A. mellifera* mtDNA early on found that even small changes in similar areas can affect how well ATP is made and how many reactive oxygen species (ROS) are made (Alburaki et al., 2023). These small differences at the molecular level could have real-world benefits. For example, colonies with healthy mitochondria might be better at fighting off parasites, especially the harmful *Varroa destructor* mite.

Recent studies indicate that it is possible to create high-throughput phenotyping systems that use both advanced metabolic assays and detailed analyses of mitochondrial DNA sequences. Respirometry to track oxygen use and metabolomic profiling are two methods that can help us understand how certain haplotypes affect the physiology of colonies. In this case, research has shown that variations in mtDNA coding regions and intergenic segments are under complicated selective pressures. Purifying selection keeps important metabolic functions going, while small mutations maybe those that change the shape of protein subunits could present small advantages in managing energy or resisting stress (Chávez-Galarza et al., 2021). For instance, if specific mtDNA haplotypes are associated with reduced ROS production during stress, this phenotype could contribute to the health of colonies.

Functional mitogenomics tries to identify mtDNA markers that can tell us about important metabolic traits. This would not only help with marker-assisted selection, which would encourage the breeding of colonies that naturally tolerate varroa and have strong metabolisms, but it would also let conservationists choose populations that are genetically likely to do well in changing environmental conditions. Conservation geneticists can acquire useful information that goes beyond just mapping genetic types by combining data from mitochondrial genome sequencing with measurements of how well a colony does. In practice, this method could make bioassays that test queens and colonies based on how well they use energy, which would help direct targeted conservation efforts and encourage more eco-friendly beekeeping methods (Alburaki et al., 2023; Chávez-Galarza et al., 2021).

30.2. Parental mtDNA Transmission in Honey Bee Hybrids

Studies on honeybee genetics have always primarily focused on the strict transmission of maternal mtDNA. This makes it easier to put together matrilineal lineages and evolutionary histories. But new research goes against this long-held belief by showing that paternal mtDNA can be passed on in rare cases, especially when different *A. mellifera subspecies* hybridize. When paternal leakage happens, children may have a mix of mtDNA from both parents. The result can make it harder to understand family trees and keep track of genetic resources in conservation efforts.

Studies have shown that when different subspecies, such as *A. m. carnica* and *A. m. capensis*, are crossed in the lab, the offspring can possess many paternal mtDNA (de Melo & Camargo, 2021). This event raises a lot of basic questions about how accurate mtDNA inheritance is and what might happen to evolution if mtDNA is passed down from both parents. For instance, could heteroplasmy make mitochondrial proteins more functionally diverse, giving them any benefits for adaptation? On the other hand, could having different mtDNA variants in the same colony lead to mitonuclear incompatibilities that make the whole colony less fit?

It's important to know how often and what effect paternal leakage has from a conservation perspective because it makes it harder to assign colonies to specific mother lines. When managed and wild populations meet, paternal mtDNA can sometimes get mixed in with the native stocks. This phenomenon makes it harder to protect different evolutionary lines because it can hide the true genetic makeup of the native stocks. More research is needed to make better molecular tests that can identify small amounts of paternal contributions and figure out how often they happen in nature. To find out if paternal leakage affects traits related to survival, like how well an organism metabolizes food or fights off diseases (de Melo & Camargo, 2021), these studies should ideally use both controlled breeding experiments and advanced sequencing methods.

Furthermore, the idea of paternal leakage makes us examine the assumptions that guide many conservation genetic studies that only look at mtDNA markers. If even a small number of people in a population show heteroplexia, conservation plans that only look at maternal markers may fail to fully capture the complexity of the genetics. To learn more about honeybee populations and their changes, we may need to study male mtDNA.

30.3. Ancient DNA and the Reconstruction of Pre-Industrial Diversity

Using ancient DNA (aDNA) techniques on museum specimens of *A. mellifera* is one of the most exciting ways to learn about the original genetic makeup of honeybee populations before humans had a big impact on them. There are many specimens from the late 1800s and early 1900s in museum collections. These specimens show genetic snapshots of populations that lived before commercial queen importation, modern beekeeping methods, and intensive breeding programs. By looking at damaged DNA from these old samples, scientists can make a record of the frequencies and diversity of mtDNA haplotypes over time. This presents them a baseline to compare with changes in modern genetics.

Scientists have been able to acquire a lot of genetic information from museum specimens (Parejo et al., 2020), thanks to recent improvements in whole-genome sequencing and ways to work with damaged DNA. For instance, looking at the mtDNA of Swiss honeybees from before the Industrial Revolution has revealed haplotypes and family groups that are now much less common or even absent in today's populations (Parejo et al., 2020). These studies have indicated that the frequency of haplotypes is changing, that bees with different genetic backgrounds are mixing more, and that there are signs of interbreeding with subspecies from outside the area. All of these things show how much human activities have changed the genetic makeup of local populations.

Comparing ancient DNA can provide valuable insights that assist conservation genetics. First, they present scientists with a historical background that helps them tell the difference between changes caused by modern farming methods and changes that happen naturally over time. Second, they can help with plans to restore or re-establish genetic diversity or ecotypes that are well-suited to the area. Should some ancient haplotypes be known to have benefits under particular local environmental conditions, these genetic markers may be given top priority in conservation breeding projects. Before industrial times, we knew the basic types of mtDNA in populations. This knowledge helps us understand how genetics have changed over time, which helps conservationists set clearer goals for saving and restoring species (Parejo et al., 2020).

Researchers will probably focus on making computer methods better for combining old and new data and finding better ways to extract and sequence DNA so that they can obtain the most information from very broken ancient DNA in the future. These methods, which combine molecular genetics with historical ecology and environmental data, could help us learn more about how *A. mellifera* has changed over time and make plans to bring back genetic diversity that has been lost.

30.4. Integration with Nuclear Genomics to Resolve Subspecies Classification

Even though mtDNA is only a small part of the genetic diversity in *A. mellifera*, it has been an important way to track maternal ancestry and phylogeographic patterns in honeybees for a long time. In contrast, nuclear genomic markers present us a much clearer picture of how populations are structured, how genes mix, and how subspecies are related to each other in terms of evolution. Scientists can now look at thousands of nuclear single nucleotide polymorphisms (SNPs) at once thanks to recent advances in high-throughput sequencing. The outcome provides them a much clearer picture of even small mixing patterns.

It is difficult to protect honeybees because there is still disagreement about how to classify them based on their physical traits, mtDNA lineages, and nuclear genomic data. For example, studies that look at mtDNA by itself sometimes identify patterns that are different from those found in nuclear markers, especially in areas where hybridization happens a lot. Studies that use both mtDNA and nuclear markers have found hidden mixing that shows the complicated history of queen trading and mating between colonies (Pinto et al., 2014). This is because mtDNA can clearly show major evolutionary groups. In places where conservation programs want to keep pure native stocks, it is essential to accurately measure how much introgression is happening with non-native subspecies.

In the future, the integration of nuclear genomic data and mitochondrial studies will impact honeybee conservation genetics by addressing classification issues. Detailed studies of the complete DNA from nuclear markers and most mitochondrial DNA sequences have started to reveal how much mixing has occurred in various regions (Parejo et al., 2016). These combined methods not only give us a deeper understanding of the general structure of the population, but they also help us identify specific genomic areas that are linked to local adaptation. This information is very useful and necessary for creating breeding programs that aim to keep the evolutionary potential of native populations. This method also lowers the risks that come with too much introgression. Using both nuclear and mitochondrial data will also help us learn more about how nuclear genes affect traits that come from mtDNA variants.

Knowing how well a colony can thrive, it's important to understand how the nuclear and mitochondrial genomes work together. In consistencies between these genomes can lead to issues that impact energy production and the colony's response to stress. Therefore, future research will heavily focus on developing bioinformatics tools and statistical models that can analyze nuclear and mitochondrial markers together to predict their impact on phenotypic traits (Pinto et al., 2014; Tanasković et al., 2021).

Finally, the fact that paternal leakage happens shows that we need to rethink the assumptions behind a lot of conservation genetic studies that only use mtDNA markers. Even if only a few individuals have heteroplasmy, conservation plans relying solely on maternal markers might underestimate a population's genetic diversity. To better understand honeybee populations and how they change over time, we may need to look more closely, including at the contributions of paternal mtDNA.

30.5. Synthesis and Future Outlook

The four research paths above represent a big shift in the field of honeybee conservation genetics. They take the field from static descriptions of mtDNA diversity to a dynamic, functionally informed, and integrative framework. Functional mitogenomics aims to connect certain mtDNA variations directly to traits that influence metabolism, like resistance to Varroa. Such an approach makes it possible to use marker-assisted selection, which could make colonies healthier and more productive. The study of paternal mtDNA transmission challenges our understanding of mitochondrial inheritance. The result makes us rethink genetic purity and population assignments in hybrid zones. By showing us the genetic makeup of honeybee populations before the Industrial Revolution, ancient DNA methods provide us a valuable look at the past. Restoration work, which aims to maintain ecotypes adapted to the area, also benefits from their assistance. Finally, combining nuclear genomic data with traditional mtDNA analyses will help answer long-standing questions about taxonomy. These developments will let conservationists use more precise and focused management strategies.

The end result of all these plans is a strong, diverse set of tools that can help *A. mellifera* address the many problems it faces in a time of rapid environmental change and genetic mixing caused by humans. For example, early studies of mtDNA diversity in the US have shown that keeping evolutionary distinct lineages is important for sustainable beekeeping (Alburaki et al., 2023), and studies of the integration of nuclear and mitochondrial markers have already demonstrated that they can help solve the genetic problems that come with managing populations (Parejo et al., 2016). The development of high-throughput sequencing technologies and advanced non-invasive sampling methods speeds up this research even more. This method ensures the swift addition of new information to conservation plans that are both adaptable and forward-looking (Madella et al., 2021).

Functional genomics, deep sequencing, and ecological modeling are all fields that work together and are likely to become the norm in conservation studies in the future. By combining detailed molecular datasets with strong phenotypic and environmental analyses, scientists will be able to guess how some genetic configurations will respond to future stressors like climate change, new diseases, and changes to habitats that are already happening. This ability to make predictions will not only help us learn more about how honeybees evolve, but it will also lead to conservation measures that protect these important pollinators' ecological function and economic viability (Ostroverkhova & Konusova, 2022; Porrini et al., 2022).

The next generation of research in conservation genetics will help make breeding programs that are designed to keep or even increase the genetic diversity of both nuclear and mitochondrial DNA. For instance, beekeepers and conservationists can use molecular data to choose which animals to breed based on mtDNA haplotypes that are strongly linked to better metabolic function or disease resistance, instead of just physical traits. Simultaneously, unraveling the mechanisms of paternal mtDNA inheritance will further complicate these breeding plans. This method will ensure that hybridization maintains the overall genetic integrity while facilitating adaptation.

Furthermore, the creation of integrated genomic frameworks will help regulatory and policy efforts to make beekeeping more environmentally friendly. These methods supply lawmakers clear and measurable data on the levels of genetic diversity and hybridization. Such information lets them set rules for queen trade, colony movement, and how to take care of conservation zones based on science. For instance, genomic monitoring allows us to modify conservation programs designed to maintain pure native stocks. Such data stops the mixing of different genetic lineages with non-native alleles by accident (Pinto et al., 2014; Tanasković et al., 2021).

Recent advances in related fields show that these new areas of research have a lot of promise. Studies that sequenced the whole genome of honeybees have already found the small genetic differences that make different populations unique. These studies have also paved the way for more comprehensive research that links these genetic differences to practical impacts (Parejo et al.,

2016). Researchers have also found that ancient DNA, even if it is badly damaged, can tell us a lot about how populations were organized in the past. This information helps us figure out how to protect species in a clear evolutionary context (Parejo et al., 2020).

As the field moves forward, it will be important for geneticists, ecologists, beekeepers, and policymakers to keep talking to each other. By working together like this, we can ensure that the advanced molecular tools and analytical methods that schools come up with are turned into useful plans that will help honeybee populations stay strong and healthy over time. Chapter 8's future directions not only show how to do academic research, but they also set the stage for a new era of evidence-based, integrative conservation that will protect *Apis mellifera* in the face of the complicated problems of the twenty-first century (Requier et al., 2019; Rortais et al., 2011).

In conclusion, the most advanced method for conserving honeybees' genetics is a convergent approach that uses functional mitogenomics to explain rare paternal mtDNA events, brings back genetic diversity from the past using ancient DNA, and combines nuclear genomic data to clear up taxonomic confusion. These research paths will help us become better at linking genotype with phenotype. These discoveries will contribute to the development of precision breeding programs, adaptive conservation strategies, and regulations that will work together to protect this species and ensure its survival (Alburaki et al., 2023; Tanasković et al., 2021). In the end, using both new genomic techniques and old-fashioned conservation methods may not only help bring back populations that were once diverse and well-suited to their environment, but it may also lead to strong beekeeping practices that can handle problems caused by people and the environment in the future.

Chapter 31

31. Paternal mtDNA Transmission in Honey Bee Hybrids

31.1. Introduction

For many years, researchers have based their studies in honeybee genetics on the idea that mitochondrial DNA (mtDNA) is passed down strictly from maternal to offspring, which has helped us figure out matrilineal lineages and evolutionary histories. However, several studies have shown that while it's usually uncommon, drones (paternal mtDNA) can also pass on their mtDNA, especially when different types of honeybees breed together. In controlled lab experiments with subspecies like *A. m. carnica* and *A. m. capensis*, researchers discovered that some paternal mtDNA mixes with the offspring, meaning both maternal and paternal mtDNA are found in the early embryo stages (Meusel & Moritz, 1993). When males pass on their genes to their offspring, it can cause mitochondrial heteroplasmy. This situation complicates phylogenetic analyses that assume strict maternal inheritance and also hinders the effectiveness of conservation and genetic resource management programs (Breton & Stewart, 2015).

31.2. Experimental Evidence of Paternal Leakage

Meusel and Moritz's groundbreaking work in 1993 gave strong proof of paternal mtDNA leakage in honeybees. The study used a method called mitochondrial DNA restriction fragment length polymorphism (RFLP) analysis with the BclI enzyme to differentiate between the mtDNA types of the two subspecies. The study achieved this by crossing *A. m. carnica*, which is typically maternal, with *A. m. capensis*, which is generally paternal. Shortly after oviposition (in newly laid eggs), researchers found paternal mtDNA at levels that constituted about 27% of the total mitochondrial DNA in newly laid eggs. This amount of leakage is unusual in most eukaryotic systems (Meusel & Moritz, 1993). However, as development goes on, the amount of paternal mtDNA drops quickly, and by the larval stage, paternal contributions are minimal or even nonexistent. Despite the ongoing decline, the presence of paternal mtDNA during early embryogenesis significantly impacts development by creating a temporary heteroplasmic state that may alter early developmental processes and the initial interactions between mitochondria and nuclei (Meusel & Moritz, 1993).

The experiment used artificially inseminated queens and a large number of eggs and larvae. This made sure that the paternal contributions were not just a result of having a small sample size but were instead a real and repeatable event. The use of special probes in filter hybridization techniques made it easier to detect even tiny amounts of paternal mtDNA (Meusel & Moritz, 1993). These methodological details strongly suggest that paternal leakage happens in honey bee hybrids and highlight the possibility of transient heteroplasmy during the early stages of development.

31.3. Evolutionary Implications

The evolutionary effects of honeybees passing on paternal mtDNA occasionally go far beyond just breaking the rules about how mothers pass on their genes. Heteroplasmy, which happens when maternal and paternal mtDNA are present at the same time, could affect how mitochondrial genes spread through natural selection. Experimental evidence shows that even a small amount of paternal mtDNA can bring new mitochondrial variations into populations, which can temporarily increase mitochondrial genetic diversity (Breton & Stewart, 2015; Meusel & Moritz, 1993). More diverse mitochondria might help colonies adapt to environmental stressors by giving them a wider range of mitochondrial phenotypes, which could make them more resilient. For instance, mitochondrial genome changes could affect bees' energy use, stress tolerance, and disease resistance. These are all essential traits for the survival and productivity of honeybee colonies.

On the other hand, adding different paternal mtDNA to a mostly maternal pool could also cause mitonuclear incompatibilities. Many mitochondrial proteins work with nuclear-encoded parts of the electron transport chain. If there are genetically different mitochondrial genomes, these important interactions could be disrupted, which could make the colony less fit overall. It appears that the evolutionary stability of mitochondrial inheritance systems depends on getting rid of paternal genomes. Thereby, when paternal leakage happens, it could mean that the mechanisms that usually enforce uniparental inheritance have stopped working (Breton & Stewart, 2015).

Intersub-specific hybridization might increase the chances of these breakdowns happening because the genetic differences between the parent lines make it harder for the cells to effectively eliminate sperm mitochondria. Because of this, when different subspecies like carnica and capensis breed, it may lead to paternal leakage, which can create a mix of mitochondria from both parents. This mix usually decreases as the organism develops, but it could still influence early development and the factors that affect it (Meusel & Moritz, 1993).

More information suggests that even if paternal mtDNA is eventually broken down or doesn't replicate with maternal mtDNA, the temporary heteroplasmic state may still have adaptive effects. If some paternal mitochondrial alleles impart an advantage in certain environments, there may be times when paternal leakage affects evolutionary paths by temporarily improving colony performance. These changes in mitochondrial diversity, even if temporary, could alter how natural selection acts on mitochondrial genomes. Such modifications could have an effect on evolution on both small and large scales (Meusel & Moritz, 1993).

31.4. Implications for Phylogenetic Inference

The assumption of strict maternal inheritance is what makes mtDNA a reliable molecular marker for phylogenetics and population genetics. However, when paternal leakage happens, this assumption is broken, and the resulting mitochondrial heteroplasmy makes it even harder to figure out how phylogenetic relationships work. Researchers have been using mtDNA markers for a long time to tell different honeybee lineages apart and figure out when they might have hybridized in the past. If even a small number of individuals exhibit heteroplasmy due to paternal leakage, the presence of recombinant or mixed mitochondrial genomes (Okwaro Louis Allan, 2019) may complicate the reconstruction of maternal lineages. This information is essential for conservation genetics because misunderstanding lineage data could lead to wrong conclusions about the genetic purity and evolutionary history of honeybee populations.

For instance, when managed and wild populations breed with each other, the occasional passing on of paternal mtDNA could make it harder to tell how "pure" the native stocks are, which would make it harder to protect unique evolutionary lineages. Hybridization events, which may still be happening in areas with mixed subspecies, make the genetic landscape even more complicated and could lead to populations where the maternal lineage is less clear. Because of this, evolutionary biologists and conservation managers need to be careful when looking at mtDNA data. They may need to combine nuclear genetic markers with mitochondrial markers to obtain a better picture of genetic structure and history (Breton & Stewart, 2015; Okwaro Louis Allan, 2019).

31.5. Conservation and Genetic Resource Management

From a conservation viewpoint, the chance of male honeybee contributions in hybrids means that strategies for managing genetic resources that mainly use mitochondrial markers need to be reconsidered. Genetic integrity is essential in conservation programs, especially those that try to protect native or endemic honeybee subspecies. Conservation plans must consider paternal contributions to fully understand the complex genetic landscape, especially since mitochondrial variation doesn't solely originate from the queen bee. The presence of paternal mtDNA in managed honeybee populations, especially in places where different honeybee subspecies mix or where non-native bees breed with local ones, could affect how we understand their evolutionary relationships (Okwaro Louis Allan, 2019).

The assumption of strict maternal inheritance is what makes mtDNA a reliable molecular marker for phylogenetics and population genetics. However, when paternal leakage happens, this assumption is broken, and the resulting mitochondrial heteroplasmy makes it even harder to figure out how phylogenetic relationships work. Researchers have been using mtDNA markers for a long time to tell different honeybee lineages apart and figure out when they might have hybridized in the past. If even a small number of people have heteroplasmy because of paternal leakage, though, the presence of recombinant or mixed mitochondrial genomes may make it harder to reconstruct maternal lineages. This information is essential for conservation genetics because misunderstanding lineage data could lead to wrong conclusions about the genetic purity and evolutionary history of honeybee populations (Okwaro Louis Allan, 2019).

For instance, when managed and wild populations breed with each other, the occasional passing on of paternal mtDNA could make it harder to tell how "pure" the native stocks are, which would make it harder to protect unique evolutionary lineages. Hybridization events, which may still be happening in areas with mixed subspecies, make the genetic landscape even more complicated and could lead to populations where the maternal lineage is less clear. Because of this, evolutionary biologists and conservation managers need to be careful when looking at mtDNA data. They may need to combine nuclear genetic markers with mitochondrial markers to obtain a better picture of genetic structure and history (Breton & Stewart, 2015; Okwaro Louis Allan, 2019).

31.6. Conservation and Genetic Resource Management

From a conservation perspective, the possibility of male honeybee hybrids sharing genes means that methods for managing genetic resources that mostly rely on mitochondrial markers need to be rethought. Genetic integrity is essential in conservation programs, especially those that try to protect native or endemic honeybee subspecies. If paternal contributions are ignored, conservation plans that assume all mitochondrial variation comes from the mother may misunderstand the true complexity of the genetic landscape. The presence of paternal mtDNA in managed honeybee populations, especially in areas where different honeybee subspecies mix or where non-native bees breed with local ones, could cause confusion about the purity of lineages and the variety of genes present (Okwaro Louis Allan, 2019).

These factors significantly influence breeding programs in the real world. Honeybee breeders often use mitochondrial markers to trace maternal lines and select desired genetic traits. If extra mitochondrial variants from fathers mix in with these lineages, the resulting mixture could make it harder to predict breeding outcomes. The colony's success may be affected by early heteroplasmy, even if most paternal mtDNA is removed later. For instance, transient heteroplasmy could impact the growth of larvae, the early immune responses of developing bees, or the metabolic efficiency of developing bees. All of these are important factors for the success of a colony in both natural and managed settings (Meusel & Moritz, 1993).

We need to take a more nuanced approach to managing genetic resources, one that looks at both mitochondrial and nuclear genes in detail. New high-throughput sequencing technologies enable the detection of low levels of paternal mtDNA contributions, which traditional RFLP analysis would not detect. Deep sequencing methods can provide us quantitative measurements of heteroplasmy across the whole mitochondrial genome. The result makes the genetic data used in conservation planning more reliable (Mandel et al., 2020).

Furthermore, better molecular tests that can determine and measure paternal mtDNA contributions are critical. Standard PCR-based methods are beneficial, but they may not be sensitive enough to pick up very low levels of paternal leakage, which could still have biologically important effects. Researchers will be able to find out the real rate of paternal leakage in wild and managed honeybee populations by using methods that combine deep sequencing with strict controls. These procedures will make it possible to make more accurate estimates of population structure, gene flow, and evolutionary paths (Meusel & Moritz, 1993; Okwaro Louis Allan, 2019).

31.7. Experimental Approaches and Methodological Considerations

Artificial insemination and detailed molecular analysis have been the main ways that scientists have been able to show that paternal mtDNA is passed on in honeybee hybrids in the lab. By carefully planning experiments that breed *A. m. carnica* with *A. m. capensis*, researchers have tracked how paternal mtDNA moves from fertilization to the early stages of embryo development. Researchers have confirmed paternal mtDNA's presence and measured its amount at various developmental stages. They did this by using diagnostic mitochondrial markers and radiolabeled hybridization techniques to make the tests more sensitive. The drop in paternal mtDNA from newly laid eggs to later larval stages shows that even though there is a lot of paternal mtDNA right after fertilization, the cells eventually get rid of these paternal genomes, leading to a consistent genetic makeup in adults (Meusel & Moritz, 1993).

One problem with these studies is that molecular assays have a limit on how low they can go in terms of detection. If paternal mtDNA is present at levels below this limit, it may not show up as much as it really is. Recent studies in this area have emphasized the necessity of high-resolution techniques capable of identifying extremely low-frequency variations. As deep sequencing methods become more widely used, they should help us better understand how paternal mtDNA influences honeybees by accurately measuring heteroplasmy at various development stages and in different tissues (Mandel et al., 2020).

Another thing to contemplate when using traditional methods is that they may sometimes confuse real paternal leakage with technical problems like contamination or the presence of nuclear mitochondrial pseudogenes (NUMTs). To solve these problems, it is important to use strict experimental controls and mitochondrial markers that are specific to each species. Researchers have made it less likely that their results will be wrong by designing studies with strict quality controls and comparing reciprocal crosses under the same experimental conditions. The evidence strongly supports the conclusion that paternal mtDNA transmission in honey bee hybrids is a real phenomenon with real biological effects (Meusel & Moritz, 1993).

This conclusion raises questions about the purity of honeybee lineages and the variety of genes present. These factors significantly influence breeding programs in the real world. Honeybee breeders often use mitochondrial markers to track maternal lines and select desired genetic traits. If extra mitochondrial variants from the paternal mix in with these lineages, the resulting mixture could make it harder to predict breeding outcomes. The colony's success may be affected by early heteroplasmy, even if most paternal mtDNA is removed later. For instance, transient heteroplasmy might affect the growth of larvae, the early immune responses of developing bees, or the metabolic efficiency of developing bees. All of these are important factors for the success of a colony in both natural and managed settings (Meusel & Moritz, 1993).

We need to take a more nuanced approach to managing genetic resources, one that looks at both mitochondrial and nuclear genes in detail. New high-throughput sequencing technologies allow us to discover small amounts of paternal mtDNA contributions that traditional RFLP analysis would miss. Deep sequencing methods can provide us quantitative measurements of heteroplasmy across the whole mitochondrial genome. This makes the genetic data used in conservation planning more reliable (Mandel et al., 2020).

Furthermore, better molecular tests that can determine and measure paternal mtDNA contributions are critical. Standard PCR-based methods are beneficial, but they may not be sensitive enough to pick up very low levels of paternal leakage, which could still have biologically important effects. Researchers will be able to find out the real rate of paternal leakage in wild and managed honeybee populations by using methods that combine deep sequencing with strict controls. These procedures will make it possible to make more accurate estimates of population structure, gene flow, and evolutionary paths (Meusel & Moritz, 1993; Okwaro Louis Allan, 2019).

It is also advantageous to look at other types of animals to better understand paternal leakage in honeybees. For instance, research on Drosophila hybrids has shown that paternal leakage can happen more often in interspecific crosses. This example illustrates how paternal contributions function in genetically distinct honeybee subspecies (Polovina et al., 2020). In the same way, studies of unusual mitochondrial inheritance in different eukaryotes support the idea that strict maternal inheritance is the norm, but deviations like paternal leakage and biparental inheritance are not new and can have big effects on evolution and ecology (Breton & Stewart, 2015).

31.8. Future Directions and Research Needs

The evidence we have right now shows that paternal mtDNA transmission in honeybee hybrids is an important factor that must be taken into account in both basic research and conservation efforts. Several important areas should be the focus of future research. First, we need to improve molecular techniques even more so that we can identify low-level paternal contributions. This is especially true for deep sequencing technologies, which are both sensitive and precise. Finding out how much paternal leakage happens in more natural populations, instead of just in hybrids that are controlled in the lab, will help us figure out if the phenomenon is just a lab artifact that happens sometimes or if it happens a lot in nature and affects honey bee population dynamics (Mandel et al., 2020; Okwaro Louis Allan, 2019).

Second, we need to quickly look into what heteroplasmy resulting from paternal leakage means for how things work. In particular, future studies should look into whether transient heteroplasmy has any measurable effects on the health of the colony, its ability to fight disease, or its metabolic efficiency. Looking into possible interactions between mitochondria and nuclei that happen when different mitochondrial genomes are present at the same time in early development is likely to help us understand the differences between positive and negative outcomes of paternal mtDNA transmission. This line of research is especially important when it comes to hybridization between different subspecies of *A. mellifera*, where genomic incompatibilities may be more obvious (Breton & Stewart, 2015; Meusel & Moritz, 1993).

Furthermore, it will be important to do comparative studies in different geographic areas and environmental conditions to learn more about how outside factors might affect the frequency of paternal leakage. For example, environmental stressors like temperature, nutrition, or even the presence of pathogens might change the cellular processes that break down mitochondria. These variables may alter the pace at which developing embryos retain paternal mtDNA. Long-term studies monitoring colonies over multiple generations could shed light on whether paternal leakage has any long-term or cumulative effects on the mitochondrial genetic landscape of honeybee population (Okwaro Louis Allan, 2019).

Furthermore, using both nuclear genetic markers and mitochondrial data in studies of conservation genetics will present us a better picture of the structure of honeybee populations. Using only mtDNA markers could make models of genetic diversity too simple, especially if paternal leakage adds hidden layers of complexity. A dual-marker approach, which uses information from both nuclear and mitochondrial genomes, will help us better understand how honeybee populations have spread to new areas, mixed with each other, and how genes have moved between and within these population (Breton & Stewart, 2015; Okwaro Louis Allan, 2019).

Lastly, the best way to learn about paternal leakage dynamics is probably through interdisciplinary approaches that combine molecular genetics, developmental biology, and ecology. Researchers from different fields, such as honeybee physiology, molecular evolution, and conservation biology, can work together to come up with controlled experiments and field studies to test ideas about the selective pressures that affect heteroplasmic individuals. These combined studies might eventually help create models that predict how paternal mtDNA leakage will influence evolution in various environments (Mandel et al., 2020; Meusel & Moritz, 1993).

31.9. In conclusion

The finding that paternal mtDNA can be passed down in honeybee hybrids is a big break from the long-held belief that only maternal mtDNA can be passed down. Paternal leakage in these insects seems to be a temporary event that happens mostly in the early stages of embryonic development and is mostly gone by later stages. However, its occurrence has important effects on evolutionary biology, phylogenetics, and conservation genetics. Meusel and Moritz's 1993 experiments indicated that paternal leakage can cause heteroplasmy, which is not just an experimental artifact but a biologically important process that could affect mitochondrial diversity, adaptive potential, and the accuracy of maternal lineage tracing (Mandel et al., 2020; Meusel & Moritz, 1993).

In terms of evolution, paternal leakage might sometimes create new mitochondrial variants that could help the colony adapt, or it could cause mitonuclear incompatibilities that hurt the colony's fitness. These different options show how important it is to do more research on how heteroplasmy affects honeybees' ability to work. Furthermore, the fact that paternal mtDNA can be passed down means that conservation genetic strategies need to be rethought, since they have always been based on the idea that maternal inheritance is strict. When managed and wild populations breed together, even rare paternal contributions can hide genetic signals and make it harder to manage evolutionary lineages (Breton & Stewart, 2015; Okwaro Louis Allan, 2019).

Paternal leakage could significantly impact the genetics of honeybee populations, making it critical to create and use advanced molecular techniques. We will need to do more research in the future using deep sequencing and dual-marker analyses, along with controlled breeding experiments, to fully understand how often paternal mtDNA is passed on and what effects it has in both wild and managed populations. In the end, these all-encompassing methods will help us learn more about the genetics of honeybee populations and lead to better conservation and breeding programs (Mandel et al., 2020; Okwaro Louis Allan, 2019).

In short, paternal leakage in honeybee hybrids goes against long-held beliefs about how mitochondria are passed down from one generation to the next. This indicates that the molecular, evolutionary, and conservation frameworks that have guided honeybee genetics research for decades need to be rethought. Researchers can obtain a more detailed and accurate picture of honeybee mitochondrial dynamics by combining information from both classical studies and new high-resolution methods. This picture will fully account for the small but important contributions of paternal mtDNA. This suggests a need to rethink the molecular, evolutionary, and conservation frameworks that have guided honeybee genetics research for decades (Breton & Stewart, 2015; Meusel & Moritz, 1993; Okwaro Louis Allan, 2019).

Chapter 32

32. Reconstructing Pre-Industrial Genetic Diversity in *Apis mellifera* Using Ancient DNA from Museum Specimens

32.1. Ancient DNA and the Reconstruction of Pre-Industrial Diversity

32.1.1. Introduction to Ancient DNA and Honeybee Conservation

Ancient DNA (aDNA) methods have become very useful for figuring out what the genetic makeup of living things was like before human activities had a big effect on it. For *A. mellifera*, taking DNA from museum specimens and studying it has given us a unique look at the genetic diversity of wild and feral honey bee populations before they were domesticated (Mikheyev et al., 2015). These museum collections, which include specimens from the late 1800s and early 1900s, are like time capsules that show how bees' genetics changed before modern beekeeping practices, a lot of queen trading, and intensive breeding programs (Parejo et al., 2020).

32.1.2. Advances in Ancient DNA Techniques

Researchers have been able to obtain high-coverage genomic data from these old samples (Mikheyev et al., 2015) by using advanced next-generation sequencing and specialized library preparation methods that work with degraded DNA. Protocols that don't use PCR amplification, for example, reduce artifacts that come from degraded samples and allow the direct sequencing of nanogram-scale aDNA. This approach preserves authentic genetic signatures that reveal the historical structure of populations (Mikheyev et al., 2015). This progress has been crucial for accurately reconstructing mitochondrial DNA (mtDNA) haplotype frequencies from pre-industrial populations. This is because mtDNA is straightforward to break down but provides high-resolution markers of maternal lineage identity (Zayed, 2009).

32.1.3. Limitations of Traditional Genetic Markers and Benefits of Whole-Genome Data

In traditional studies of honeybees, researchers often used markers like the tRNA-leu-cox2 intergenic region to distinguish between different evolutionary lines. But looking at these small pieces of DNA alone doesn't show the full range of historical genetic diversity (D. S. M. Henriques, 2018). The introduction of whole mitogenome sequencing and the addition of nuclear genomic data have greatly improved resolution. This technique has made it possible to observe changes in haplotype frequencies and admixture events that have happened over the past century (Parejo et al., 2020). Comparing genomes from museum specimens to those from modern bees has consistently shown a big drop in mitochondrial haplotype diversity in modern populations. The decrease is a sign of historical bottlenecks caused by things like the introduction of *Varroa destructor* and other human-made stressors (D. S. M. Henriques, 2018).

32.1.4. Mitochondrial vs. Nuclear Diversity: Mechanisms and Patterns

One of the most intriguing things that ancient DNA studies have found is that mitochondrial diversity has shrunk a lot, but nuclear genetic diversity has stayed relatively stable overall. Queen bees mate with more than one male, which leads to a lot of nuclear recombination and genetic mixing within colonies (Mikheyev et al., 2015; Parejo et al., 2020). This process is primarily the reason for the observed difference. The phenomenon shows that selective pressures affect maternally inherited genomes differently than biparentally inherited genomes. This shows how important it is to look at both genomic compartments to attain a full picture of population dynamics (Mikheyev et al., 2015).

32.1.5. Insights from Historical Museum Specimens

For example, in-depth studies of museum specimens from Switzerland have shown that mtDNA haplotype frequencies have changed over time in ways that are not seen in current domestic stocks. Researchers have found that specimens from before the Industrial Revolution had a wider range of mtDNA haplotypes.

Many of these haplotypes are now either lost or much less common in modern populations (D. Henriques, Browne, et al., 2018; Parejo et al., 2020). These kinds of reconstructions not only show when and how big demographic bottlenecks were, but they also show the hidden history of introgression events, which are when non-native subspecies or lineages have been added to native gene pools over time (Parejo et al., 2020).

32.1.6. Methodological Rigor and Bioinformatics for *(aDNA)*

There are many different ways that methodological advances have made these insights possible. Strict rules for extracting aDNA from museum specimens take into account common types of damage, like cytosine deamination, which can cause sequencing mistakes if not fixed correctly (Mikheyev et al., 2015). Researchers have created computational pipelines that use strong variant calling tools and strict filtering criteria to tell the difference between real genetic polymorphisms and artifacts that happen when DNA breaks down (Card et al., 2021). These integrated bioinformatic methods are needed to obtain accurate estimates of historical effective population sizes and genetic differentiation indices (D. S. M. Henriques, 2018) (Espregueira Themudo et al., 2020) and to create reliable data sets.

32.1.7. Conservation Applications of Ancient DNA

Firstly, utilizing ancient DNA (aDNA) from museum specimens to reconstruct genetic diversity prior to the industrial age significantly impacts conservation genetics. First, these studies present us a historical baseline that helps us tell the difference between natural evolutionary processes and recent changes caused by human actions. By comparing ancient genetic data to modern samples, researchers can determine the extent of genetic material loss over the past century. Researchers can also identify the lost genetic lineages (D. Henriques, Browne, et al., 2018; Parejo et al., 2020). This kind of information is essential for setting conservation goals and guiding management practices that aim to protect or restore lost genetic diversity.

Secondly, studying ancient DNA can help identify locally adapted ecotypes that may have genetic variations that make them more resistant to modern stressors like parasites, pesticides, and habitat loss (D. S. M. Henriques, 2018). For example, if certain mtDNA haplotypes found in old samples are connected to stronger immune systems or better ability to handle stress, these traits can be prioritized in breeding programs aimed at restoring native populations. *A. mellifera* once had a wide range of mitochondrial lineages in the wild, demonstrating the significant genetic diversity that could be useful for conservation efforts (D. Henriques, Browne, et al., 2018; Parejo et al., 2020).

32.1.8. Long-Term Evolutionary Perspectives

Furthermore, combining old and new genomic data not only helps us understand what happened in the past, but it also helps us make predictions about how resilient we will be in the future. Several studies using museum DNA analyses have revealed a higher level of local genetic continuity in wild honeybee populations than previously believed (D. Henriques, Browne, et al., 2018; Mikheyev et al., 2015). Museum DNA analyses have shown in several studies that wild honeybee populations have a higher level of local genetic continuity than was previously thought. They have also shown that recent admixture events are mostly due to human-mediated introductions rather than natural gene flow (Mikheyev et al., 2015; Parejo et al., 2020).

Another important result of studying ancient DNA is the discovery of mito-nuclear discordance in some historical specimens. For instance, studies have indicated that museum samples had mitochondrial haplotypes that are usually linked to one lineage, but the nuclear genome showed that the samples came from a different lineage (Parejo et al., 2020). These patterns suggest that introgressive hybridization events happened before modern interventions became common. These differences make it even harder to figure out how admixture processes have changed the genetic makeup of honeybee populations today. They might also show hidden genetic reservoirs that have helped species adapt over time (Mikheyev et al., 2015).

32.1.9. Policy, Restoration, and Future Techniques

Using ancient DNA to recreate genetic diversity before the industrial era presents us a strong way to judge how well current conservation policies are working. To figure out how much introgression and loss of native haplotypes there has been, legislative efforts to protect native subspecies, like those that have been implemented in some parts of Europe, depend on accurate assessments of historical genetic baselines (Taurisano et al., 2024). When comparing ancient DNA data with modern genetic surveys in some areas, it has been found that while modern admixture has increased overall heterozygosity in some managed populations, important indigenous mitochondrial haplotypes have been greatly reduced or replaced (Espregueira Themudo et al., 2020; Taurisano et al., 2024). These results show how important it is to put conservation actions that focus on protecting the unique genetic heritage found in museum specimens at the top of the list.

Researchers are now putting more effort into improving aDNA extraction methods so that they can obtain the most high-quality DNA from museum collections, which often have very broken samples. Better DNA repair enzymes, better buffer conditions, and targeted enrichment strategies are all helping to make sequencing ancient honeybee DNA more successful (Mikheyev et al., 2015). At the same time, the creation of more advanced bioinformatics pipelines that can accurately model and remediate the damage patterns that are naturally present in aDNA is making it easier for us to put together complete genomic profiles from these old materials (Card et al., 2021).

32.1.10. Modeling Genetic Change and Selection Over Time

Combining these new technologies with larger sets of ecological and historical data could help us understand the evolutionary history of *A. mellifera* on a continental scale. Researchers can create detailed models that illustrate how environmental factors have influenced genetic diversity since industrialization by looking at changes in mtDNA haplotype frequencies in museum specimens alongside historical records of beekeeping, habitat changes, and the spread of invasive parasites. These kinds of models not only help us understand how genetic change happens, but they also help us guess what will happen in the future under different management or environmental conditions.

One intriguing thing about museum-based aDNA studies is that they can detect signs of selection that have affected honeybee populations over time. Genome-wide scans of ancient specimens have found hundreds of genes that show signs of natural selection. Disease resistance, developmental timing, and behavioral traits that promote colony health are associated with many of these genes (Mikheyev et al., 2015; Parejo et al., 2020). Finding these adaptive loci in populations that lived before the Industrial Revolution is especially useful because it presents conservation breeding programs with clear goals. Conservationists can try to bring back or strengthen these beneficial genetic variants in modern populations that have lost genetic diversity by choosing alleles that were once common in the wild and probably gave advantages in the local environment (D. Henriques, Browne, et al., 2018; Parejo et al., 2020).

32.1.11. Taxonomy, Gene Flow, and Lineage Complexity

Rebuilding ancient genetic diversity also provides us a chance to look at the taxonomic and evolutionary links between different *A. mellifera* lineages again. Historical genetic data have shown that some groups that were thought to be genetically similar actually had a complicated mix of mitochondrial and nuclear ancestries (Mikheyev et al., 2015; Parejo et al., 2020). These results go against what most people perceive as how different some subspecies are and show how gene flow has changed over time in the evolutionary history of honeybees. Detailed reconstructions that combine genomic data from the past and the present help with both conservation planning and a better basic understanding of how honeybees have changed over time (Espregueira Themudo et al., 2020; D. S. M. Henriques, 2018).

32.1.12. Restoration Goals and Policy Evaluation

There is a lot of potential for ancient DNA studies to help with conservation work. One useful way to use historical genetic baselines is to set goals for restoration. By measuring the genetic change between pre-industrial specimens and modern populations, researchers can determine how much current gene pools have changed from their historical states (Parejo et al., 2020). This kind of information is essential for creating breeding programs that try to bring back genetic diversity that has been lost, especially in areas where native honeybee stocks are at risk from genetic homogenization and hybridization with non-native strains (Parejo et al., 2020; Taurisano et al., 2024).

Ancient DNA can also aid in conservation efforts by monitoring the effectiveness of laws designed to safeguard native honeybee subspecies. For example, some parts of Europe have passed laws to protect native lineages like *A. m. ligustica* by controlling the introduction of foreign queens (Taurisano et al., 2024). Conservationists can see if these laws are protecting the native genetic heritage by examining the mitochondrial haplotype distributions of museum specimens, which show the genetic makeup before these laws were in place, and comparing them to the genetic makeup of current populations (Espregueira Themudo et al., 2020; Taurisano et al., 2024).

32.1.13. Integrative Approaches and the Future of Honeybee Genomics

In the future, research on ancient DNA from honeybee museum specimens will probably benefit from more collaboration between different fields. Combining molecular genetic data with historical records, climate reconstructions, and studies of land use can help us obtain a full picture of how changes in the environment and human activities have affected honeybee diversity over the past 150 years (Panziera et al., 2022; Schwartz et al., 2007). These kinds of integrative methods help us learn more about how evolution has worked in the past and provide us important information about how honeybee populations might react to future problems like climate change, new diseases, and habitat fragmentation (Espregueira Themudo et al., 2020).

32.1.14. Conclusion: Looking Backward to Move Forward

In short, using ancient DNA techniques on museum specimens of *A. mellifera* has greatly improved our understanding of genetic diversity before the Industrial Revolution. These methods present us detailed pictures of mitochondrial haplotype frequencies, show how far back in history admixture and introgression have happened, and tell us the difference between natural evolutionary processes and those caused by modern beekeeping practices (Mikheyev et al., 2015; Parejo et al., 2020). These studies have important implications for conservation because they set baseline genetic diversity metrics that can help with restoration and management plans that aim to protect local adaptations and the overall evolutionary resilience of honeybee populations (Espregueira Themudo et al., 2020; D. S. M. Henriques, 2018).

It is certain that future research will focus on improving both bioinformatics methods so that they can collect as much useful genetic data as possible from samples that have been badly damaged. Better extraction methods and more accurate damage-repair methods should provide us better ancient DNA. At the same time, better computer tools will help us combine these historical datasets with modern genomic profiles (Card et al., 2021; Mikheyev et al., 2015). Researchers will be able to create improved models that illustrate how *A. mellifera* populations have changed over time and how to maintain their genetic integrity during periods of rapid environmental change (Card et al., 2021).

32.1.15. The Broader Impact of aDNA in Honeybee Conservation

As conservation efforts rely increasingly on genomic resources, using ancient DNA to recreate pre-industrial diversity will be a key part of making decisions and taking action. Finding and studying old genetic information not only helps identify lost or declining haplotypes, but it also plays a crucial role in assessing how well conservation efforts are working to lessen the impacts of mixing species and losing genetic diversity (Parejo et al., 2020; Taurisano et al., 2024). Conservation biologists and managers can create breeding programs, plans to reintroduce species, and strategies to protect habitats that honor the natural history of local honeybee populations by using both current genetic data and historical information (Espregueira Themudo et al., 2020; D. S. M. Henriques, 2018).

In the end, combining ancient DNA data with modern genomic surveys has changed the way we view how honeybees evolved. We can now follow the lineage-specific effects of important historical events, such as the introduction of invasive parasites like *Varroa destructor* and the later hybridization with commercial stocks by humans. We can also see how these events have changed the genetic structure of *Apis mellifera* populations (Mikheyev et al., 2015; Parejo et al., 2020). This kind of knowledge is not only intriguing from an academic perspective, but it is also necessary to protect the ecosystem services that these pollinators provide, which are still crucial for global biodiversity conservation and agricultural production (Espregueira Themudo et al., 2020; D. S. M. Henriques, 2018).

32.1.16. Final Thoughts: Preserving an Evolutionary Legacy

In conclusion, studies of ancient DNA from museum specimens of *A. mellifera* have changed our understanding of genetic diversity before the industrial revolution in a way that can never be undone. These studies provide a standard against which we can measure the genetic effects of modern beekeeping and environmental changes. These reconstructions have shown significant changes in the frequencies of mitochondrial haplotypes, found cases of mito-nuclear discordance, and demonstrated how much human activity has changed the genetic makeup of honey bee populations (Mikheyev et al., 2015; Parejo et al., 2020). The ongoing improvement of aDNA methods and the combination of historical genomic data with modern biological and environmental data offer tremendous hope for guiding future conservation efforts aimed at protecting one of our planet's most important pollinators' evolutionary potential and ecological functions (Espregueira Themudo et al., 2020; D. S. M. Henriques, 2018).

Because of these new discoveries, it is important that conservation plans for *A. mellifera* use information from studies of ancient DNA. Scientists and policymakers can work together to ensure that honeybee populations survive and thrive in the face of new and ongoing challenges. They can do this by restoring lost genetic diversity and strengthening the genetic integrity of indigenous stocks. The historical perspective offered by museum specimens thus provides a critical foundation for informed conservation management, ensuring that the evolutionary legacy captured in centuries-old specimens continues to shape the destiny of modern honeybees (Parejo et al., 2020; Taurisano et al., 2024).

In the end, using ancient DNA to recreate the diversity of pre-industrial times is more than just a scientific project. It is also a necessary step toward understanding and reducing the serious ecological and economic effects of genetic erosion in honeybees. As we stand at the crossroads of the past and the present, aDNA studies remind us that history holds the keys to sustainable management and conservation. These lessons are both timeless and urgently needed in an age of rapid environmental change (Panziera et al., 2022; Schwartz et al., 2007).

Chapter 33

33. Integration with Nuclear Genomics to Resolve Subspecies Classification

33.1. Limitations of Mitochondrial DNA in Honey Bee Subspecies Classification

Researchers have long used mitochondrial DNA (mtDNA) markers to trace maternal ancestry and broad phylogeographic patterns in the study of *A. mellifera subspecies* classification. However, mtDNA is only a small part of the total genetic information available in honeybees and only shows a single, uniparental mode of inheritance (Buswell et al., 2023). Nuclear genomic markers, on the other hand, are passed down from both parents and include a wide range of genetic diversity that encompasses both historical and recent admixture events. This versatility makes them a strong tool for defining population structure and finding subspecies boundaries (Buswell et al., 2023; Eimanifar, Brooks, et al., 2018).

33.2. Advancements in Sequencing Technologies and SNP Detection

Researchers can now identify even the smallest signs of introgression and admixture that traditional mtDNA methods may miss (D. S. M. Henriques, 2018), thanks to high-throughput sequencing technologies like whole-genome sequencing (WGS) and genotyping-by-sequencing (GBS). These technologies allow them to look at thousands of single nucleotide polymorphisms (SNPs) across the nuclear genome. This combined method addresses a major issue in honeybee conservation: the mismatch between physical traits, mtDNA lineage classifications, and nuclear genomic data, particularly in regions where queen trading and inter-colony mating have led to significant hybridization (Buswell, 2022; Eimanifar, Brooks, et al., 2018).

33.3. Nuclear SNPs Reveal Fine-Scale Population Structure and Admixture

Nuclear SNP data have been very useful for finding fine-scale population structure. They have also been able to measure how much gene flow there is between native M-lineage subspecies and the commercial C-lineage bees that were brought in (Eimanifar, Brooks, et al., 2018; D. Henriques, Parejo, et al., 2018). Studies using smaller SNP panels from whole-genome resequencing have gotten assignment accuracies of over 98%, even with only a few ancestry-informative markers (AIMs). This study shows how nuclear markers can be used as cheap, high-throughput diagnostic tools for regularly checking genetic purity and introgression in conservation programs (D. S. M. Henriques, 2018).

33.4. Mitonuclear Discordance and the Need for Dual-Genome Approaches

Recent studies have indicated that there is a consistent pattern of disagreement between mtDNA and nuclear markers. For example, mtDNA analyses may show a relatively homogeneous maternal lineage that hides significant levels of biparental gene flow shown by nuclear SNP data (Buswell et al., 2023; Eimanifar, Kimball, et al., 2018). There are several reasons for this difference between mtDNA and nuclear markers, including gene flow that favors one sex, past areas where different species mixed, and the random gain or loss of maternal genetic types. These all demonstrate the need for a dual-genome approach to fully comprehend the evolutionary history of honey bee subspecies (Chávez-Galarza et al., 2021; D. Henriques, Browne, et al., 2018).

33.5. Uncovering Cryptic Introgression Through Integrative Genomics

Recent studies using integrative genomics have combined detailed whole-genome resequencing of nuclear markers with nearly complete mitochondrial genomes. This has provided the most detailed view yet of how subspecies are related, uncovering both hidden gene mixing and small-scale mixing patterns across whole populations (D. Henriques, Browne, et al., 2018; Tihelka et al., 2020). These kinds of studies show that mitochondrial markers, like the tRNAleu-cox2 intergenic region, are still very useful for tracing maternal lineage and broad-scale evolutionary lineages. However, they often don't show the subtle spatial dynamics of hybridization that nuclear SNP analyses do (D. Henriques, Browne, et al., 2018; Muñoz & De la Rúa, 2021).

33.6. Conservation Applications: Monitoring Introgression and Local Adaptation

For example, in areas where conservation programs want to protect native stocks of *A. m. mellifera,* nuclear SNP genotyping is crucial for management decisions because it shows how much C-lineage introgression has happened. These analyses reveal clear signs of gene mixing that mtDNA data alone would miss, helping managers create better breeding and removal strategies to protect local adaptations (Buswell et al., 2023; D. Henriques, Browne, et al., 2018). Furthermore, nuclear genomic data can show specific genomic regions that are being selected for, which are often linked to traits like disease resistance, foraging efficiency, and adaptability to changing climates. These are all important traits for managing honeybee populations in a way that is sustainable in the face of rapid environmental change (D. Henriques, Browne, et al., 2018).

33.7. Mitonuclear Co-evolution and Colony Fitness

Combining nuclear and mitochondrial datasets also opens up new areas of research into mitonuclear co-evolution. This is an important part of honeybee biology because the coordinated evolution of mitochondrial and nuclear genes affects how well a colony does and how it responds to stress (Wragg et al., 2017). Differences between mtDNA and nuclear SNP data can show that the two types of DNA don't work well together, which could mean less energy production or weaker stress responses. This study shows how important it is to look at both types of data together to attain a full picture of colony health and resilience (Eimanifar, Kimball, et al., 2018; Wragg et al., 2017).

33.8. Bioinformatics and Statistical Tools for Integrated Genomic Analysis

The development of bioinformatic tools and statistical models that can merge nuclear SNP and mtDNA data has simplified our ability to predict how these two parts of the genome influence physical traits. Researchers attain a more complete picture of a person's genetics by using integrative methods like local ancestry inference and multilocus admixture analyses. These methods take into account both ancient divergence and recent gene flow (D. Henriques, Browne, et al., 2018). This integrative analysis is necessary to distinguish the difference between real introgression events and incomplete lineage sorting. This difference is crucial for understanding the evolutionary history of honey bee subspecies and coming up with conservation plans (Eimanifar, Kimball, et al., 2018; Tanasković et al., 2021).

33.9. Environmental Correlations and Adaptive Potential

Furthermore, using nuclear markers to classify subspecies presents with us a better idea of how honeybees adapt to different environments. Genome-wide scans that examine nuclear SNP data and environmental factors have found specific areas in the genome linked to how well honeybees adjust to local weather, diseases, and pesticide exposure (D. Henriques, Browne, et al., 2018; Muñoz & De la Rúa, 2021). These results enable conservation programs to implement methods that protect genetic integrity and maintain the adaptability necessary for native populations to survive long-term amid global challenges (Buswell et al., 2023; Soltani, 2023).

33.10. Development of Diagnostic Tools for Genetic Monitoring

It is clear that combining nuclear genomic data has real-world benefits, as seen in the creation of reduced SNP assays that are specifically designed to determine introgression in honeybee populations. For example, reduced panels with as few as 48 high-quality SNPs have been shown to correctly assign individuals to their original subspecies with amazing accuracy. Conservation programs rely on these panels for routine genetic monitoring and to control genetic pollution (D. Henriques, Browne, et al., 2018). These diagnostic tools help beekeepers and conservationists make smart choices about breeding, raising queens, and moving stocks by making testing quick and cheap. This process helps protect unique genetic lineages that are at risk of being mixed with other species by humans (Eimanifar, Brooks, et al., 2018; D. Henriques, Parejo, et al., 2018).

33.11. Predictive Modeling and Future Population Dynamics

The combination of nuclear and mitochondrial genomics is not only helping to clear up current taxonomic confusion caused by introgression, but it is also paving the way for predictive modeling of population dynamics. Researchers can now use large genomic datasets to model different gene flow scenarios and see how management actions might change the genetic structure of honeybee populations over time. These kinds of predictive models are very advantageous for figuring out what will happen if hybridization keeps happening and for testing ways to stop the loss of native genetic variation (D. Henriques, Browne, et al., 2018; Wragg et al., 2017).

33.12. Dual Perspectives on Honey Bee Evolution

Furthermore, integrated genomic approaches help us learn more about the evolutionary processes that have led to the diversity of honeybees. Human activity, such as queen trading and migratory beekeeping, has caused more recent and ongoing gene flow events, as revealed by nuclear markers. Mitochondrial DNA, on the other hand, often shows signals from ancient divergence events. This two-sided view not only makes it easier to understand where different subspecies came from in terms of evolution, but it also makes it easier to see clear conservation units that are important for protecting populations that are well-suited to their environment and have evolutionary significance (Eimanifar, Brooks, et al., 2018; D. Henriques, Browne, et al., 2018).

33.13. Preserving Adaptive Traits Amidst Hybridization

One important part of this integrated approach is that it can detect small genetic signs of local adaptation that are often hidden by a lot of mixing. Conservation geneticists can identify adaptive variants that confer organisms an edge in certain environmental conditions by looking for genomic regions that are under selection. For example, they can isolate variants that make organisms resistant to pathogens like *Varroa destructor* or tolerant of certain types of weather. Once these adaptive loci have been reliably found, they can be watched in populations to verify that breeding programs don't accidentally get rid of traits that evolved in that area and are important for colony survival (Buswell et al., 2023; D. Henriques, Browne, et al., 2018).

33.14. Conclusion: Toward Holistic Conservation Strategies

In short, combining nuclear genomics with traditional mitochondrial markers is a big step forward in figuring out how to classify honey bee subspecies. This holistic genomic approach goes beyond the problems of mtDNA-only studies by giving us a full picture of how genetic variation is passed down from both parents, how admixture works, and how to locate genomic regions that are locally adaptive. It also deals with the natural differences between different genetic systems, which helps with accurate taxonomic classification and the creation of specific conservation plans. The powerful combination of high-throughput sequencing technologies and advanced statistical tools that can analyze nuclear and mitochondrial data together will be even more useful for future research that aims to fully understand the complex evolutionary history and adaptation of honeybees (D. Henriques et al., 2021; D. Henriques, Browne, et al., 2018).

Adding nuclear genomics to subspecies classification strengthens our scientific knowledge and has significant impacts on the protection and management of *A. mellifera*. This integrative approach shows the full extent of introgression and the genomic basis of adaptive divergence. It provides important evidence for creating conservation programs that protect genetic diversity and help honeybee populations around the world survive in the long term (Buswell et al., 2023; D. Henriques, Browne, et al., 2018). Researchers and conservationists should continue to use and improve these combined methods to guarantee that we fully understand both nuclear and mitochondrial aspects, which are important for protecting the evolutionary history and ecological health of one of our planet's main pollinators (D. Henriques, Browne, et al., 2018).

Chapter 34

34. Toward a Genomic Renaissance in Honey Bee Conservation: Integrative Strategies for Adaptive Management, Precision Breeding, and Policy Innovation

34.1. Introduction

Honey bee conservation is undergoing a significant change right now. It is moving from static descriptions of mitochondrial diversity to a framework that includes functional, ecological, and evolutionary aspects. This topic brings together four related areas of research that could entirely change how we perceive the genetics of *A. mellifera*. These include functional mitogenomics, which connects mitochondrial DNA (mtDNA) variants to important metabolic and adaptive traits; reevaluating mitochondrial inheritance by looking at unusual paternal mtDNA transmission; the use of ancient DNA (aDNA) methods to recreate historical genetic landscapes and help with modern restoration efforts; and finally, combining large nuclear genomic datasets with traditional mtDNA analyses to clear up long-standing taxonomic uncertainties. Together, these methods make up the foundation of an adaptive and predictive framework for programs to protect honeybees and breed them with precision. In the end, such techniques will help policymakers make decisions based on facts to protect one of the most important pollinators on Earth (Eimanifar, Kimball, et al., 2018).

34.2. Current state-of-the-art in honeybee conservation genetics

34.2.1. The diversity of mitochondrial DNA and functional mitogenomics

Studies of honeybee genetics have mostly focused on mtDNA diversity in the past. This is because the mitochondrial genome is passed down from mother to child, has a high mutation rate, and doesn't mix well. Early research indicated that there were big differences in mtDNA haplotype diversity between subspecies like *A. m. capensis* and *A. m. scutellata* exhibit significant differences in mtDNA haplotype diversity. *m. scutellata*, even though it was challenging to determine genetic markers that could tell them apart (Eimanifar,

Kimball, et al., 2018). Functional mitogenomics is a growing field that promises to go beyond these static descriptions by directly linking certain mtDNA variants to traits that affect how well the body works, like better disease resistance or better energy metabolism. For instance, some mtDNA haplotypes are becoming increasingly linked to resistance to varroosis and other energy shortages in bee colonies. This approach is the first step toward marker-assisted selection that can make colonies healthier and more productive (Panziera et al., 2022).

34.2.2. Nuclear–Mitochondrial Data Integration

At the same time, combining nuclear genomic data with traditional mtDNA analyses has become an important way to navigate around the problems that come with only maternal inheritance. Nuclear markers, such as microsatellites and genome-wide single nucleotide polymorphisms (SNPs), have helped us understand how honey bee populations interact, how they adapt to new environments, and how they maintain their adaptive potential (D. Henriques, Parejo, et al., 2018; Wragg et al., 2017). When used together, these methods help clear up taxonomic confusion that can happen when only using mtDNA. They also present a fuller picture of honeybee genetic diversity, population structure, and gene flow (Ostroverkhova & Konusova, 2022).

34.2.3. Paternal Mitochondrial Transmission

Recent studies have started to question long-held beliefs about how mtDNA is passed down from one parent (the queen bee). New studies have shown that paternal mitochondrial transmission is rare in insects. The evidence suggests that paternal mtDNA may play a bigger role in the genetic makeup of populations than was previously thought (Kelemen & Rehan, 2021). This new idea of biparental or paternal mtDNA inheritance makes genetic studies in honeybees even more complicated and suggests that we need to rethink how we identify subspecies and protect them (Kenzy Ivveth Peña Carrillo, 2021).

34.2.4. Ancient DNA and Genetic Restoration

Researchers can use ancient DNA methods to learn a lot about the genetic diversity of honeybees in the past. This lets them recreate genetic landscapes from before the industrial era and see how human-made changes have changed allele frequencies over time (Espregueira Themudo et al., 2020). Scientists can identify lost alleles or ancestral gene combinations that used to support local adaptations by taking DNA from museum specimens and archaeological remains and studying them. These historical baselines are crucial for genetic restoration projects that want to bring back lost biodiversity to modern populations. These improvements will make the colonies more adaptable and stronger overall (Irving-Pease et al., 2019).

34.2.5. Future Directions in Honey Bee Conservation Genetics

Improving Functional Mitogenomics The goal of functional mitogenomics is to go beyond just listing different types of mtDNA by linking specific mitochondrial variants to specific functional outcomes. Mitochondrial genes affect many traits, including metabolism, thermal regulation, immune system function, and responses to oxidative stress. Next-generation sequencing and bioinformatics have advanced significantly, enabling them to identify potential mtDNA polymorphisms that could be beneficial in specific environmental or pathogenic conditions (Tihelka et al., 2020). For example, high-throughput sequencing has shown small differences in the order of genes, the way tRNA is arranged, and the way codons are used. These differences could lead to significant changes in how cells breathe and how much energy they produce. In the context of climate change, these kinds of functional insights can help choose mitochondrial haplotypes that improve fitness when it's hot or when food is scarce. The evidence supports marker-assisted selection methods that are specific to these adaptive responses (Pearman et al., 2024).

34.3. Exploring Paternal mtDNA Contributions

It has always been thought that honeybees only acquire their mtDNA from their mothers (maternals). However, research indicating the potential for paternal mtDNA contributions provides a fresh perspective on the inheritance of traits. To figure out how often paternal mitochondrial transmission happens, how it works, and what it does to cells, scientists will need to come up with new ways to design experiments and identify sensitive ways to detect it. Finding paternal mtDNA and figuring out how it might interact with the maternally derived mitochondrial genome could help us identify new sources of genetic variation within a population that could affect the colony's fitness, especially in places where hybridization or gene flow is strong (Kelemen & Rehan, 2021; Kenzy Ivveth Peña Carrillo, 2021).

34.4. Integrating Ancient DNA into Contemporary Frameworks

By looking at ancient DNA, conservation geneticists can compare the genetic makeup of the past with that of the present. The significant differences between museum specimens and modern samples show how genetic bottlenecks and introgression events have changed populations over time (Espregueira Themudo et al., 2020; Pearman et al., 2024). Using aDNA methods, scientists can recreate lost genetic diversity and come up with plans to bring back useful alleles that may have been lost when commercial beekeeping became more popular. As sequencing technologies develop better and more ancient specimens become available for study, these efforts will create important datasets that help us learn more about long-term population dynamics and show how important it is to protect heritage genetic lineages so that they can adapt in the future (Parejo et al., 2020).

34.5. Constructing an Integrated Genomic Framework

The best hope for the future of honeybee conservation genetics is to fully combine genome-wide data. Researchers can obtain a much clearer picture of population structure, evolutionary paths, and adaptive potential by combining functional mitogenomics with nuclear genomic analyses and adding historical aDNA data to the mix. To make this integration work, we need to create standardized tools and methods that let us combine data from different genetic sources. Such an approach will make it possible to run strong marker-assisted selection programs that can tell the difference between adaptive and neutral genetic variation (D. Henriques, Parejo, et al., 2018; Wragg et al., 2017).

In technical terms, the future workflow may include using whole-genome sequencing and high-density SNP arrays to figure out how populations are different and identify genes that could help them adapt. You can then compare these genetic markers with mitochondrial genomic data to figure out cases of mito-nuclear discordance that are common in mixed populations. When combined with cutting-edge bioinformatics and machine learning techniques, predictive models can predict how people will react to environmental changes, like climate change. In the end, these combined models will help conservationists come up with ways to maintain genetic diversity and reproductive fitness in both managed and wild populations (Ostroverkhova & Konusova, 2022).

34.6. Breeding Programs and Policy Implications

One of the most concrete results of combining genomic research will be the creation of precision breeding programs that aim to improve the health and productivity of colonies. Beekeepers will be able to use special selection methods to improve the genetics of their bees by identifying mtDNA haplotypes that are closely associated with excellent health traits or strong resistance to diseases. This method surpasses traditional selective breeding, which focuses solely on phenotypic traits. It uses molecular benchmarks to ensure that the organism will last a long time and be able to adapt (Panziera et al., 2022).

As genomic datasets grow, they will also present policymakers measurable numbers for genetic diversity and introgression rates. These kinds of metrics are important for making rules about queen trade, migratory beekeeping, and setting aside areas for conservation. One way to use genomic data to make policy changes is to create protected genetic reserves where native honeybee lineages are kept in their purest form. Such reserves would help reduce the negative effects of human gene flow (Pearman et al., 2024).

34.7. Synthesis and Future Outlook

The combined effects of these research paths are likely to lead to a strong set of tools for dealing with the difficult problems that honeybee populations are facing. Functional mitogenomics will directly link genetic variations in mtDNA to important traits like metabolism, temperature tolerance, and resistance to diseases. Such analysis not only helps us learn more about how evolution works, but it also presents us with useful information for precision breeding and conservation programs (Eimanifar, Kimball, et al., 2018; Tihelka et al., 2020).

Simultaneously, emerging evidence of paternal mitochondrial transmission has the potential to challenge established notions about the genetic inheritance of honeybees. Putting these results into conservation genetic frameworks will help us learn more about genetic admixture, especially in hybrid zones where managed and wild populations mix. To measure how paternal mtDNA affects overall genetic diversity, we need more advanced sequencing methods and careful population-level analyses (Kelemen & Rehan, 2021).

Studies of ancient DNA act as a time bridge, putting modern genetic research in a historical context that shows how honeybee populations have reacted to natural and human-made stressors over time. These studies indicate that significant changes in genetics have happened because of things like industrial beekeeping, habitat fragmentation, and the introduction of non-native lineages. Conservationists can create better restoration plans that aim to restore lost genetic diversity and improve the ability of existing populations to adapt when they obtain historical baselines (Espregueira Themudo et al., 2020; Parejo et al., 2020).

Putting together nuclear and mitochondrial genomic data isn't just an academic exercise; it's a useful way to clear up taxonomic confusion that can get in the way of conservation efforts. In areas where people of different races mix a lot, a combined genomic approach lets scientists determine different subspecies and evolutionarily important units. This technique has immediate uses in policy and management because decisions about bringing in queens, moving colonies, and protecting habitats must be based on a correct understanding of genetic structure (Ostroverkhova & Konusova, 2022; Wragg et al., 2017).

Collaboration between different fields will shape the future of honeybee conservation genetics. Combining genomics with ecological modeling and climate projections will not only help us predict how populations will react to changes in the environment, but it will also help us determine areas where genetic resilience is strongest. For example, using models like Gradient Forests and Generalized Dissimilarity Modeling (GDM) can help us make predictions about genetic turnover in space and point out possible refugia that could hold genetic diversity in the future (Pearman et al., 2024).

Technological advances will keep this field moving forward. High-throughput sequencing platforms, coupled with improved bioinformatics pipelines and non-invasive sampling methods, will enable real-time monitoring of honeybee genetic diversity. In a time of rapid environmental change, quick monitoring is essential because it allows for quick adaptive management responses. These new technologies will facilitate precision breeding by utilizing both nuclear and mitochondrial markers to maintain a balanced genetic portfolio capable of withstanding both biotic and abiotic stressors (D. Henriques, Parejo, et al., 2018; Parejo et al., 2020).

In the future, combining these different types of research will lead to a flexible and changing conservation plan. Breeding programs that utilize functional mitogenomics data can identify and disseminate mtDNA haplotypes associated with beneficial traits, such as Varroa resistance and improved thermal stability. Simultaneously, these programs can detect any unwanted paternal mtDNA leakage that may compromise genetic purity. These kinds of programs promise to improve the ecological and economic health of honeybee populations, making sure that they stay important pollinators in both natural and agricultural ecosystems (Panziera et al., 2022; Pearman et al., 2024).

It will be crucial to have policy frameworks based on integrated genomic data to manage honeybee populations in a way that lasts. Regulators will be able to set strict rules for importing queens and breeding them in captivity if they can acquire measurable information about genetic diversity from combining nuclear and mitochondrial datasets. Such an approach will help protect native genotypes from invasive gene flow from commercial stocks. These kinds of policies are important for lowering the risk of genetic homogenization, which can make it harder for local populations to adapt (Espregueira Themudo et al., 2020; Pearman et al., 2024). To turn advanced genomic knowledge into real-world conservation actions, researchers from different fields need to keep working together. Geneticists, ecologists, apiculturists, and policymakers need to work together to create monitoring programs that combine high-resolution genomic data with information about phenotypes, climate, and ecosystems. We can only hope to make strong, predictive models of how honeybees adapt that can help with future restoration and management efforts by encouraging these kinds of partnerships (Pearman et al., 2024).

34.8. Practical Implications for Precision Breeding and Adaptive Management

In practice, the four research areas covered in this synthesis come together to make a powerful set of tools for adaptive management and precision breeding. Functional mitogenomics makes it possible to identify mitochondrial haplotypes that are closely linked to beneficial physiological traits, like better energy metabolism or more resistance to common pathogens. Beekeepers can use these genetic markers to choose colonies that perform better and have a strong genetic background that will help them adapt over time. At the same time, figuring out how paternal mtDNA is passed on will improve these breeding plans by making sure that the spread of beneficial traits doesn't accidentally bring in harmful genetic elements (Eimanifar, Kimball, et al., 2018; Kenzy Ivveth Peña Carrillo, 2021).

Furthermore, using ancient DNA analyses presents genetic restoration efforts with an evolutionary context that they can use as a guide. By looking at the genetic profiles of ancient and modern species, conservationists can find out where genetic diversity has been lost and take specific steps to bring back alleles that provide species the ability to adapt. These kinds of restoration efforts are especially important because the environment is changing quickly and new threats are popping up, like new pathogens and extreme weather (Espregueira Themudo et al., 2020; Pearman et al., 2024).

An integrated genomic framework that combines data from both nuclear and mitochondrial sources is the first step toward a complete understanding of population structure and gene flow. This, in turn, helps to plan conservation strategies that are specific to the area. For instance, high-resolution spatial genetic analyses could identify areas as genetic refugia that should be prioritized for protection or the creation of protected breeding areas. Genomic monitoring could then be used to update regulatory guidelines on the fly, making sure that management strategies stay flexible in the face of environmental and human-made pressures (Pearman et al., 2024; Wragg et al., 2017).

34.9. Toward an Interdisciplinary

Future Based on Data Collaborating across fields is crucial for the future of honeybee conservation genetics. Bringing together functional mitogenomics, high-throughput nuclear sequencing, advanced ecological modeling, and ancient DNA techniques will create a detailed set of data. This complete dataset will help researchers not only guess how evolution will respond to changing environments, but it will also help them make targeted conservation policies. In this new way of thinking, everyone involved, from academic researchers to government agencies, will benefit from having a single, high-resolution view of the genetic health and adaptability of honeybees (Panziera et al., 2022; Pearman et al., 2024).

Interdisciplinary methods are already showing their value in related fields. For example, whole-genome sequencing studies have found small genetic markers that show where populations end and predict how they will respond to changes in the environment. These kinds of advances show us exactly how to use integrated genomic analysis on honeybees. Such findings will help us come up with conservation plans that are both scientifically sound and doable (D. Henriques, Parejo, et al., 2018; Parejo et al., 2020).

As the field grows, we'll be able to better predict genetic turnover in different climate scenarios thanks to new analytical tools like machine learning algorithms and better general circulation models. Using tools like Generalized Dissimilarity Modeling (GDM) and Gradient Forest analyses will help us find the main environmental factors that affect genetic diversity. The results will help us come up with adaptive management strategies that are accurate in time and space (Pearman et al., 2024).

34.10. Conclusion

In summary, the future of honeybee conservation genetics will combine different methods, including functional mitogenomics, studies of paternal mitochondria, ancient DNA research, and both nuclear and mitochondrial genomic data. All of these research paths together hold the promise of closing the gap between static genetic descriptions and dynamic, predictive models of adaptation. The framework that comes out of this work will not only allow for precise breeding programs that make sure honeybee populations stay healthy, but it will also give policymakers the information they need to put in place strong conservation and management plans in the face of environmental changes that have never happened before (Eimanifar, Kimball, et al., 2018; Pearman et al., 2024).

Modern genomic techniques promise to improve our understanding of honeybee biology, lead to new developments in marker-assisted selection, and ultimately protect the genetic integrity of populations against both natural and human-made threats. Combining modern genomic methods with traditional conservation methods may be the best way to make beekeeping strong and long-lasting in the twenty-first century (Espregueira Themudo et al., 2020; Panziera et al., 2022).

Combining these new research areas represents not only a significant advancement in honeybee conservation genetics but also a significant shift in our understanding of biodiversity protection in the modern world. The future of *A. mellifera* conservation looks bright thanks to more interdisciplinary research, better data integration, and a dedication to adaptive management. These efforts will lead to healthier colonies, more adaptive potential, and a scientifically sound policy framework that will protect these important pollinators for generations to come (Pearman et al., 2024; Wragg et al., 2017).

Combining new genomic technologies with old ecological knowledge provides us a plan for a conservation strategy that can adapt and stay strong. The field can ensure that new genomic research leads to actions that protect honeybees, a key species for both global biodiversity and food security, by encouraging geneticists, ecologists, beekeepers, and policymakers to work together (Ostroverkhova & Konusova, 2022; Pearman et al., 2024).

In conclusion, this synthesis shows a convergent path that will lead to a new era in the genetics of honeybee conservation. The next step in protecting honeybee diversity is to use functional mitogenomics, look into the complicated ways that paternal mitochondrial DNA is passed down, bring back genetic variation from the past using ancient DNA, and combine nuclear and mitochondrial data in a way that works. As these strategies improve and are included in management plans and policies based on evidence, they will create a complete conservation method that helps restore and keep genetic diversity and promotes sustainable beekeeping practices during a time of rapid global changes (D. Henriques, Parejo, et al., 2018; Pearman et al., 2024). To help honeybees survive in a world of environmental uncertainty and human-made pressures, we need to use a multi-layered approach that includes advanced molecular techniques, data integration from different fields, and making decisions based on facts (Parejo et al., 2020; Pearman et al., 2024).

35. References

Abdulla, B. S., Rashed, R. J., Hamasalih, R. M., Ahmed, S. H., & Majid, A. O.2023). MOLECULAR DIVERSITY AND PHYLOGENETIC ANALYSIS OF SOME IRAQI HONEY BEE POPULATION INFERRED FROM MITOCHONDRIAL DNA GENE SEGMENTSCOI AND 16S rDNA). *Applied Ecology and Environmental Research*, *21* (1), 229–242. https://doi.org/10.15666/aeer/2101_229242

Abicht, A., Scharf, F., Kleinle, S., Schön, U., Holinski-Feder, E., Horvath, R., Benet-Pagès, A., & Diebold, I.2018). Mitochondrial and nuclear disease panelMito-aND-Panel): Combined sequencing of mitochondrial and nuclear DNA by a cost-effective and sensitive NGS-based method. *Molecular Genetics \& Genomic Medicine*, *6* (6), 1188–1198. https://doi.org/https://doi.org/10.1002/mgg3.500

Abou-Shaara, H. F.2019). Utilizing bioinformatics to detect genetic similarities between African honey bee subspecies. *Journal of Genetics*, *98* (4), 96. https://doi.org/10.1007/s12041-019-1145-7

Abou-Shaara, H. F., Al-Ghamdi, A. A., Khan, K. A., & Al-Kahtani, S. N.2021). Genetic network analysis between *Apis mellifera* subspecies based on mtDNA argues the purity of specimens from North Africa, the Levant and Saudi Arabia. *Saudi Journal of Biological Sciences*, *28* (5), 2718–2725. https://doi.org/https://doi.org/10.1016/j.sjbs.2021.03.032

Abou-Shaara, H. F., Syed Abbas, A., AL-Kahtani, S. N., Taha, E.-K. A., Ali Khan, K., Jamal, Z. A., Alhumaidi Alotaibi, M., Ahmad, B., Ahmad Khan, N., Qamer, S., Ishtiaq Anjum, S., Khan, S., Hossam Mahmoud, A., Mohammed, O. B., & Gamal El Den Nasser, M.2021). Exploring the non-coding regions in the mtDNA of some honey bee species and subspecies. *Saudi Journal of Biological Sciences*, *28* (1), 204–209. https://doi.org/https://doi.org/10.1016/j.sjbs.2020.09.047

Adams, C. I. M., Knapp, M., Gemmell, N. J., Jeunen, G. J., Bunce, M., Lamare, M. D., & Taylor, H. R.2019). Beyond biodiversity: Can environmental DNAeDNA) cut it as a population genetics tool? *Genes*, *10* (3), 1–19. https://doi.org/10.3390/genes10030192

Adema, C. M.2021). Sticky problems: extraction of nucleic acids from molluscs. *Philosophical Transactions of the Royal Society B: Biological Sciences*, *376* (1825), 20200162. https://doi.org/10.1098/rstb.2020.0162

Adolph, R.2016). *DETERMINING LOW LEVELS OF AFRICANIZATION IN UNMANAGED HONEY BEE COLONIES USING THREE DIAGNOSTIC TECHNIQUES*. 1–23.

Ajao, A. M., Nneji, L. M., Adeola, A. C., Oladipo, S. O., Ayoola, A. O., Wang, Y.-Y., Adeniyi, A. V, & Olademeji, Y. U.2021). Genetic diversity and population structure of the native Western African honeybeeApis mellifera adansonii Latreille, 1804) in Nigeria based on mitochondrial COI sequences. *Zoologischer Anzeiger*, *293*, 17–25. https://doi.org/https://doi.org/10.1016/j.jcz.2021.05.007

Akyildiz, İ., Yetimoğlu, E., Raday, S., Erdem, Ö., Acar, S., Yilmaz, Ö., Uzunöner, D., Düz, G., & Damarli, E.2022). Development of a novel pretreatment protocol for the efficient isolation and enrichment of honey proteome using pine honey and the hypopharyngeal glands of Apis mellifera L. *Journal of Food Measurement and Characterization*, *16*. https://doi.org/10.1007/s11694-022-01380-8

Al-Dhafer, H. M., Balaji, R., Abdel-Dayem, M. S., Rasool, I., Mohamed, A., & Palanisamy, S.2025). Simple DNA extraction for museum beetle specimens to unlock genetic data from historical collections. *MethodsX*, *14*, 103236. https://doi.org/https://doi.org/10.1016/j.mex.2025.103236

Al-Ghamdi, A. A., Nuru, A., Khanbash, M. S., & Smith, D. R.2013). Geographical distribution and population variation of Apis mellifera jemenitica Ruttner. *Journal of Apicultural Research*, *52* (3), 124–133. https://doi.org/10.3896/IBRA.1.52.3.03

Alattal, Y., & Algamdi, A.2022). Assessment of genetic variation in Apis mellifera jemeniticaHymenoptera: Apidae) based on mitochondrial Cytochrome Oxidase Subunit II and III. *PLoS ONE*, *17* (3 March), 1–10. https://doi.org/10.1371/journal.pone.0265454

Alattal, Y., Alsharhi, D. M., Al-Ghamdi, A., ALFAIFY, S., Migdadi, H., & Ansari, M.2014). Characterization of the native honey bee subspecies in Saudi

Arabia using the mtDNA COI–COII intergenic region and morphometric characteristics. *Bulletin of Insectology, 67.*

Alburaki, M., Bertrand, B., & Legout, H.2013). A fifth major genetic group among honeybees revealed in Syria. *BMC Genetics, 14*, 117.

Alburaki, M., Madella, S., Lopez, J., Bouga, M., Chen, Y., & vanEngelsdorp, D.2023). Honey bee populations of the USA display restrictions in their mtDNA haplotype diversity. *Frontiers in Genetics, Volume 13.* https://doi.org/10.3389/fgene.2022.1092121

Alburaki, M., Moulin, S., Legout, H., Alburaki, A., & Garnery, L.2011). Mitochondrial structure of Eastern honeybee populations from Syria, Lebanon and Iraq. *Apidologie, 42* (628).

Ali, N., Rampazzo, R. de C. P., Costa, A. D. T., & Krieger, M. A.2017). Current Nucleic Acid Extraction Methods and Their Implications to Point-of-Care Diagnostics. *BioMed Research International, 2017* (1), 9306564. https://doi.org/https://doi.org/10.1155/2017/9306564

Alqahtani, F.2020). Algorithms for Mitochondrial Genome Assembly and Haplogroup Assignment from Low-Coverage Whole-Genome Sequencing Data. *Doctoral Dissertations.*
https://opencommons.uconn.edu/dissertations/2416

Alsharhi, M., Al-Ghamdi, A., Al-Garadi, M. A., & Alburaki, M.2025). Genetic diversity and novel haplotypes of Apis mellifera jemenitica on the Arabian Peninsula: insights from mtDNA markers. *Frontiers in Genetics, Volume 16.* https://doi.org/10.3389/fgene.2025.1532988

Andermann, T., Torres Jiménez, M. F., Matos-Maraví, P., Batista, R., Blanco-Pastor, J. L., Gustafsson, A. L. S., Kistler, L., Liberal, I. M., Oxelman, B., Bacon, C. D., & Antonelli, A.2020). A Guide to Carrying Out a Phylogenomic Target Sequence Capture Project. *Frontiers in Genetics, Volume 10.* https://doi.org/10.3389/fgene.2019.01407

Andrews, K. R., De Barba, M., Russello, M. A., & Waits, L. P.2021). Advances in Using Non-invasive, Archival, and Environmental Samples for Population Genomic Studies. In P. A. Hohenlohe & O. P. RajoraEds.),

*Population Genomics: Wildlife*pp. 63–99). Springer International Publishing. https://doi.org/10.1007/13836_2018_45

Angelique, M., & Diversité, T.2017). *mellifera dans les îles du sud-ouest de l ' océan Indien To cite this version : Diversité génétique et phylogéographie de l ' abeille Apis mellifera dans les îles du Sud-Ouest de l ' océan Indien.*

Arias, M. C., Rinderer, T. E., & Sheppard, W. S.2006). Further characterization of honey bees from the Iberian Peninsula by allozyme, morphometric and mtDNA haplotype analyses. *J. Apic. Res*, *45*, 188–196.

Arias, M. C., & Sheppard, W. S.1996). Molecular Phylogenetics of Honey Bee SubspeciesApis melliferaL.) Inferred from Mitochondrial DNA Sequence. *Molecular Phylogenetics and Evolution*, *5* (3), 557–566. https://doi.org/https://doi.org/10.1006/mpev.1996.0050

Arias, M. C., & Sheppard, W. S.2006). Phylogenetic relationships of honey beesHymenoptera: Apinae: Apini) inferred from nuclear and mitochondrial DNA sequence data. *Mol. Phylogenet. Evol*, *37*, 25–35.

Ariede, R. B., Freitas, M. V, Hata, M. E., Matrochirico-Filho, V. A., Utsunomia, R., Mendonça, F. F., Foresti, F., Porto-Foresti, F., & Hashimoto, D. T.2018). Developing population genetics microsatellite markers from metagenomic shotgun next generation sequencing data X Barton. Developing population genetics microsatellite markers from metagenomic shotgun next generation sequencing data. Unknown journal, 2021. *Molecular Biology Reports*, *45* (1), 9–18. https://doi.org/10.1007/s11033-017-4134-z

Atoui, A., & El Khoury, A.2017). PCR-RFLP for Aspergillus Species. In A. Moretti & A. SuscaEds.), *Mycotoxigenic Fungi: Methods and Protocols*pp. 313–320). Springer New York. https://doi.org/10.1007/978-1-4939-6707-0_20

Attardo, G. M., Abd-Alla, A. M. M., Acosta-Serrano, A., Allen, J. E., Bateta, R., Benoit, J. B., Bourtzis, K., Caers, J., Caljon, G., Christensen, M. B., Farrow, D. W., Friedrich, M., Hua-Van, A., Jennings, E. C., Larkin, D. M., Lawson, D., Lehane, M. J., Lenis, V. P., Lowy-Gallego, E., … Aksoy, S.2019). Comparative genomic analysis of six Glossina genomes, vectors of African trypanosomes. *Genome Biology*, *20* (1), 187. https://doi.org/10.1186/s13059-019-1768-2

Avila, E., Graebin, P., Chemale, G., Freitas, J., Kahmann, A., & Alho, C. S.2019). Full mtDNA genome sequencing of Brazilian admixed populations: A forensic-focused evaluation of a MPS application as an alternative to Sanger sequencing methods. *Forensic Science International: Genetics*, *42*, 154–164. https://doi.org/10.1016/j.fsigen.2019.07.004

Baele, G., & Lemey, P.2013). Bayesian evolutionary model testing in the phylogenomics era: matching model complexity with computational efficiency. *Bioinformatics*, *29* (16), 1970–1979. https://doi.org/10.1093/bioinformatics/btt340

Balkanska, R., Stefanova, K., Stoikova-Grigorova, R., & Manolova, V.2018). Preliminary Study of Dna Extraction From Bulgarian Honeys and Its Amplification By Pcr for Botanical Identification. *Food and Health*, *4* (3), 194–201. https://doi.org/10.3153/fh18020

Barbosa, S., Hendricks, S. A., Funk, W. C., Rajora, O. P., & Hohenlohe, P. A.2021). Wildlife Population Genomics: Applications and Approaches. In P. A. Hohenlohe & O. P. RajoraEds.), *Population Genomics: Wildlife*pp. 3–59). Springer International Publishing. https://doi.org/10.1007/13836_2020_83

Barrow, L. N., Bigelow, A. T., Phillips, C. A., & Lemmon, E. M.2015). Phylogeographic inference using Bayesian model comparison across a fragmented chorus frog species complex. *Molecular Ecology*, *24* (18), 4739–4758. https://doi.org/https://doi.org/10.1111/mec.13343

Battle, S. L., Puiu, D., mtDNA Working Group, Topm., Verlouw, J., Broer, L., Boerwinkle, E., Taylor, K. D., Rotter, J. I., Rich, S. S., Grove, M. L., Pankratz, N., Fetterman, J. L., Liu, C., & Arking, D. E.2022). A bioinformatics pipeline for estimating mitochondrial DNA copy number and heteroplasmy levels from whole genome sequencing data. *NAR Genomics and Bioinformatics*, *4* (2), lqac034. https://doi.org/10.1093/nargab/lqac034

Bayerri, C. R.2019). *Does antiretroviral therapy accelerate the speed of physiological ageing through the clonal expansion of mitochondrial DNA mutations? September*.

Beall, S. G., Cantera, J., Diaz, M. H., Winchell, J. M., Lillis, L., White, H., Kalnoky, M., Gallarda, J., & Boyle, D. S.2019). Performance and workflow assessment of six nucleic acid extraction technologies for use in resource limited settings. *PLOS ONE*, *14* (4), e0215753. https://doi.org/10.1371/journal.pone.0215753

Behura, S. K.2007). Analysis of Nuclear Copies of Mitochondrial Sequences in HoneybeeApis mellifera) Genome. *Molecular Biology and Evolution*, *24* (7), 1492–1505. https://doi.org/10.1093/molbev/msm068

Benčič, A., Toplak, N., Koren, S., Košir, A. B., Milavec, M., Tomič, V., Lužnik, D., & Dreo, T.2024). Metrological evaluation of DNA extraction method effects on the bacterial microbiome and resistome in sputum. *MSystems*, *9* (9), e00735-24. https://doi.org/10.1128/msystems.00735-24

Bertrand, B.2013). *Analyse de la diversité moléculaire de populations d'abeilles de la lignée ouest-méditerranéenneApis mellifera mellifera) dans le but de la conservation*. https://theses.hal.science/tel-00951288

Bertrand, B., Alburaki, M., Legout, H., Moulin, S., Mougel, F., & Garnery, L.2015). MtDNA COI-COII marker and drone congregation area: An efficient method to establish and monitor honeybeeApis mellifera L.) conservation centres. *Molecular Ecology Resources*, *15* (3), 673–683. https://doi.org/https://doi.org/10.1111/1755-0998.12339

Blair, C., & Ané, C.2019). Phylogenetic Trees and Networks Can Serve as Powerful and Complementary Approaches for Analysis of Genomic Data. *Systematic Biology*, *69* (3), 593–601. https://doi.org/10.1093/sysbio/syz056

Boardman, L.2019). The complete mitochondrial genome of Apis mellifera unicolorInsecta: Hymenoptera: Apidae), the Malagasy honey bee. *Mitochondrial DNA Part B*, *4*, 3286–3287.

Bogožalec Košir, A., Lužnik, D., Tomič, V., & Milavec, M.2023). Evaluation of DNA Extraction Methods for Reliable Quantification of Acinetobacter baumannii, Klebsiella pneumoniae, and Pseudomonas aeruginosa. In *Biosensors*Vol. 13, Issue 4). https://doi.org/10.3390/bios13040463

Borland, E. M., & Kading, R. C.2021). Modernizing the Toolkit for Arthropod Bloodmeal Identification. *Insects*, *12* (1). https://doi.org/10.3390/insects12010037

Bossert, S., Murray, E. A., Pauly, A., Chernyshov, K., Brady, S. G., & Danforth, B. N.2020). Gene Tree Estimation Error with Ultraconserved Elements: An Empirical Study on Pseudapis Bees. *Systematic Biology*, *70* (4), 803–821. https://doi.org/10.1093/sysbio/syaa097

Bouga, M.2011). A review of methods for discrimination of honey bee populations as applied to European beekeeping. *J. Apic. Res*, *50*, 51–84.

Bovo, S., Ribani, A., Utzeri, V. J., Schiavo, G., Bertolini, F., & Fontanesi, L.2018). Shotgun metagenomics of honey DNA: Evaluation of a methodological approach to describe a multi-kingdom honey bee derived environmental DNA signature. *PLOS ONE*, *13* (10), 1–19. https://doi.org/10.1371/journal.pone.0205575

Bovo, S., Utzeri, V. J., Ribani, A., Cabbri, R., & Fontanesi, L.2020). Shotgun sequencing of honey DNA can describe honey bee derived environmental signatures and the honey bee hologenome complexity. *Scientific Reports*, *10* (1), 9279. https://doi.org/10.1038/s41598-020-66127-1

Bovo, S., Utzeri, V. J., Ribani, A., Taurisano, V., Schiavo, G., & Fontanesi, L.2022). A genotyping by sequencing approach can disclose Apis mellifera population genomic information contained in honey environmental DNA. *Scientific Reports*, *12* (1), 19541. https://doi.org/10.1038/s41598-022-24101-z

Breton, S., & Stewart, D. T.2015). Atypical mitochondrial inheritance patterns in eukaryotes. *Genome*, *58* (10), 423–431. https://doi.org/10.1139/gen-2015-0090

Bris, C., Goudenege, D., Desquiret-Dumas, V., Charif, M., Colin, E., Bonneau, D., Amati-Bonneau, P., Lenaers, G., Reynier, P., & Procaccio, V.2018). Bioinformatics Tools and Databases to Assess the Pathogenicity of Mitochondrial DNA Variants in the Field of Next Generation Sequencing. *Frontiers in Genetics*, *Volume 9*-. https://doi.org/10.3389/fgene.2018.00632

Bubnič, J., Mole, K., Prešern, J., & Moškrič, A.2020). Non-Destructive Genotyping of Honeybee Queens to Support Selection and Breeding. In *Insects*Vol. 11, Issue 12). https://doi.org/10.3390/insects11120896

Buswell, V. G.2022). *A Phenotypic and Genomic Investigation of the Identity and Variation in the European Dark Bee Apis mellifera mellifera) A Phenotypic and Genomic Investigation of the Identity and Variation in the European Dark Bee Apis mellifera mellifera) Awarding .*

Buswell, V. G., Ellis, J. S., Huml, J. V., Wragg, D., Barnett, M. W., Brown, A., Group, T. S. B. A. C. S., & Knight, M. E.2023). When One's Not Enough: Colony Pool-Seq Outperforms Individual-Based Methods for Assessing Introgression in Apis mellifera mellifera. *Insects*, *14* (5). https://doi.org/10.3390/insects14050421

Calfee, E., Agra, M. N., Palacio, M. A., Ramírez, S. R., & Coop, G.2020). Selection and hybridization shaped the rapid spread of African honey bee ancestry in the americas. *PLoS Genetics*, *16* (10), 1–59. https://doi.org/10.1371/journal.pgen.1009038

Card, D. C., Shapiro, B., Giribet, G., Moritz, C., & Edwards, S. V.2021). Museum Genomics. *Annual Review of Genetics*, *55* (Volume 55, 2021), 633–659. https://doi.org/https://doi.org/10.1146/annurev-genet-071719-020506

Carina Ferreira Matias, H.2021). *Universidade do Minho Escola de Psicologia*. *5*, 1–30.

Carr, S. M.2023). Multiple mitogenomes indicate Things Fall Apart with Out of Africa or Asia hypotheses for the phylogeographic evolution of Honey BeesApis mellifera). *Scientific Reports*, *13* (1), 9386. https://doi.org/10.1038/s41598-023-35937-4

Carroll, C. J., Brilhante, V., & Suomalainen, A.2014). Next-generation sequencing for mitochondrial disorders. *British Journal of Pharmacology*, *171* (8), 1837–1853. https://doi.org/https://doi.org/10.1111/bph.12469

Chahbar, N., Muñoz, I., Dall'Olio, R., De la Rúa, P., Serrano, J., & Doumandji, S.2013). Population structure of North African honey bees is influenced by both biological and anthropogenic factors. *Journal of Insect Conservation*, *17* (2), 385–392. https://doi.org/10.1007/s10841-012-9520-1

Chapman, N. C., Harpur, B. A., Lim, J., Rinderer, T. E., Allsopp, M. H., Zayed, A., & Oldroyd, B. P.2015). A SNP test to identify Africanized honeybees via proportion of "African" ancestry. *Molecular Ecology Resources*, *15* (6), 1346–1355. https://doi.org/10.1111/1755-0998.12411

Chávez-Galarza, J., Garnery, L., Henriques, D., Neves, C. J., Loucif-Ayad, W., Jonhston, J. S., & Pinto, M. A.2017). Mitochondrial DNA variation of Apis mellifera iberiensis: further insights from a large-scale study using sequence data of the tRNAleu-cox2 intergenic region. *Apidologie*, *48* (4), 533–544. https://doi.org/10.1007/s13592-017-0498-2

Chávez-Galarza, J., Henriques, D., Johnston, J. S., Carneiro, M., Rufino, J., Patton, J. C., & Pinto, M. A.2015). Revisiting the Iberian honey beeApis mellifera iberiensis) contact zone: Maternal and genome-wide nuclear variations provide support for secondary contact from historical refugia. *Mol. Ecol*, *24*, 2973–2992.

Chávez-Galarza, J., López-Montañez, R., Jiménez, A., Ferro-Mauricio, R., Oré, J., Medina, S., Rea, R., & Vásquez, H.2021). Mitochondrial DNA Variation in Peruvian Honey BeeApis mellifera L.) Populations Using the tRNAleu-cox2 Intergenic Region. In *Insects*Vol. 12, Issue 7). https://doi.org/10.3390/insects12070641

Chen, C.2015). Genomic analyses reveal demographic history and temperate adaptation of the newly discovered honey bee subspecies Apis mellifera sinisxinyuan n. ssp. *Mol. Biol. Evol*, *33*, 1337–1348.

Chen, C., Parejo, M., Momeni, J., Langa, J., Nielsen, R. O., Shi, W., Contributors, S. W. D., Vingborg, R., Kryger, P., Bouga, M., Estonba, A., & Meixner, M.2022). Population Structure and Diversity in European Honey BeesApismellifera L.)-An Empirical Comparison of Pool and Individual Whole-Genome Sequencing. *Genes*, *13* (2). https://doi.org/10.3390/genes13020182

Chen, F., Ma, H., Ma, C., Zhang, H., Zhao, M., Meng, Y., Wei, H., & Ma, L.2016). Sequencing and characterization of mitochondrial DNA genome for Brama japonicaPerciformes: Bramidae) with phylogenetic consideration. *Biochemical Systematics and Ecology*, *68*, 109–118. https://doi.org/https://doi.org/10.1016/j.bse.2016.06.012

Chen, R., Aldred, M. A., Xu, W., Zein, J., Bazeley, P., Comhair, S. A. A., Meyers, D. A., Bleecker, E. R., Liu, C., Erzurum, S. C., & Hu, B.2021).

Comparison of whole genome sequencing and targeted sequencing for mitochondrial DNA. *Mitochondrion*, *58*, 303–310. https://doi.org/https://doi.org/10.1016/j.mito.2021.01.006

Chibani Bahi Amar, A., Tabet Aoul, N., Fridi, R., Vignal, A., & Canale-Tabet, K.2024). New COI-COII mtDNA Region Haplotypes in the Endemic Honey Bees Apis mellifera intermissa and Apis mellifera sahariensisHymenoptera: Apidae) in Algeria. *Insects*, *15* (7). https://doi.org/10.3390/insects15070549

Chua, P. Y. S., Bourlat, S. J., Ferguson, C., Korlevic, P., Zhao, L., Ekrem, T., Meier, R., & Lawniczak, M. K. N.2023). Future of DNA-based insect monitoring. *Trends in Genetics*, *39* (7), 531–544. https://doi.org/10.1016/j.tig.2023.02.012

Cilia, G., Flaminio, S., & Quaranta, M.2022). A novel and non-invasive method for DNA extraction from dry bee specimens. *Scientific Reports*, *12* (1), 11679. https://doi.org/10.1038/s41598-022-15595-8

Cleary, D.2022). *Mitochondrial DNA Diversity, Parasite and Pathogen Occurrence, and a Potential Disease Vector in Managed and Unmanaged Honey Bee, Apis mellifera L …*. https://scholarworks.uark.edu/etd/4655

Condamine, F. L., NABHOLZ, B., CLAMENS, A.-L., DUPUIS, J. R., & SPERLING, F. A. H.2018). Mitochondrial phylogenomics, the origin of swallowtail butterflies, and the impact of the number of clocks in Bayesian molecular dating. *Systematic Entomology*, *43* (3), 460–480. https://doi.org/https://doi.org/10.1111/syen.12284

Cornuet, J. M., & Garnery, L.1991). Mitochondrial DNA variability in honeybees and its phylogeographic implications. *Apidologie*, *22* (6), 627–642. https://doi.org/10.1051/apido:19910606

Corrales, C., Poczai, P., Mulcahy, D., Mackenzie-Dodds, J., Haring, E., Krukenhauser, L., Martín, M. P., Sanmartín, I., Ruiz, Y., & Diéguez-Uribeondo, J.2023). Genomic Characterisation. *Biodiversity Biobanking*, 145.

Coulibaly, K. A. S., Majeed, M. Z., Chen, C., Yeo, K., Shi, W., & Ma, C.-S.2019). Insights into the Maternal Ancestry of Côte d'Ivoire Honeybees Using the Intergenic Region COI-COII. *Insects*, *10* (4). https://doi.org/10.3390/insects10040090

Cridland, J. M., Tsutsui, N. D., & Ramírez, S. R.2017). The complex demographic history and evolutionary origin of the Western honey bee, Apis mellifera. *Genome Biol. Evol*, *9*, 457–472.

Dahn, H. A., Mountcastle, J., Balacco, J., Winkler, S., Bista, I., Schmitt, A. D., Pettersson, O. V., Formenti, G., Oliver, K., Smith, M., Tan, W., Kraus, A., Mac, S., Komoroske, L. M., Lama, T., Crawford, A. J., Murphy, R. W., Brown, S., Scott, A. F., … Fedrigo, O.2022). Benchmarking ultra-high molecular weight DNA preservation methods for long-read and long-range sequencing. *GigaScience*, *11*. https://doi.org/10.1093/gigascience/giac068

Dalla-Costa, L. M., Morello, L. G., Conte, D., Pereira, L. A., Palmeiro, J. K., Ambrosio, A., Cardozo, D., Krieger, M. A., & Raboni, S. M.2017). Comparison of DNA extraction methods used to detect bacterial and yeast DNA from spiked whole blood by real-time PCR. *Journal of Microbiological Methods*, *140*, 61–66.
https://doi.org/https://doi.org/10.1016/j.mimet.2017.06.020

Dames, S., Chou, L.-S., Xiao, Y., Wayman, T., Stocks, J., Singleton, M., Eilbeck, K., & Mao, R.2013). The Development of Next-Generation Sequencing Assays for the Mitochondrial Genome and 108 Nuclear Genes Associated with Mitochondrial Disorders. *The Journal of Molecular Diagnostics*, *15* (4), 526–534. https://doi.org/10.1016/j.jmoldx.2013.03.005

Darby, B., Bryant, R., Keller, A., Jochim, M., Moe, J., Schreiner, Z., Pratt, C., Euliss Jr., N. H., Park, M., Simmons, R., & Otto, C.2020). Molecular sequencing and morphological identification reveal similar patterns in native bee communities across public and private grasslands of eastern North Dakota. *PLOS ONE*, *15* (1), 1–22. https://doi.org/10.1371/journal.pone.0227918

de la Cruz-Ramos, J. M., Hernández-Triana, L. M., García-De la Peña, C., González-Álvarez, V. H., Weger-Lucarelli, J., Siller-Rodríguez, Q. K., Sánchez Rámos, F. J., Rodríguez, A. D., & Ortega-Morales, A. I.2019). Comparison of two DNA extraction methods from larvae, pupae, and adults of <em>Aedes aegypti</em>. *Heliyon*, *5* (10).
https://doi.org/10.1016/j.heliyon.2019.e02660

De la Rúa, P., Jaffé, R., Dall'Olio, R., Muñoz, I., & Serrano, J.2009). Biodiversity, conservation and current threats to European honeybees. *Apidologie*, *40* (3), 263–284. https://doi.org/10.1051/apido/2009027

de Melo, K. P., & Camargo, M.2021). Mechanisms for sperm mitochondrial removal in embryos. *Biochimica et Biophysica ActaBBA) - Molecular Cell Research, 1868* (2), 118916. https://doi.org/https://doi.org/10.1016/j.bbamcr.2020.118916

Deiner, K., Bik, H. M., Mächler, E., Seymour, M., Lacoursière-Roussel, A., Altermatt, F., Creer, S., Bista, I., Lodge, D. M., & de Vere, N.2017). Environmental DNA metabarcoding: Transforming how we survey animal and plant communities. *Mol. Ecol*, 5872–5895. https://doi.org/doi: 10.1111/mec.14350

Delaney, D. A., Meixner, M. D., Schiff, N. M., & Sheppard, W. S.2009). Genetic Characterization of Commercial Honey BeeHymenoptera: Apidae) Populations in the United States by Using Mitochondrial and Microsatellite Markers. *Annals of the Entomological Society of America, 102* (4), 666–673. https://doi.org/10.1603/008.102.0411

Desalle, R., Schierwater, B., & Hadrys, H.2017). *MtDNA : The small workhorse of evolutionary studies*. 873–887.

Didelot, A., Kotsopoulos, S., Lupo, A., Pekin, D., Li, X., Atochin, I., Srinivasan, P., Zhong, Q., Olson, J., Link, D., Laurent-Puig, P., Blons, H., Hutchison, J., & Valerie, T. A.2013). Multiplex Picoliter-Droplet Digital PCR for Quantitative Assessment of DNA Integrity in Clinical Samples. *Clinical Chemistry, 59*. https://doi.org/10.1373/clinchem.2012.193409

Dimbiarimanga, R. A., Herisolo, R., Claude, J., Jens, R. A., Henriette, R., Ravaomanarivo, R., & Harivelo, L.2024). *MORPHOLOGICAL DIVERSITY AND HYGIENIC BEHAVIOR OF APIS MELLIFERA UNICOLOR. 2*, 20–28.

Doctor, G.2009). *Thank God for all his blessings My family for continuous and unconditional support.*

Dogantzis, K. A., Tiwari, T., Conflitti, I. M., Dey, A., Patch, H. M., Muli, E. M., Garnery, L., Whitfield, C. W., Stolle, E., Alqarni, A. S., Allsopp, M. H., & Zayed, A.2021). Thrice out of Asia and the adaptive radiation of the western honey bee. *Science Advances, 7* (49), 1–10. https://doi.org/10.1126/sciadv.abj2151

Dogantzis, K. A., & Zayed, A.2019). Recent advances in population and quantitative genomics of honey bees. *Curr. Opinion Insect Sci*, *31*, 93–98.

Donato, L., Scimone, C., Alibrandi, S., Pitruzzella, A., Scalia, F., D'Angelo, R., & Sidoti, A.2020). Possible A2E Mutagenic Effects on RPE Mitochondrial DNA from Innovative RNA-Seq Bioinformatics Pipeline. *Antioxidants*, *9* (11). https://doi.org/10.3390/antiox9111158

Dong, Z., Wang, Y., Li, C., Li, L., & Men, X.2021). Mitochondrial DNA as a Molecular Marker in Insect Ecology: Current Status and Future Prospects. *Annals of the Entomological Society of America*, *114* (4), 470–476. https://doi.org/10.1093/aesa/saab020

Donthu, R., Marcelino, J. A. P., Giordano, R., Tao, Y., Weber, E., Avalos, A., Band, M., Akraiko, T., Chen, S.-C., Reyes, M. P., Hao, H., Ortiz-Alvarado, Y., Cuff, C. A., Claudio, E. P., Soto-Adames, F., Smith-Pardo, A. H., Meikle, W. G., Evans, J. D., Giray, T., … Huang, Z.2024). HBeeID: a molecular tool that identifies honey bee subspecies from different geographic populations. *BMC Bioinformatics*, *25* (1), 278. https://doi.org/10.1186/s12859-024-05776-9

Düttmann, C., Flores, B., Sheleby-Elías, J., Castillo, G., Rodriguez, D., Maggi, M., & Demedio, J.2022). Africanized honeybee populationApis mellifera L.) in Nicaragua: Forewing length and mitotype lineages. *PloS One*, *17* (4), e0267600. https://doi.org/10.1371/journal.pone.0267600

Eimanifar, A., Brooks, S. A., Bustamante, T., & Ellis, J. D.2018). Population genomics and morphometric assignment of western honey beesApis mellifera L.) in the Republic of South Africa. *BMC Genomics*, *19* (1), 615. https://doi.org/10.1186/s12864-018-4998-x

Eimanifar, A., Kimball, R. T., Braun, E. L., & Ellis, J. D.2018). Mitochondrial genome diversity and population structure of two western honey bee subspecies in the Republic of South Africa. *Scientific Reports*, *8* (1), 1333. https://doi.org/10.1038/s41598-018-19759-3

Eimanifar, A., Pieplow, J. T., Asem, A., & Ellis, J. D.2020). Genetic diversity and population structure of two subspecies of western honey bees\textit{Apis mellifera} L.) in the Republic of South Africa as revealed by microsatellite genotyping. *PeerJ*, *8*, e8280. https://doi.org/10.7717/peerj.8280

Eimanifar, A., Rebecca, T. K., Edward, L. B., Dahi, M. M., Nizar, H., Stefan, F., Bernd, G., & and Ellis, J. D.2017). The complete mitochondrial genome of the Egyptian honey bee, Apis mellifera lamarckiiInsecta: Hymenoptera: Apidae). *Mitochondrial DNA Part B*, *2* (1), 270–272. https://doi.org/10.1080/23802359.2017.1325343

Eimanifar, A., T Kimball, R., L Braun, E., Fuchs, S., Grünewald, B., & Ellis, J. D.2017a). The complete mitochondrial genome of Apis mellifera medaInsecta: Hymenoptera: Apidae). *Mitochondrial DNA. Part B, Resources*, *2* (1), 268–269. https://doi.org/10.1080/23802359.2017.1325342

Eimanifar, A., T Kimball, R., L Braun, E., Fuchs, S., Grünewald, B., & Ellis, J. D.2017b). The complete mitochondrial genome of an east African honey bee, Apis mellifera monticola SmithInsecta: Hymenoptera: Apidae). *Mitochondrial DNA. Part B, Resources*, *2* (2), 589–590. https://doi.org/10.1080/23802359.2017.1372722

El-Seedi, H. R., El-Wahed, A. A. A., Zhao, C., Saeed, A., Zou, X., Guo, Z., Hegazi, A. G., Shehata, A. A., El-Seedi, H. H. R., Algethami, A. F., Al Naggar, Y., Agamy, N. F., Rateb, M. E., Ramadan, M. F. A., Khalifa, S. A. M., & Wang, K.2022). A Spotlight on the Egyptian HoneybeeApis mellifera lamarckii). *Animals : An Open Access Journal from MDPI*, *12* (20). https://doi.org/10.3390/ani12202749

Ellis, J. S., Soland-Reckeweg, G., Buswell, V. G., Huml, J. V., Brown, A., & Knight, M. E.2018). Introgression in native populations of Apis mellifera mellifera L: implications for conservation. *Journal of Insect Conservation*, *22* (3), 377–390. https://doi.org/10.1007/s10841-018-0067-7

Engel, M. S., Rasmussen, C., & Gonzalez, V. H.2021). Bees. In C. K. StarrEd.), *Encyclopedia of Social Insects*pp. 93–109). Springer International Publishing. https://doi.org/10.1007/978-3-030-28102-1_14

Erwin, D., Erickson, I., Delwiche, M., Colwell, F., Strap, J., & Crawford, R.2005). Diversity of Oxygenase Genes from Methane- and Ammonia-Oxidizing Bacteria in the Eastern Snake River Plain Aquifer. *Applied and Environmental Microbiology*, *71*, 2016–2025. https://doi.org/10.1128/AEM.71.4.2016-2025.2005

Espregueira Themudo, G., Rey-Iglesia, A., Robles Tascón, L., Bruun Jensen, A., da Fonseca, R. R., & Campos, P. F.2020). Declining genetic diversity of European honeybees along the twentieth century. *Scientific Reports*, *10* (1), 10520. https://doi.org/10.1038/s41598-020-67370-2

Evans, A.2022). *Development of a Sample Preparation Platform for Bacterial and Viral Pathogens by. March.*

Evans, J. D., Ryan S, S., Yan Ping, C., Giles, B., Robert S, C., Pilar, D. la R., Joachim R, de M., Sylvain, F., Leonard, F., Laurent, G., Elke, G., Sebastian, G., Antje, J., Robert, K., Dawn, L., Cheng Man, L., Robin F A, M., Ryszard, M., Irene, M., & and Pinto, M. A.2013). Standard methods for molecular research in Apis mellifera. *Journal of Apicultural Research*, *52* (4), 1–54. https://doi.org/10.3896/IBRA.1.52.4.11

Fan, P., Fjeldså, J., Liu, X., Dong, Y., Chang, Y., Qu, Y., Song, G., & Lei, F.2021). An approach for estimating haplotype diversity from sequences with unequal lengths. *Methods in Ecology and Evolution*, *12* (9), 1658–1667. https://doi.org/https://doi.org/10.1111/2041-210X.13643

Felczykowska, A., Krajewska, A., Zielińska, S., & Łos, J. M.2015). Sampling, metadata and DNA extraction - Important steps in metagenomic studies. *Acta Biochimica Polonica*, *62* (1), 151–160. https://doi.org/10.18388/abp.2014_916

Ferrari, R. R., Ricardo, P. C., Dias, F. C., Araújo, N. S., Soares, D. O., Zhou, Q.-S., Zhu, C.-D., Coutinho, L. L., Arias, M. C., & Batista, T. M.2024). *The complete genome of Tetragonisca angustulaApidae: Meliponini), the most commonly reared stingless bee in Brazil.* https://doi.org/10.21203/rs.3.rs-3833242/v1

Ferrari, R. R., Ricardo, P. C., Dias, F. C., de Souza Araujo, N., Soares, D. O., Zhou, Q.-S., Zhu, C.-D., Coutinho, L. L., Arias, M. C., & Batista, T. M.2024). The nuclear and mitochondrial genome assemblies of Tetragonisca angustulaApidae: Meliponini), a tiny yet remarkable pollinator in the Neotropics. *BMC Genomics*, *25* (1), 587. https://doi.org/10.1186/s12864-024-10502-z

Ferreira, H.2017). *Diversity patterns of honey beeApis mellifera L.) populations from the Archipelago of the Azores: insights from mtDNA and wing geometric morphometrics*. https://api.semanticscholar.org/CorpusID:91260644

Fisher, K. E., Bradbury, S. P., & Coates, B. S.2020). Prediction of mitochondrial genome-wide variation through sequencing of mitochondrion-enriched extracts. *Scientific Reports*, *10* (1), 19123. https://doi.org/10.1038/s41598-020-76088-0

Formenti, G., Rhie, A., Balacco, J., Haase, B., Mountcastle, J., Fedrigo, O., Brown, S., Capodiferro, M. R., Al-Ajli, F. O., Ambrosini, R., Houde, P., Koren, S., Oliver, K., Smith, M., Skelton, J., Betteridge, E., Dolucan, J., Corton, C., Bista, I., … Consortium, T. V. G. P.2021). Complete vertebrate mitogenomes reveal widespread repeats and gene duplications. *Genome Biology*, *22* (1), 120. https://doi.org/10.1186/s13059-021-02336-9

Forouzan, E., Maleki, M. S. M., Karkhane, A. A., & Yakhchali, B.2017). Evaluation of nine popular de novo assemblers in microbial genome assembly. *Journal of Microbiological Methods*, *143*, 32–37. https://doi.org/https://doi.org/10.1016/j.mimet.2017.09.008

Franck, P., Garnery, L., Loiseau, A., Oldroyd, B. P., Hepburn, H. R., Solignac, M., & Cornuet, J.-M.2001). Genetic diversity of the honeybee in Africa: microsatellite and mitochondrial data. *Heredity*, *86* (4), 420–430. https://doi.org/10.1046/j.1365-2540.2001.00842.x

Françoso, E., de Souza Araujo, N., Ricardo, P. C., Santos, P. K. F., Zuntini, A. R., & Arias, M. C.2020). Evolutionary perspectives on bee mtDNA from mito-OMICS analyses of a solitary species. *Apidologie*, *51* (4), 531–544. https://doi.org/10.1007/s13592-020-00740-x

Frazier, M., Muli, E., & Patch, H.2024). Ecology and Management of African Honey BeesApis mellifera L.). *Annual Review of Entomology*, *69*, 439–453. https://doi.org/10.1146/annurev-ento-020823-095359

Gajić, B., Stevanović, J., Radulović, Ž., Kulišić, Z., Vejnović, B., Glavinić, U., & Stanimirović, Z.2016). Haplotype identification and detection of mitochondrial DNA heteroplasmy in Varroa destructor mites using ARMS and

PCR–RFLP methods. *Experimental and Applied Acarology*, *70* (3), 287–297. https://doi.org/10.1007/s10493-016-0086-6

Galanis, A., Vardakas, P., Reczko, M., Harokopos, V., Hatzis, P., Skoulakis, E. M. C., Pavlopoulos, G. A., & Patalano, S.2022). Bee foraging preferences, microbiota and pathogens revealed by direct shotgun metagenomics of honey. *Molecular Ecology Resources*, *22* (7), 2506–2523. https://doi.org/https://doi.org/10.1111/1755-0998.13626

Galindo-Cardona, A., Acevedo-Gonzalez, J. P., Rivera-Marchand, B., & Giray, T.2013). Genetic structure of the gentle Africanized honey bee populationgAHB) in Puerto Rico. *BMC Genetics*, *14* (1), 65. https://doi.org/10.1186/1471-2156-14-65

Galuszynski, N. C., & Potts, A. J.2020a). Application of High Resolution Melt analysisHRM) for screening haplotype variation in a non-model plant genus: Cyclopiahoneybush). *PeerJ*, *2020* (3). https://doi.org/10.7717/peerj.9187

Galuszynski, N. C., & Potts, A. J.2020b). Application of High Resolution Melt analysisHRM) for screening haplotype variation in non-model plants: a case study of HoneybushCyclopia Vent.). *BioRxiv*. https://doi.org/10.1101/2020.02.05.921080

García-Giménez, J. L., Marta, S.-C., Trygve O., T., Carlos, R.-M., Lorena, P.-C., Pablo, L., & and Pallardó, F. V.2017). Epigenetic biomarkers: Current strategies and future challenges for their use in the clinical laboratory. *Critical Reviews in Clinical Laboratory Sciences*, *54* (7–8), 529–550. https://doi.org/10.1080/10408363.2017.1410520

García-Olivares, V., Muñoz-Barrera, A., Lorenzo-Salazar, J. M., Zaragoza-Trello, C., Rubio-Rodríguez, L. A., Díaz-de Usera, A., Jáspez, D., Iñigo-Campos, A., González-Montelongo, R., & Flores, C.2021). A benchmarking of human mitochondrial DNA haplogroup classifiers from whole-genome and whole-exome sequence data. *Scientific Reports*, *11* (1). https://doi.org/10.1038/s41598-021-99895-5

García-Sandoval, R.2014). Why some clades have low bootstrap frequencies and high Bayesian posterior probabilities. *Israel Journal of Ecology and Evolution*, *60* (1), 41–44. https://doi.org/10.1080/15659801.2014.937900

Garnery, L., Franck, P., Baudry, E., Vautrin, D., Cornuet, J.-M., & Solignac, M.1998). Genetic diversity of the west European honey beeApis mellifera mellifera and A. m. iberica) II. Microsatellite loci. *Genetics Selection Evolution*, *30* (1), S49. https://doi.org/10.1186/1297-9686-30-S1-S49

Garrick, R. C., Collins, B. D., Yi, R. N., Dyer, R. J., & Hyseni, C.2015). Identification of Eastern United States Reticulitermes Termite Species via PCR-RFLP, Assessed Using Training and Test Data. *Insects*, *6* (2), 524–537. https://doi.org/10.3390/insects6020524

Gautam, A.2022). DNA Isolation by Chelex Method. In *DNA and RNA Isolation Techniques for Non-Experts*pp. 79–84). Springer International Publishing. https://doi.org/10.1007/978-3-030-94230-4_10

Ghasemi, M., Mohseni, M., Fattahi, Z., Edizadeh, M., Beheshtian, M., Keshavarzi, F., Jalalvand, K., Omrani, M., Khanbazi, A., Riazalhosseini, Y., Akbari, M. R., Kahrizi, K., & Najmabadi, H.2025). Haplogroup Structure and Genetic Variation Analyses of Mitochondrial Genome SNPs in the Iranian Population. *Arch Iran Med*, *28* (3), 140–148. https://doi.org/10.34172/aim.33639

Gonçalves, L. T., Françoso, E., & Deprá, M.2023). Mitochondrial phylogenomics of bumblebees, BombusHymenoptera: Apidae): a tale of structural variation, shifts in selection constraints, and tree discordance. *Zoological Journal of the Linnean Society*, *202* (1), zlad178. https://doi.org/10.1093/zoolinnean/zlad178

Göze, İ., & Özdil, F.2023). Genetic Diversity of Kırklareli Honey BeeApis mellifera L.) Populations in Thrace Region of Turkey: Identification of Mitochondrial COI and ND5 Gene Regions. *Tekirdağ Ziraat Fakültesi Dergisi*. https://api.semanticscholar.org/CorpusID:266329946

Grass, A. A., Hosie, C., & McDowall, I. L.2016). The complete mitochondrial genome and phylogenetic position of the critically endangered Trinidad Piping Guan, Pipile pipile synonym Aburria pipileAves: Galliformes). *Mitochondrial DNA. Part B, Resources*, *1* (1), 649–650. https://doi.org/10.1080/23802359.2016.1219622

Greenslit, C.2018). *Population Genetics and Bumble Bee Conservation : Saving Species by Thinking Small.*

Grozinger, C. M., & Zayed, A.2020). Improving bee health through genomics. *Nat. Rev. Genet*, *21*, 277–291.

Gueuning, M.2019). *Assessing wild bee diversity using next generation sequencing*. Université de Neuchâtel, Faculté des sciences.

Gujarati, D., & Porter, D.2010). *DNA BARCODING AND MORPHOMETRICS OF STINGLESS BEESAPIDAE: MELIPONINAE) IN THREE SELECTED FORESTS IN KENYA*. *January*.

Haddad, N. J.2015). Draft genome sequence of the Algerian bee Apis mellifera intermissa. *Genom. Data*, *4*, 24–25.

Haddad, N. J., Meixner, M. D., Fuchs, S., Migdadi, H. M., Garnery, L., & Sheppard, W. S.2009). Mitochondrial DNA support for genetic reserves of Apis mellifera/syriaca in Jordan. *Journal of Apicultural Research*, *48*, 19–22. https://api.semanticscholar.org/CorpusID:85384790

Haddad, N., Mahmud Batainh, A., Suleiman Migdadi, O., Saini, D., Krishnamurthy, V., Parameswaran, S., & Alhamuri, Z.2016). Next generation sequencing of Apis mellifera syriaca identifies genes for Varroa resistance and beneficial bee keeping traits. *Insect Science*, *23* (4), 579–590. https://doi.org/https://doi.org/10.1111/1744-7917.12205

Hafeez, M. A., Shivaramaiah, S., Dorsey, K. M., Ogedengbe, M. E., El-Sherry, S., Whale, J., Cobean, J., & Barta, J. R.2015). Simultaneous identification and DNA barcoding of six Eimeria species infecting turkeys using PCR primers targeting the mitochondrial cytochrome c oxidase subunit ImtCOI) locus. *Parasitology Research*, *114* (5), 1761–1768. https://doi.org/10.1007/s00436-015-4361-y

Hailu, T. G., D'Alvise, P., & Hasselmann, M.2021). Disentangling Ethiopian Honey BeeApis mellifera) Populations Based on Standard Morphometric and Genetic Analyses. *Insects*, *12* (3). https://doi.org/10.3390/insects12030193

Hailu, T. G., D'Alvise, P., Tofilski, A., Fuchs, S., Greiling, J., Rosenkranz, P., & Hasselmann, M.2020). Insights into Ethiopian honey bee diversity based on wing geomorphometric and mitochondrial DNA analyses. *Apidologie*, *51*, 1182–1198.

Han, F., Wallberg, A., & Webster, M. T.2012). From where did the Western honeybeeApis mellifera) originate? *Ecology and Evolution, 2* (8), 1949–1957. https://doi.org/https://doi.org/10.1002/ece3.312

Harman, A.2021). Beeswax. In C. K. StarrEd.), *Encyclopedia of Social Insects*pp. 115–117). Springer International Publishing. https://doi.org/10.1007/978-3-030-28102-1_173

Harvey, N. R., Albury, C. L., Stuart, S., Benton, M. C., Eccles, D. A., Connell, J. R., Sutherland, H. G., Allcock, R. J. N., Lea, R. A., Haupt, L. M., & Griffiths, L. R.2019). Ion torrent high throughput mitochondrial genome sequencingHTMGS). *PLOS ONE, 14* (11), 1–13. https://doi.org/10.1371/journal.pone.0224847

Hassett, J., Browne, K. A., McCormack, G. P., Moore, E., Society, N. I. H. B., Soland, G., & Geary, M.2018). A significant pure population of the dark European honey beeApis mellifera mellifera) remains in Ireland. *Journal of Apicultural Research, 57* (3), 337–350. https://doi.org/10.1080/00218839.2018.1433949

Hassler, G. W., Magee, A. F., Zhang, Z., Baele, G., Lemey, P., Ji, X., Fourment, M., & Suchard, M. A.2023). Data Integration in Bayesian Phylogenetics. *Annual Review of Statistics and Its Application, 10* (Volume 10, 2023), 353–377. https://doi.org/https://doi.org/10.1146/annurev-statistics-033021-112532

Hayes, C. N., Nakahara, H., Ono, A., Tsuge, M., & Oka, S.2024). From Omics to Multi-Omics: A Review of Advantages and Tradeoffs. *Genes, 15* (12). https://doi.org/10.3390/genes15121551

Hazır, C., & Bock, C. H.2019). Success of DNA extraction and PCR amplification from dry pinned sand beesAndrena spp. Fabricius, 1775) using newly-designed primers TT - İğnelenerek kurutulmuş kum arılarındanAndrena spp. Fabricius, 1775) DNA eldesi ve yeni tasarlanmış primerler kull. *Turkish Journal of Entomology, 43* (1), 79–96. https://doi.org/10.16970/entoted.460948

He, B., Su, T., Wu, Y., Xu, J., & Huang, D.2018). Phylogenetic analysis of the mitochondrial genomes in beesHymenoptera: Apoidea: Anthophila). *PLOS ONE, 13* (8), 1–15. https://doi.org/10.1371/journal.pone.0202187

Henriques, D., Browne, K. A., Barnett, M. W., Parejo, M., Kryger, P., Freeman, T. C., Muñoz, I., Garnery, L., Highet, F., Jonhston, J. S., McCormack, G. P., & Pinto, M. A.2018). High sample throughput genotyping for estimating C-lineage introgression in the dark honeybee: An accurate and cost-effective SNP-based tool. *Sci. Rep, 8* (8552).

Henriques, D., Chávez-Galarza, J. C., Quaresma, A., Neves, C. J., Lopes, A. R., Costa, C., Costa, F. O., Rufino, J., & Pinto, M. A.2019). From the popular tRNAleu-COX2 intergenic region to the mitogenome: insights from diverse honey bee populations of Europe and North Africa. *Apidologie*, *50*, 215–229. https://api.semanticscholar.org/CorpusID:71148716

Henriques, D., Costa, C., Rufino, J., & Pinto, M. A.2022). The mitochondrial genome of Apis mellifera siciliana. *Mitochondrial DNA. Part B, Resources*, *7* (5), 828–830. https://doi.org/10.1080/23802359.2022.2073844

Henriques, D., Lopes, A. R., Chejanovsky, N., Dalmon, A., Higes, M., Jabal-Uriel, C., Le Conte, Y., Reyes-Carreño, M., Soroker, V., Martín-Hernández, R., & Pinto, M. A.2021). A SNP assay for assessing diversity in immune genes in the honey beeApis mellifera L.). *Scientific Reports*, *11* (1), 15317. https://doi.org/10.1038/s41598-021-94833-x

Henriques, D., Parejo, M., Lopes, A. R., & Pinto, M. A.2019). Mitochondrial SNP markers to monitor evolutionary lineage ancestry in Apis melliferamellifera conservation programs. *Apidologie*, *50* (4), 538–541. https://doi.org/10.1007/s13592-019-00666-z

Henriques, D., Parejo, M., Vignal, A., Wragg, D., Wallberg, A., Webster, M. T., & Pinto, M. A.2018). Developing reduced SNP assays from whole-genome sequence data to estimate introgression in an organism with complex genetic patterns, the Iberian honeybeeApis mellifera iberiensis). *Evol. Appl*, *11*, 1270–1282.

Henriques, D. S. M.2018). *Employment of whole genome resequencing to reveal the evolutionary history and to develop molecular tools for Western European honey beesApis mellifera subspecies). 52* (10), 1176–1184.

Hereward, J. P., Smith, T. J., Gloag, R., Brookes, D. R., & Walter, G. H.2025). Reassessing Hybridisation in Australian Tetragonula Stingless Bees Using

Multiple Genetic Markers. *Ecology and Evolution*, *15* (2), 1–31. https://doi.org/10.1002/ece3.70912

Holloway, B. A., Tarver, M. R., & Rinderer, T. E.2013). An economical and effective high-throughput DNA extraction protocol for molecular marker analysis in honeybees. *Entomologia Experimentalis et Applicata*, *148* (2), 196–200. https://doi.org/https://doi.org/10.1111/eea.12090

Hu, P., Lu, Z.-X., Haddad, N., Noureddine, A., Loucif-Ayad, W., Wang, Y.-Z., Zhao, R.-B., Zhang, A.-L., Guan, X., Zhang, H.-X., & Niu, H.2016). Complete mitochondrial genome of the Algerian honeybee, Apis mellifera intermissaHymenoptera: Apidae). *Mitochondrial DNA Part A*, *27* (3), 1791–1792. https://doi.org/10.3109/19401736.2014.963815

Huang, T.2011). Next Generation Sequencing to Characterize Mitochondrial Genomic DNA Heteroplasmy. *Current Protocols in Human Genetics*, *71* (1), 19.8.1-19.8.12. https://doi.org/https://doi.org/10.1002/0471142905.hg1908s71

Ilyasov, R. A., Lee, M. –., Takahashi, J. –., Kwon, H. W., & Nikolenko, A. G.2020). A revision of subspecies structure of western honey bee Apis mellifera. *Saudi J. Biol. Sci*, *27*, 3615–3621.

Ilyasov, R. A., Nikolenko, A., Tuktarov, V. R., Goto, K., Takahashi, J., & Kwon, H. W.2019). Comparative analysis of mitochondrial genomes of the honey bee subspecies A. m. caucasica and A. m. carpathica and refinement of their evolutionary lineages. *Journal of Apicultural Research*, *58*, 567–579. https://api.semanticscholar.org/CorpusID:196653745

Ilyasov, R. A., Poskryakov, A. V, & Nikolenko, A. G.2016). Seven genes of mitochondrial genome enabling differentiation of honeybee subspecies Apis mellifera. *Russian Journal of Genetics*, *52* (10), 1062–1070. https://doi.org/10.1134/S1022795416090064

Indahsari, A. N., & Purwanto, H.2024). Detection of Entomological Origin of Honey Sold in Indonesia Based on 16S rRNA Gene Analysis. *Journal of Tropical Biodiversity and Biotechnology*, *9* (3), 1–10. https://doi.org/10.22146/jtbb.86256

Irving-Pease, E. K., Ryan, H., Jamieson, A., Dimopoulos, E. A., Larson, G., & Frantz, L. A. F.2019). Paleogenomics of Animal Domestication. In C. Lindqvist & O. P. RajoraEds.), *Paleogenomics: Genome-Scale Analysis of Ancient DNA*pp. 225–272). Springer International Publishing. https://doi.org/10.1007/13836_2018_55

J¨unemann, S.2022). Quality is a Myth - Assessing and Addressing Errors in Sequencing Data Sebastian J ¨ unemann. *Faculty of Technology, Bielefeld University, Germany For.*

Jacques, F., Bolivar, P., Pietras, K., & Hammarlund, E. U.2023). Roadmap to the study of gene and protein phylogeny and evolution—A practical guide. *PLOS ONE, 18* (2), 1–35. https://doi.org/10.1371/journal.pone.0279597

Jain, R. K., Sharma, A., Lalwani, J., Chaurasia, D., & Perumal, N.2024). Impact of relative humidity on SARS-CoV-2 RNA extraction using Nextractor automated extraction system. *Journal of Biological Methods, 11* (2), e99010012. https://doi.org/10.14440/jbm.2024.0001

Jain, S. A., de Jesus, F. T., Marchioro, G. M., & de Araújo, E. D.2013). Extraction of DNA from honey and its amplification by PCR for botanical identification. *Food Science and Technology, 33* (4), 753–756. https://doi.org/10.1590/S0101-20612013000400022

Jangra, S., & Ghosh, A.2022). Rapid and zero-cost DNA extraction from soft-bodied insects for routine PCR-based applications. *PLOS ONE, 17* (7), e0271312. https://doi.org/10.1371/journal.pone.0271312

Jenke, N., & Gruenstaeudl, M.2022). *Depth and evenness of sequence coverage are associated with assembly quality , genome structure , and choice of sequencing platform in archived plastid genomes. 2009*, 1–17.

Jensen, A. B., Palmer, K. A., Boomsma, J. J., & Pedersen, B. V.2005). Varying degrees of Apis mellifera ligustica introgression in protected populations of the black honeybee, Apis mellifera mellifera, in northwest Europe. *Mol Ecol, 14*, 93–106.

Ji, Y.2021). The geographical origin, refugia, and diversification of honey beesApis spp.) based on biogeography and niche modeling. *Apidologie, 52* (2), 367–377. https://doi.org/10.1007/s13592-020-00826-6

Jiang, L., Zhang, M., Deng, L., Xu, Z., Shi, H., Jia, X., Lai, Z., Ruan, Q., & Chen, W.2020). Characteristics of the mitochondrial genome of Rana omeimontis and related species in Ranidae: Gene rearrangements and phylogenetic relationships. *Ecology and Evolution*, *10* (23), 12817–12837. https://doi.org/https://doi.org/10.1002/ece3.6824

Joseph, N., Bahtiar, N., Mahmud, F., Hamid, K. A., Raman, R., Chee, H. Y., & Nordin, S. A.2022). Comparison of Automated and Manual Viral Nucleic Acid Extraction Kits for Covid-19 Detection Using qRT-PCR. *Malaysian Journal of Medicine and Health Sciences*, *18* (1), 14–19.

Kadri, S. M., Harpur, B. A., Orsi, R. O., & Zayed, A.2016). A variant reference data set for the Africanized honeybee, Apis mellifera. *Scientific Data*, *3* (1), 160097. https://doi.org/10.1038/sdata.2016.97

Kahnt, B., Gerth, M., Paxton, R. J., Bleidorn, C., & Husemann, M.2015). The complete mitochondrial genome of the endemic and highly specialized South African bee species Rediviva intermixtaHymenoptera: Melittidae), with a comparison with other bee mitogenomes. *Biological Journal of the Linnean Society*, *116* (4), 940–953. https://doi.org/10.1111/bij.12627

Kahya, Y.2020). HRM Analysis of Spermathecal Contents to Determine the Origin of Drones that Inseminated Honey Bee Queens. *Journal of Apicultural Science*, *64* (2), 241–249. https://doi.org/10.2478/jas-2020-0018

Kaltenpoth, M., Showers Corneli, P., Dunn, D. M., Weiss, R. B., Strohm, E., & Seger, J.2012). Accelerated Evolution of Mitochondrial but Not Nuclear Genomes of Hymenoptera: New Evidence from Crabronid Wasps. *PLOS ONE*, *7* (3), 1–13. https://doi.org/10.1371/journal.pone.0032826

Kandemir, I., Kence, M., Sheppard, W. S., & Kence, A.2006). Mitochondrial DNA variation in honey beeApis mellifera L.) populations from Turkey. *J. Apic. Res*, *45*, 33–38.

Karuppaiah, V., Gadge, A. S., Shirsat, D. V, Soumia, P. S., Mainkar, P., Kumar, S., Jaiswal, D. K., & Mahajan, V.2023). The complete mitochondrial genome of the Indian dammer bee, Tetragonula iridipennis, and the phylogenomics of Meliponini. *Frontiers in Ecology and Evolution*, *Volume 11*. https://doi.org/10.3389/fevo.2023.1171242

Kaskinova, M. D., Gaifullina, L. R., & Saltykova, E. S.2023). Haplotypes of the tRNAleu-COII mtDNA Region in Russian Apis mellifera Populations. *Animals*, *13* (14). https://doi.org/10.3390/ani13142394

Keane, A.2020). *Computing a Yeast Tree of Life. December.*

Kek, S. P., Chin, N. L., Tan, S. W., Yusof, Y. A., & Chua, L. S.2017). Molecular identification of honey entomological origin based on bee mitochondrial 16S rRNA and COI gene sequences. *Food Control*, *78*, 150–159. https://doi.org/https://doi.org/10.1016/j.foodcont.2017.02.025

Kek, S. P., Chin, N. L., Tan, S. W., Yusof, Y. A., & Chua, L. S.2018). Comparison of DNA extraction methods for entomological origin identification of honey using simple additive weighting method. *International Journal of Food Science and Technology*, *53* (11), 2490–2499. https://doi.org/10.1111/ijfs.13840

Kekeçoğlu, M., Bir, S., Acar, M. K., & Ilyasov, R.2024). Honeybee biodiversity based on five mitochondrial genes and new evidence for different sublineages in Türkiye. *Turkish Journal of Veterinary and Animal Sciences*, *48* (6). https://doi.org/10.55730/1300-0128.4362

Kelemen, E. P., & Rehan, S. M.2021). Conservation insights from wild bee genetic studies: Geographic differences, susceptibility to inbreeding, and signs of local adaptation. *Evolutionary Applications*, *14* (6), 1485–1496. https://doi.org/https://doi.org/10.1111/eva.13221

Kelomey, A., Paraïso, A., SINA, H., Legout, H., Adjanohoun, A., Garnery, L., & Baba-Moussa, L.2018). Genetic Variability of the Mitochondrial DNA in HoneybeesApis mellifera L.) from Benin. *Journal of Agricultural Science and Technology*, *2017*. https://doi.org/10.17265/2161-6256/2017.08.006

Kenzy Ivveth Peña Carrillo.2021). *The ongoing speciation ina neotropical ant species complex.*

Khalid, A., Imran, M., Ali, A., Muzammil, S., Badar, M., Hayat, S., & Malik, I. R.2022). Molecular MarkerPcr-Rflp) Assisted Identification of Meat Species By Mitochondrial Cytochrome C Oxidase Subunit ICoi) Gene. *Journal of Animal and Plant Sciences*, *32* (6), 1724–1730. https://doi.org/10.36899/JAPS.2022.6.0580

Khedidji, H., Abderrahmani, K., Oulebsir-Mohandkaci, H., Ladjali-Mohammedi, K., & Mohammedi, A.2022). Effects of Pollen Deprivation in Groups of TellianApis mellifera intermissa) and SaharanApis mellifera sahariensis) Honey Bees under Controlled Conditions. *Insects*, *13* (8). https://doi.org/10.3390/insects13080727

Kim, J. S., Ah Rha, W., Min Jee, K., Keon Hee, L., & and Kim, I.2019). Single-nucleotide polymorphism markers in mitochondrial genomes for identifying Varroa destructor-resistant and -susceptible strains of Apis melliferaHymenoptera: Apidae). *Mitochondrial DNA Part A*, *30* (3), 477–489. https://doi.org/10.1080/24701394.2018.1551385

Kim, J. S., Kim, M. J., Kim, H.-K., Vung, N. N., & Kim, I.2019). Development of single nucleotide polymorphism markers specific to Apis melliferaHymenoptera: Apidae) line displaying high hygienic behavior against Varroa destructor, an ectoparasitic mite. *Journal of Asia-Pacific Entomology*, *22* (4), 1031–1039. https://doi.org/https://doi.org/10.1016/j.aspen.2019.08.005

Kjærstad, G., Majaneva, M., Falahati-Anbaran, M., Topstad, L., Finstad, A. G., Arnekleiv, J. V., & Ekrem, T.2022). DNA Metabarcoding of Preservative Ethanol Reveals Changes in Invertebrate Community Composition Following Rotenone Treatment. *Frontiers in Environmental Science*, *Volume 10*. https://doi.org/10.3389/fenvs.2022.872739

Knoll, A., Langová, L., Přidal, A., & Urban, T.2024). Haplotype Diversity in mtDNA of Honeybee in the Czech Republic Confirms Complete Replacement of Autochthonous Population with the C Lineage. *Insects*, *15* (7). https://doi.org/10.3390/insects15070495

Koeniger, N., Koeniger, G., & Smith, D. R.2011). Phylogeny of the genus Apis. In H. R. Hepburn & S. E. RadloffEds.), *Honeybees of Asia*. Springer-Verlag, Berlin.

Kong, S., Sánchez-Pacheco, S. J., & Murphy, R.2023). Median-Joining Networks and Bayesian Phylogenies Often Do Not Tell the Same Story. *Bulletin of the Society of Systematic Biologists*, *2* (1), 1–13. https://doi.org/10.18061/bssb.v2i1.9625

Kowalczyk, M., Staniszewski, A., Kamińska, K., Domaradzki, P., & Horecka, B.2021). Advantages, Possibilities, and Limitations of Mitochondrial DNA

Analysis in Molecular Identification. *Folia BiologicaPoland), 69* (3), 101–111. https://doi.org/10.3409/fb_69-3.12

Kuang, J., Yan, X., Genders, A. J., Granata, C., & Bishop, D. J.2018). An overview of technical considerations when using quantitative real-time PCR analysis of gene expression in human exercise research. *PLOS ONE, 13* (5), e0196438. https://doi.org/10.1371/journal.pone.0196438

Lado, J.2023). *Assessment of DNA Concentration and Degradation from Extracted Bumble Bee Thorax Muscle 2023*. 11–13.

Lalhmangaihi, R., Ghatak, S., Laha, R., Gurusubramanian, G., & Kumar, N. S.2014). Protocol for optimal quality and quantity pollen DNA isolation from honey samples. *Journal of Biomolecular Techniques : JBT*, *25* (4), 92–95. https://doi.org/10.7171/jbt.14-2504-001

Law, G., da Silva, C. R. B., Vlasich-Brennan, I., Taylor, B. A., Harpur, B. A., Heard, T., Nacko, S., Riegler, M., Dorey, J. B., Stevens, M. I., Lo, N., & Gloag, R.2024). Gene Flow Between Populations With Highly Divergent Mitogenomes in the Australian Stingless Bee, Tetragonula hockingsi. *Ecology and Evolution*, *14* (11), e70475.
https://doi.org/https://doi.org/10.1002/ece3.70475

Lefebre, R., Broeckx, B. J. G., De Smet, L., Peelman, L., & de Graaf, D. C.2024). Population-wide modelling reveals prospects of marker-assisted selection for parasitic mite resistance in honey bees. *Scientific Reports*, *14* (1), 7866. https://doi.org/10.1038/s41598-024-58596-5

Lefebre, R., De Smet, L., Tehel, A., Paxton, R. J., Bossuyt, E., Verbeke, W., van Dooremalen, C., Ulgezen, Z. N., van den Bosch, T., Schaafsma, F., Valkenburg, D.-J., Dall'Olio, R., Alaux, C., Dezmirean, D. S., Giurgiu, A. I., Capela, N., Simões, S., Sousa, J. P., Bencsik, M., … de Graaf, D. C.2024). Allele Frequencies of Genetic Variants Associated with Varroa Drone Brood ResistanceDBR) in Apis mellifera Subspecies across the European Continent. *Insects*, *15* (6). https://doi.org/10.3390/insects15060419

Legati, A., Zanetti, N., Nasca, A., Peron, C., Lamperti, C., Lamantea, E., & Ghezzi, D.2021). Current and New Next-Generation Sequencing Approaches to Study Mitochondrial DNA. *The Journal of Molecular Diagnostics*, *23* (6), 732–741. https://doi.org/10.1016/j.jmoldx.2021.03.002

Legesse, G., & Getu, E.2022). Geographical and population diversity of African honeybees, Apismellifera L. (Hymenoptera: Apidae): A review. *SINET: Ethiopian Journal of Science*, *45* (2), 235–248.

Li, H., Chen, X., Yang, Y., Xu, J., Gu, J., Fu, J., Qian, X., Zhang, S., Wu, J., & Liu, K.2011). Development and genetic mapping of microsatellite markers from whole genome shotgun sequences in Brassica oleracea. *Molecular Breeding*, *28* (4), 585–596. https://doi.org/10.1007/s11032-010-9509-y

Li, J., Zeng, C., Yan, G., & He, L.2019). Characterization of the mitochondrial genome of an ancient amphipod Halice sp. MT-2017Pardaliscidae) from 10,908 m in the Mariana Trench. *Scientific Reports*, *9* (1), 2610. https://doi.org/10.1038/s41598-019-38735-z

Li, Y.-R., Wang, Z.-W., Corlett, R. T., & Yu, W.-B.2024). Comparative analyses of mitogenomes in the social bees with insights into evolution of long inverted repeats in the Meliponini. *Zoological Research*, *45* (1), 160–175. https://doi.org/10.24272/j.issn.2095-8137.2023.169

Linck, E.2019). *Gene flow and models of avian speciation in tropical mountains.*

Lindberg, M. R., Schmedes, S. E., Hewitt, F. C., Haas, J. L., Ternus, K. L., Kadavy, D. R., & Budowle, B.2016). A Comparison and Integration of MiSeq and MinION Platforms for Sequencing Single Source and Mixed Mitochondrial Genomes. *PLOS ONE*, *11* (12), 1–16. https://doi.org/10.1371/journal.pone.0167600

Lindgren, J., Hatch, A., Hourdez, S., Seid, C., & Rouse, G.2019). Phylogeny and Biogeography of BranchipolynoePolynoidae, Phyllodocida, Aciculata, Annelida), with Descriptions of Five New Species from Methane Seeps and Hydrothermal Vents. *Diversity*, *11*, 153. https://doi.org/10.3390/d11090153

Litvinoff, L., Menescardi, F., Porrini, L., Russo, R., Liendo, M. C., Nucci, A., Lusarreta, E., Ventura, R., Espasadin, L., Monmany-Garzia, A. C., Scannapieco, A. C., & Galindo-Cardona, A.2023). Morphometric and genetic characterization as tools for selection of Apis melliferaHymenoptera: Apidae) stocks in an area of natural hybridization in Argentina. *Frontiers in Insect Science*, *Volume 2*-. https://doi.org/10.3389/finsc.2022.1073999

Liu Bsc, M.2021). *Using DNA metabarcoding to assess beetle biodiversity for ecologically sustainable forest management. March.*

Liu, Y.-Y., & Harbison, S.2018). A review of bioinformatic methods for forensic DNA analyses. *Forensic Science International: Genetics, 33*, 117–128. https://doi.org/10.1016/j.fsigen.2017.12.005

López-Uribe, M. M., Soro, A., & Jha, S.2017). Conservation genetics of bees: advances in the application of molecular tools to guide bee pollinator conservation. *Conservation Genetics, 18* (3), 501–506. https://doi.org/10.1007/s10592-017-0975-1

Lu, H., He, B., Hao, Y., Zhou, Z., Su, C., & Huang, D.2021). Comparative Mitogenomic Analysis of Two Cuckoo BeesApoidea: Anthophila: Megachilidae) with Phylogenetic Implications. *Insects, 12* (1). https://doi.org/10.3390/insects12010029

Lyons, E. A., Scheible, M. K., Sturk-Andreaggi, K., Irwin, J. A., & Just, R. S.2013). A high-throughput Sanger strategy for human mitochondrial genome sequencing. *BMC Genomics, 14* (1), 881. https://doi.org/10.1186/1471-2164-14-881

Ma, C., Hu, R., Costa, C., & Li, J.2022). Genetic Drift and Purifying Selection Shaped Mitochondrial Genome Variation in the High Royal Jelly-Producing Honeybee StrainApis mellifera ligustica). *Frontiers in Genetics, 13*, 835967. https://doi.org/10.3389/fgene.2022.835967

Ma, J., Du, C., Zhou, C., Sheng, Y., Fan, Z., Yue, B., & Zhang, X.2017). Complete mitochondrial genomes of two blattid cockroaches, Periplaneta australasiae and Neostylopyga rhombifolia, and phylogenetic relationships within the Blattaria. *PLOS ONE, 12* (5), e0177162. https://doi.org/10.1371/journal.pone.0177162

Ma, K., Li, H., Cao, Y., Zhao, X., Liu, W., & Zhao, X.2016). Haplotype diversity in mitochondrial genome in a Chinese Han population. *Journal of Human Genetics, 61* (10), 903–906. https://doi.org/10.1038/jhg.2016.74

Macken, W. L., Vandrovcova, J., Hanna, M. G., & Pitceathly, R. D. S.2021). Applying genomic and transcriptomic advances to mitochondrial medicine.

Nature Reviews Neurology, *17* (4), 215–230. https://doi.org/10.1038/s41582-021-00455-2

Madella, S., Grubbs, K., & Alburaki, M.2021). Non-invasive genotyping of honey bee queens apis mellifera l.: Transition of the drai mtdna coi-coii test to in silico. *Insects*, *12* (1), 1–11. https://doi.org/10.3390/insects12010019

Mahmud, S., Biswas, S., Afrose, S., Mita, M. A., Hasan, M. R., Shimu, M. S. S., Paul, G. K., Chung, S., Saleh, M. A., Alshehri, S., Ghoneim, M. M., Alruwaily, M., & Kim, B.2022). Use of Next-Generation Sequencing for Identifying Mitochondrial Disorders. *Current Issues in Molecular Biology*, *44* (3), 1127–1148. https://doi.org/10.3390/cimb44030074

Malik, M., Archana, A., Praveen, P., Neela, K., & Vammi, T. K. S.2021). *DNA Pro fi ling in Forensic Science : A Review.*

Mandel, J. R., Ramsey, A. J., Holley, J. M., Scott, V. A., Mody, D., & Abbot, P.2020). Disentangling Complex Inher

itance Patterns of Plant Organellar Genomes: An Example From Carrot. *Journal of Heredity*, *111* (6), 531–538. https://doi.org/10.1093/jhered/esaa037

Maria Cristina Arias, Rute Magalhães Brito, Flávio de Oliveira Francisco, Geraldo Moretto, Favízia Freitas de Oliveira, Daniela Silvestre, & Walter Steven Sheppard.2006). Molecular markers as a tool for population and evolutionary studies of stingless bees. *Apidologie*, *37* (2), 259–274. https://doi.org/10.1051/apido:2006021

Meixner, M. D., Pinto, M. A., Bouga, M., Kryger, P., Ivanova, E., & Fuchs, S.2013). Standard methods for characterising subspecies and ecotypes of Apis mellifera. *Journal of Apicultural Research*, *52* (4). https://doi.org/10.3896/IBRA.1.52.4.05

Meusel, M. S., & Moritz, R. F. A.1993). Transfer of paternal mitochondrial DNA during fertilization of honeybeeApis mellifera L.) eggs. *Current Genetics*, *24* (6), 539–543. https://doi.org/10.1007/BF00351719

Miguel, I., Iriondo, M., Garnery, L., Sheppard, W. S., & Estonba, A.2007). Gene flow within the M evolutionary lineage of Apis mellifera: role of the Pyrenees, isolation by distance and post-glacial re-colonization routes in the western Europe. *Apidologie*, *38* (2), 141–155.

https://doi.org/10.1051/apido:2007007

Miguel, I., Lionel, G., Mikel, I., Michel, B., Carmen, M., W, S. S., & and Estonba, A.2015). Origin, evolution and conservation of the honey bees from La Palma IslandCanary Islands): molecular and morphological data. *Journal of Apicultural Research*, *54* (5), 427–440. https://doi.org/10.1080/00218839.2016.1180017

Mikheyev, A. S., Tin, M. M. Y., Arora, J., & Seeley, T. D.2015). Museum samples reveal rapid evolution by wild honey bees exposed to a novel parasite. *Nature Communications*, *6* (1), 7991. https://doi.org/10.1038/ncomms8991

Miller, M. A., Pfeiffer, W., & Schwartz, T.2010). Creating the CIPRES Science Gateway for inference of large phylogenetic trees. *2010 Gateway Computing Environments WorkshopGCE)*, 1–8. https://doi.org/10.1109/GCE.2010.5676129

Modaber, M., Nazemi Rafie, J., & Rajabi-Maham, H.2019). Population genetic structure of native Iranian population of Apis mellifera meda based on intergenic region and COX2 gene of mtDNA. *Insectes Sociaux*, *66* (3), 413–424. https://doi.org/10.1007/s00040-019-00701-3

Mohamadzade Namin, S., Ghosh, S., & Jung, C.2023). Honey Quality Control: Review of Methodologies for Determining Entomological Origin. *Molecules*, *28* (10). https://doi.org/10.3390/molecules28104232

Mohamadzade Namin, S., Yeasmin, F., Choi, H. W., & Jung, C.2022). DNA-Based Method for Traceability and Authentication of Apis cerana and A. dorsata HoneyHymenoptera: Apidae), Using the NADH dehydrogenase 2 Gene. *Foods*, *11* (7). https://doi.org/10.3390/foods11070928

Momeni, J., Parejo, M., Nielsen, R. O., Langa, J., Montes, I., Papoutsis, L., Farajzadeh, L., Bendixen, C., Căuia, E., Charrière, J.-D., Coffey, M. F., Costa, C., Dall'Olio, R., De la Rúa, P., Drazic, M. M., Filipi, J., Galea, T., Golubovski, M., Gregorc, A., … Estonba, A.2021). Authoritative subspecies diagnosis tool for European honey bees based on ancestry informative SNPs. *BMC Genomics*, *22* (1), 101. https://doi.org/10.1186/s12864-021-07379-7

MOREIRA, S. A., HORGAN, G. F., MURRAY, E. T., & KAKOULI-DUARTE, T.2013). BumblebeeHymenoptera: Apidae) sample storage for a posteriori molecular studies: Interactions between sample storage and DNA-extraction techniques. *EJE, 110* (3), 419–425. https://www.eje.cz/artkey/eje-201303-0005.php

Moškrič, A., Marinč, A., Ferk, P., Leskošek, B., Mosbech, M.-B., Bunikis, I., Pettersson, O. V., Soler, L., & Prešern, J.2022). The Carniolan Honeybee from Slovenia-A Complete and Annotated Mitochondrial Genome with Comparisons to Closely Related Apis mellifera Subspecies. *Insects, 13* (5). https://doi.org/10.3390/insects13050403

Moškrič, A., Mole, K., & Prešern, J.2021). EPIC markers of the genus Apis as diagnostic tools for detection of honey fraud. *Food Control, 121*, 107634. https://doi.org/https://doi.org/10.1016/j.foodcont.2020.107634

Moškrič, A., Pavlin, A., Mole, K., Marinč, A., Bubnič, J., Opara, A., Kovačić, M., Puškadija, Z., Uzunov, A., Andonov, S., Dahle, B., & Prešern, J.2023). Cutting corners: The impact of storage and DNA extraction on quality and quantity of DNA in honeybeeApis mellifera) spermatheca. *Frontiers in Physiology, Volume 14*. https://doi.org/10.3389/fphys.2023.1139269

Moya-Moraga, M. R., & Pérez-Ruíz, C.2022). Application of MaxEnt Modeling and HRM Analysis to Support the Conservation and Domestication of Gevuina avellana Mol. in Central Chile. *Plants, 11* (20). https://doi.org/10.3390/plants11202803

Muñoz, I., Dall'Olio, R., Lodesani, M., & Rúa, P. D. La.2009). Population genetic structure of coastal Croatian honeybeesApis mellifera carnica). *Apidologie, 40* (6), 617–626. https://doi.org/10.1051/apido/2009041

Muñoz, I., Dall'Olio, R., Lodesani, M., & la Rúa, P.2014). Estimating introgression in Apis mellifera siciliana populations: are the conservation islands really effective? *Insect Conservation and Diversity, 7* (6), 563–571. https://doi.org/https://doi.org/10.1111/icad.12092

Muñoz, I., & De la Rúa, P.2021). Wide genetic diversity in Old World honey bees threaten by introgression. *Apidologie, 52* (1), 200–217. https://doi.org/10.1007/s13592-020-00810-0

Muñoz, I., Henriques, D., Jara, L., Johnston, J. S., Chávez-Galarza, J., Rúa, P. D. La, & Pinto, M. A.2017). SNPs selected by information content outperform randomly selected microsatellite loci for delineating genetic identification and introgression in the endangered dark European honeybeeApis mellifera mellifera). *Mol. Ecol. Resour*, *17*, 783–795.

Muñoz, I., Henriques, D., Johnston, J. S., Chávez-Galarza, J., Kryger, P., & Pinto, M. A.2015). Reduced SNP panels for genetic identification and introgression analysis in the dark honey beeApis mellifera mellifera). *PLOS ONE*, *10*, 124365.

Nakagawa, I., Maeda, M., Chikano, M., Okuyama, H., Murray, R., & Takahashi, J.-I.2018). The complete mitochondrial genome of the yellow coloured honeybee Apis mellifera Insecta: Hymenoptera: Apidae) of New Zealand. *Mitochondrial DNA. Part B, Resources*, *3* (1), 66–67. https://doi.org/10.1080/23802359.2017.1422401

Nanetti, A., Cilia, G., Bonetto, M., Tesoriero, D., & Garrido, C.2019). Beekeeping together within Agriculture. *46th Apimondia International Apicultural Congress*, 232.

Nascimento, F. F., Reis, M. dos, & Yang, Z.2017). A biologist's guide to Bayesian phylogenetic analysis. *Nature Ecology & Evolution*, *1* (10), 1446–1454. https://doi.org/10.1038/s41559-017-0280-x

Neumann, P., Jay D, E., Jeff S, P., Christian W W, P., Marc O, S., Gina, T., & and Ellis, J. D.2013). Standard methods for small hive beetle research. *Journal of Apicultural Research*, *52* (4), 1–32. https://doi.org/10.3896/IBRA.1.52.4.19

Nikolikj, I.2017). *Identification of berry species and cultivars by High Resolution MeltingHRM) analysis*. https://doi.org/10.25365/thesis.49124

Nikulin, S. L., Hesketh-Best, P. J., Mckeown, D. A., Spivak, M., & Schroeder, D. C.2024). A semi-automated and high-throughput approach for the detection of honey bee viruses in bee samples. *PLOS ONE*, *19* (3), e0297623. https://doi.org/10.1371/journal.pone.0297623

Njeri, N & Njeri, N.2024). *Discrimination of Hypotrigona speciesApidae: Meliponinae) in Kenya using nest architecture, cephalic secretions and molecular tools.*

Offornedo, Q., Menkir, A., Babalola, D., & Gedil, M.2022). Developing and deploying an efficient genotyping workflow for accelerating maize improvement in developing countries. *Gates Open Research*, *6*, 1–31. https://doi.org/10.12688/gatesopenres.13338.3

Okwaro Louis Allan.2019). GENETIC CHARACTERIZATION OF HONEYBEEApis mellifera) IN COMO ROS ISLANDS USING MITOCHONDRIAL AND MICROSATELLITE MARK ERS. *SustainabilitySwitzerland)*, *11* (1), 1–14. http://scioteca.caf.com/bitstream/handle/123456789/1091/RED2017-Eng-8ene.pdf?sequence=12&isAllowed=y%0Ahttp://dx.doi.org/10.1016/j.regsciurbeco.2008.06.005%0Ahttps://www.researchgate.net/publication/305320484_SISTEM_PEMBETUNGAN_TERPUSAT_STRATEGI_MELESTARI

Olate-Olave, V. R., Verde, M., Vallejos, L., Perez Raymonda, L., Cortese, M. C., & Doorn, M.2021). Bee Health and Productivity in Apis mellifera, a Consequence of Multiple Factors. *Veterinary Sciences*, *8* (5). https://doi.org/10.3390/vetsci8050076

Oleksa, A., Chybicki, I., Tofilski, A., & Burczyk, J.2011). Nuclear and mitochondrial patterns of introgression into native dark beesApis mellifera mellifera) in Poland. *Journal of Apicultural Research*, *50* (2), 116–129. https://doi.org/10.3896/IBRA.1.50.2.03

Oleksa, A., Kusza, S., & Tofilski, A.2021). Mitochondrial DNA suggests the introduction of honeybees of African ancestry to East-Central Europe. *Insects*, *12* (5), 410.

Oleksa, A., & Tofilski, A.2015). Wing geometric morphometrics and microsatellite analysis provide similar discrimination of honey bee subspecies. *Apidologie*, *46* (1), 4960.

Orsini, M., Cuccuru, G., Uva, P., & Fotia, G.2016). Bacterial Genomic Data Analysis in the Next-Generation Sequencing Era. *Methods in Molecular BiologyClifton, N.J.)*, *1415*, 407–422. https://doi.org/10.1007/978-1-4939-3572-7_21

Ostroverkhova, N. V., & Konusova, O. L.2022). SOME PROBLEMS OF IDENTIFICATION OF HONEYBEE SUBSPECIES AND THEIR SOLUTION ON THE EXAMPLE OF STUDYING THE Apis mellifera IN SIBERIA. *Sel'skokhozyaistvennaya Biologiya*, *57* (2), 283–303. https://doi.org/10.15389/agrobiology.2022.2.283eng

Özdil, F., Fakhri, B., Meydan, H., Yıldız, M. A., & Hall, H. G.2009). Mitochondrial DNA Variation in the CoxI–CoxII Intergenic Region among Turkish and Iranian Honey BeesApis mellifera L.). *Biochemical Genetics*, *47* (9), 717–721. https://doi.org/10.1007/s10528-009-9269-8

Pacheco, M. A., Cepeda, A. S., Bernotienė, R., Lotta, I. A., Matta, N. E., Valkiūnas, G., & Escalante, A. A.2018). Primers targeting mitochondrial genes of avian haemosporidians: PCR detection and differential DNA amplification of parasites belonging to different genera. *International Journal for Parasitology*, *48* (8), 657–670. https://doi.org/https://doi.org/10.1016/j.ijpara.2018.02.003

Palmer, M. R., Smith, D. R., & Kaftanoglu, O.2000). Brief communication. Turkish honeybees: genetic variation and evidence for a fourth lineage of Apis mellifera mtDNA. *Journal of Heredity*, *91* (1), 42–46. https://doi.org/10.1093/jhered/91.1.42

Panziera, D., Requier, F., Chantawannakul, P., Pirk, C. W. W., & Blacquière, T.2022). The Diversity Decline in Wild and Managed Honey Bee Populations Urges for an Integrated Conservation Approach. *Frontiers in Ecology and Evolution*, *10* (March), 1–7. https://doi.org/10.3389/fevo.2022.767950

Parejo, M., Wragg, D., Gauthier, L., Vignal, A., Neumann, P., & Neuditschko, M.2016). Using whole-genome sequence information to foster conservation efforts for the european dark honey bee, Apis mellifera mellifera. *Front. Ecol. Evol*, *4* (140).

Parejo, M., Wragg, D., Henriques, D., Charrière, J.-D., & Estonba, A.2020). Digging into the Genomic Past of Swiss Honey Bees by Whole-Genome Sequencing Museum Specimens. *Genome Biology and Evolution*, *12* (12), 2535–2551. https://doi.org/10.1093/gbe/evaa188

Pathiraja, D., Cho, J., Kim, J., & Choi, I.-G.2023). Metabarcoding of eDNA for tracking the floral and geographical origins of bee honey. *Food Research International*, *164*, 112413. https://doi.org/https://doi.org/10.1016/j.foodres.2022.112413

Pearman, P. B., Broennimann, O., Aavik, T., Albayrak, T., Alves, P. C., Aravanopoulos, F. A., Bertola, L. D., Biedrzycka, A., Buzan, E., Cubric-Curik, V., Djan, M., Fedorca, A., Fuentes-Pardo, A. P., Fussi, B., Godoy, J. A.,

Gugerli, F., Hoban, S., Holderegger, R., Hvilsom, C., … Bruford, M.2024). Monitoring of species' genetic diversity in Europe varies greatly and overlooks potential climate change impacts. *Nature Ecology & Evolution*, *8* (2), 267–281. https://doi.org/10.1038/s41559-023-02260-0

Peck, M. A., Sturk-Andreaggi, K., Thomas, J. T., Oliver, R. S., Barritt-Ross, S., & Marshall, C.2018). Developmental validation of a Nextera XT mitogenome Illumina MiSeq sequencing method for high-quality samples. *Forensic Science International: Genetics*, *34*, 25–36. https://doi.org/10.1016/j.fsigen.2018.01.004

Pierre Franck, Lionel Garnery, Michel Solignac, & Jean-Marie Cornuet.2000). Molecular confirmation of a fourth lineage in honeybees from the Near East. *Apidologie*, *31* (2), 167–180. https://doi.org/10.1051/apido:2000114

Pinto, M. A., Henriques, D., Chávez-Galarza, J., Kryger, P., Garnery, L., van der Zee, R., Dahle, B., Soland-Reckeweg, G., la Rúa, P. D., Dall'Olio, R., Carreck, N. L., & Johnston, J. S.2014). Genetic integrity of the Dark European honey beeApis mellifera mellifera) from protected populations: A genome-wide assessment using SNPs and mtDNA sequence data. *J. Apic. Res*, *53*, 269–278.

Polovina, E.-S., Parakatselaki, M.-E., & Ladoukakis, E. D.2020). Paternal leakage of mitochondrial DNA and maternal inheritance of heteroplasmy in Drosophila hybrids. *Scientific Reports*, *10* (1), 2599. https://doi.org/10.1038/s41598-020-59194-x

Porrini, L. P., Brasesco, C., Maggi, M., Eguaras, M. J., & Quintana, S.2021). High-resolutionmtDNA) melting analysis for simple and efficient characterization of Africanized honey bee Apis melliferaHymenoptera:Apidae). *Genetica*, *149* (5–6), 343–350. https://doi.org/10.1007/s10709-021-00139-1

Porrini, L. P., Quintana, S., Brasesco, C., Maggi, M. D., Porrini, M. P., Garrido, M. P., Fernández Iriarte, P. J., & Eguaras, M. J.2022). Current genetic diversity of managed and commercially produced Apis mellifera colonies in Argentina inferred by wing geometric morphometrics and COI-COII mtDNA locus. *Apidologie*, *53* (5), 61. https://doi.org/10.1007/s13592-022-00970-1

Pratas, D., Toppinen, M., Pyöriä, L., Hedman, K., Sajantila, A., & Perdomo, M. F.2020). A hybrid pipeline for reconstruction and analysis of viral genomes at multi-organ level. *GigaScience*, *9* (8), giaa086. https://doi.org/10.1093/gigascience/giaa086

Qiu, L., Dong, J., Li, X., Parey, S. H., Tan, K., Orr, M., Majeed, A., Zhang, X., Luo, S., Zhou, X., Zhu, C., Ji, T., Niu, Q., Liu, S., & Zhou, X.2023). Defining honeybee subspecies in an evolutionary context warrants strategized conservation. *Zoological Research*, *44* (3), 483–493. https://doi.org/10.24272/j.issn.2095-8137.2022.414

Quero, F. J., Aidelberg, G., Vielfaure, H., de Kermadec, Y. H., Cazaux, S., Pandi, A., Pascual-Garrigos, A., Arce, A., Sakyi, S., Gaudenz, U., Federici, F., Molloy, J. C., & Lindner, A. B.2024). qByte: Open-source isothermal fluorimeter for democratizing analysis of nucleic acids, proteins and cells. *BioRxiv*, 2024.11.03.621723. https://doi.org/10.1101/2024.11.03.621723

Rahimi, A., Mirmoayedi, A., Kahrizi, D., Zarei, L., & Jamali, S.2016). Genetic diversity of Iranian honey beeApis mellifera meda Skorikow, 1829) populations based on ISSR markers. *Cellular and Molecular BiologyNoisy-Le-Grand, France)*, *62* (4), 53–58.

Rahimi, A., Mirmoayedi, A., Kahrizi, D., Zarei, L., & Jamali, S.2018). Genetic Variation in Iranian Honey bees, Apis mellifera meda Skorikow, 1829,Hymenoptera: Apidae) Inferred from PCR-RFLP Analysis of two mtDNA Gene SegmentsCOI and 16S rDNA). *Sociobiology*. https://api.semanticscholar.org/CorpusID:92573546

Raju, C. M., Elpa, D. P., & Urban, P. L.2024). Automation and Computerization ofBio)sensing Systems. *ACS Sensors*, *9* (3), 1033–1048. https://doi.org/10.1021/acssensors.3c01887

Razzaq, A. detection of genetic changes in the fungal endophyte E. festucae A. during adaptation to a new perennial ryegrass host.2020). *Bioinformatic detection of genetic changes in the fungal endophyte Epichloë festucae AR37 during adaptation to a new perennial ryegrass host*.

Requier, F., Garnery, L., Kohl, P. L., Njovu, H. K., Pirk, C. W. W., Crewe, R. M., & Steffan-Dewenter, I.2019). The Conservation of Native Honey Bees Is

Crucial. *Trends in Ecology & Evolution*, *34* (9), 789–798. https://doi.org/10.1016/j.tree.2019.04.008

Ribani, A., Taurisano, V., Utzeri, V. J., & Fontanesi, L.2022). Honey Environmental DNA Can Be Used to Detect and Monitor Honey Bee Pests: Development of Methods Useful to Identify Aethina tumida and Galleria mellonella Infestations. *Veterinary Sciences*, *9* (5). https://doi.org/10.3390/vetsci9050213

Ribani, A., Utzeri, V. J., Taurisano, V., Galuppi, R., & Fontanesi, L.2021). Analysis of honey environmental DNA indicates that the honey beeApis mellifera L.) trypanosome parasite Lotmaria passim is widespread in the apiaries of the North of Italy. *Journal of Invertebrate Pathology*, *184*, 107628. https://doi.org/https://doi.org/10.1016/j.jip.2021.107628

Rishan, S. T.2024). Development and Application of Environmental DNA eDNA) Techniques to Detect Rough-Footed and Yellow Mud Turtles : A Seasonal Field Study to Detect Turtle eDNA. *ScholarWorks @ UTRGV*.

Rishan, S. T., Kline, R. J., & Rahman, M. S.2025). Development and validation of rapid eDNA detection method for yellow mud turtle, Kinosternon flavescens: a field study in South Texas, USA. *Frontiers in Conservation Science*, *Volume 6*-. https://doi.org/10.3389/fcosc.2025.1431348

Rohde, A. T., Strange, J. P., Tobin, K. B., Mock, K. E., & Branstetter, M. G.2025). Genome-wide markers test the status of two putative species of North American bumble bees. *Conservation Genetics*, *26* (3), 429–448. https://doi.org/10.1007/s10592-025-01674-6

Rojas-Sandoval, J.2022). Apis mellifera capensisCape honeybee). *CABI Compendium*. https://doi.org/10.1079/cabicompendium.88902422

Ronquist, F., Teslenko, M., van der Mark, P., Ayres, D. L., Darling, A., Höhna, S., Larget, B., Liu, L., Suchard, M. A., & Huelsenbeck, J. P.2012). MrBayes 3.2: efficient Bayesian phylogenetic inference and model choice across a large model space. *Syst. Biol*, *61*, 539–542. https://doi.org/10.1093/sysbio/sys029.

Rortais, A., Arnold, G., Alburaki, M., Legout, H., & Garnery, L.2011). Review of the DraICOI-COII test for the conservation of the black honeybeeApis mellifera mellifera). *Conservation Genetics Resources, 3* (2), 383–391. https://doi.org/10.1007/s12686-010-9351-x

Ruppert, K. M., Davis, D. R., Rahman, M. S., & Kline, R. J.2022). Development and assessment of an environmental DNAeDNA) assay for a cryptic SirenAmphibia: Sirenidae). *Environmental Advances, 7,* 100163. https://doi.org/https://doi.org/10.1016/j.envadv.2021.100163

Sarhaddi Avval, M. S. A., Fazeli-Nasab, B. F., Pahlavan, A. P., & Mirzaei, A. R. M.2024). Investigating the Phylogenetic Relationships of the Iranian Honey Bee with Other Bee Breeds Worldwide Using the Cytochrome Oxidase I Region. *Gene, Cell and Tissue, 12* (1), e157608. https://doi.org/10.5812/gct-157608

Schmidt, J., Opgenoorth, L., Mao, K., Baniya, C. B., & Hofmann, S.2023). Molecular phylogeny of mega-diverse Carabus attests late Miocene evolution of alpine environments in the Himalayan–Tibetan Orogen. *Scientific Reports, 13* (1). https://doi.org/10.1038/s41598-023-38999-6

Schwartz, M. K., Luikart, G., & Waples, R. S.2007). Genetic monitoring as a promising tool for conservation and management. *Trends in Ecology & Evolution, 22* (1), 25–33.
https://doi.org/https://doi.org/10.1016/j.tree.2006.08.009

Sewe, S. O., Silva, G., Sicat, P., Seal, S. E., & Visendi, P.2022). Trimming and Validation of Illumina Short Reads Using Trimmomatic, Trinity Assembly, and Assessment of RNA-Seq Data. In D. EdwardsEd.), *Plant Bioinformatics: Methods and Protocols*pp. 211–232). Springer US.
https://doi.org/10.1007/978-1-0716-2067-0_11

Shen, R., Fan, J.-B., Campbell, D., Chang, W., Chen, J. C., Doucet, D., Yeakley, J. M., Bibikova, M., Garcia, E. W., McBride, C., Steemers, F., Garcia, F., Kermani, B. G., Gunderson, K. L., & Oliphant, A. R.2005). High-throughput SNP genotyping on universal bead arrays. *Mutation Research, 573 1-2*, 70–82. https://api.semanticscholar.org/CorpusID:10239733

Smith, C. D., Cornman, R. S., Fike, J. A., Kraus, J. M., Oyler-McCance, S. J., Givens, C. E., Hladik, M. L., Vandever, M. W., Kolpin, D. W., & Smalling, K. L.2024). Comparing modern identification methods for wild bees:

Metabarcoding and image-based morphological taxonomic assignment. *PLOS ONE*, *19* (4), e0301474. https://doi.org/10.1371/journal.pone.0301474

Soares, S., Amaral, J. S., Oliveira, M. B. P. P., & Mafra, I.2015). Improving DNA isolation from honey for the botanical origin identification. *Food Control*, *48*, 130–136. https://doi.org/10.1016/j.foodcont.2014.02.035

Soares, S., Grazina, L., Mafra, I., Costa, J., Pinto, M. A., Duc, H. P., Oliveira, M. B. P. P., & Amaral, J. S.2018). Novel diagnostic tools for AsianApis cerana) and EuropeanApis mellifera) honey authentication. *Food Research International*, *105*, 686–693. https://doi.org/https://doi.org/10.1016/j.foodres.2017.11.081

Soares, S., Grazina, L., Mafra, I., Costa, J., Pinto, M. A., Oliveira, M. B. P. P., & Amaral, J. S.2019). Towards honey authentication: Differentiation of Apis mellifera subspecies in European honeys based on mitochondrial DNA markers. *Food Chemistry*, *283*, 294–301. https://doi.org/https://doi.org/10.1016/j.foodchem.2018.12.119

Soares, S., Rodrigues, F., & Delerue-Matos, C.2023). Towards DNA-Based Methods Analysis for Honey: An Update. In *Molecules*Vol. 28, Issue 5). https://doi.org/10.3390/molecules28052106

Soltani, C.2023). *Development of a cost-effective SNP tool to detect genetic pollution in honey bee spermatheca content.*

Song, F., Li, H., Jiang, P., Zhou, X., Liu, J., Sun, C., Vogler, A. P., & Cai, W.2016). Capturing the Phylogeny of Holometabola with Mitochondrial Genome Data and Bayesian Site-Heterogeneous Mixture Models. *Genome Biology and Evolution*, *8* (5), 1411–1426. https://doi.org/10.1093/gbe/evw086

Song, S.-N., Tang, P., Wei, S.-J., & Chen, X.-X.2016). Comparative and phylogenetic analysis of the mitochondrial genomes in basal hymenopterans. *Scientific Reports*, *6* (1), 20972. https://doi.org/10.1038/srep20972

Straeuli, R.2022). *Phylogeny of Emperor moths and phylogeography of Gonimbrasia belina in Southern Africa*. *December*, 23.

Syromyatnikov, M. Y., Borodachev, A. V, Kokina, A. V, & Popov, V. N.2018). A Molecular Method for the Identification of Honey Bee Subspecies Used by Beekeepers in Russia. *Insects*, *9* (1). https://doi.org/10.3390/insects9010010

Szalanski, A. L., Tripodi, A. D., Trammel, C. E., & Downey, D.2016). Mitochondrial DNA genetic diversity of honey bees, Apis mellifera, in Hawaii. *Apidologie*, *47* (5), 679–687. https://doi.org/10.1007/s13592-015-0416-4

Talavera, G., & Vila, R.2011). What is the phylogenetic signal limit from mitogenomes? The reconciliation between mitochondrial and nuclear data in the Insecta class phylogeny. *BMC Evolutionary Biology*, *11* (1), 315. https://doi.org/10.1186/1471-2148-11-315

Tan, H.-W., Liu, G.-H., Dong, X., Lin, R.-Q., Song, H.-Q., Huang, S.-Y., Yuan, Z.-G., Zhao, G.-H., & Zhu, X.-Q.2011). The Complete Mitochondrial Genome of the Asiatic Cavity-Nesting Honeybee Apis ceranaHymenoptera: Apidae). *PLOS ONE*, *6* (8), e23008. https://doi.org/10.1371/journal.pone.0023008

Tanasković, M., Erić, P., Patenković, A., Erić, K., Mihajlović, M., Tanasić, V., Stanisavljević, L., & Davidović, S.2021). MtDNA Analysis Indicates Human-Induced Temporal Changes of Serbian Honey Bees Diversity. In *Insects*Vol. 12, Issue 9). https://doi.org/10.3390/insects12090767

Taurisano, V., Ribani, A., Sami, D., Nelson Johnson, K. E., Schiavo, G., Utzeri, V. J., Bovo, S., & Fontanesi, L.2024). Distribution of honey bee mitochondrial DNA haplotypes in an Italian region where a legislative act is protecting the Apis mellifera ligustica subspecies. *Scientific Reports*, *14* (1), 20583. https://doi.org/10.1038/s41598-024-71233-5

Taylor, C. R., Kiesler, K. M., Sturk-Andreaggi, K., Ring, J. D., Parson, W., Schanfield, M., Vallone, P. M., & Marshall, C.2020). Platinum-Quality Mitogenome Haplotypes from United States Populations. *Genes*, *11* (11). https://doi.org/10.3390/genes11111290

Taylor, M., & Sheppard, W.2014). *Old World honey beeApis mellifera) populations: A genetic resource for U.S. honey bee breeding.*

Techer, M. A., Clémencet, J., Simiand, C., Preeaduth, S., Azali, H. A., Reynaud, B., & Hélène, D.2017). Large-scale mitochondrial DNA analysis of native honey bee Apis mellifera populations reveals a new African subgroup private to the South West Indian Ocean islands. *BMC Genetics*, *18* (1), 1–21. https://doi.org/10.1186/s12863-017-0520-8

Techer, M. A., Clémencet, J., Simiand, C., Turpin, P., Garnery, L., Reynaud, B., & Delatte, H.2017). Genetic diversity and differentiation among insular honey bee populations in the southwest Indian Ocean likely reflect old geographical isolation and modern introductions. *PloS One*, *12* (12), e0189234. https://doi.org/10.1371/journal.pone.0189234

Tihelka, E., Cai, C., Pisani, D., & Donoghue, P. C. J.2020). Mitochondrial genomes illuminate the evolutionary history of the Western honey beeApis mellifera). *Scientific Reports*, *10* (1), 14515. https://doi.org/10.1038/s41598-020-71393-0

Torricelli, M., Pierboni, E., Tovo, G. R., Curcio, L., & Rondini, C.2016). In-house Validation of a DNA Extraction Protocol from Honey and Bee Pollen and Analysis in Fast Real-Time PCR of Commercial Honey Samples Using a Knowledge-Based Approach. *Food Analytical Methods*, *9*, 3439–3450. https://api.semanticscholar.org/CorpusID:89205120

Tozkar, C.2020). Genetic structure of honey beeApis mellifera linnaeus, 1758) subspecies based on trnaleu-cox2 and nd5 regions of mtdna. *Applied Ecology and Environmental Research*, *18* (2), 2269–2284. https://doi.org/10.15666/aeer/1802_22692284

Utzeri, V. J., Ribani, A., Taurisano, V., Banqué, C. H. I., & Fontanesi, L.2021). Distribution of the main apis mellifera mitochondrial dna lineages in italy assessed using an environmental dna approach. *Insects*, *12* (7). https://doi.org/10.3390/insects12070620

Utzeri, V. J., Ribani, A., Taurisano, V., & Fontanesi, L.2022). Entomological authentication of honey based on a DNA method that distinguishes Apis mellifera mitochondrial C mitotypes: Application to honey produced by A. m. ligustica and A. m. carnica. *Food Control*, *134*, 108713. https://doi.org/https://doi.org/10.1016/j.foodcont.2021.108713

Utzeri, V. J., Schiavo, G., Ribani, A., Bertolini, F., Bovo, S., & Fontanesi, L.2019). A next generation sequencing approach for targeted Varroa destructorAcari: Varroidae) mitochondrial DNA analysis based on honey derived environmental DNA. *Journal of Invertebrate Pathology*, *161*, 47–53. https://doi.org/https://doi.org/10.1016/j.jip.2019.01.005

Utzeri, V. J., Schiavo, G., Ribani, A., Tinarelli, S., Bertolini, F., Bovo, S., & Fontanesi, L.2018). Entomological signatures in honey: an environmental DNA metabarcoding approach can disclose information on plant-sucking insects in agricultural and forest landscapes. *Scientific Reports*, *8* (1), 9996. https://doi.org/10.1038/s41598-018-27933-w

Veisi, H., Nazemi-Rafie, J., Azizi, A., & Rajabi-Maham, H.2020). Genetic characteristics of the Iranian honey bee, Apis mellifera meda, based on mitochondrial genes of ND4, ND4L and ND6 and their internal transcribed spacers. *Insectes Sociaux*, *67* (3), 439–448. https://doi.org/10.1007/s00040-020-00769-2

Vences, M., Patmanidis, S., Schmidt, J.-C., Matschiner, M., Miralles, A., & Renner, S. S.2024). Hapsolutely: a user-friendly tool integrating haplotype phasing, network construction, and haploweb calculation. *Bioinformatics Advances*, *4* (1), vbae083. https://doi.org/10.1093/bioadv/vbae083

Vesterlund, S.-R., Sorvari, J., & Vasemägi, A.2014). Molecular identification of cryptic bumblebee species from degraded samples using PCR–RFLP approach. *Molecular Ecology Resources*, *14* (1), 122–126. https://doi.org/https://doi.org/10.1111/1755-0998.12168

Von, V., & Hardulak, L. A.2020). *Development and Improvement of Next Generation Sequencing Pipelines for Mixed and Bulk Samples of German Fauna*. https://semspub.epa.gov/work/05/249512.pdf%0Ahttps://edoc.ub.uni-muenchen.de/27000/

Wallinger, C., Staudacher, K., Sint, D., Thalinger, B., Oehm, J., Juen, A., & Traugott, M.2017). Evaluation of an automated protocol for efficient and reliable DNA extraction of dietary samples. *Ecology and Evolution*, *7* (16), 6382–6389. https://doi.org/https://doi.org/10.1002/ece3.3197

Wang, A. R., Kim, J. S., Kim, M. J., Kim, H.-K., Choi, Y. S., & Kim, I.2018). Comparative description of mitochondrial genomes of the honey bee

ApisHymenoptera: Apidae): four new genome sequences and Apis phylogeny using whole genomes and individual genes. *Journal of Apicultural Research*, *57* (4), 484–503. https://doi.org/10.1080/00218839.2018.1494885

White, O. W., Hall, A., Price, B. W., Williams, S. T., & Clark, M. D.2025). A Snakemake Toolkit for the Batch Assembly, Annotation and Phylogenetic Analysis of Mitochondrial Genomes and Ribosomal Genes From Genome Skims of Museum Collections. *Molecular Ecology Resources*, *25* (1), e14036. https://doi.org/https://doi.org/10.1111/1755-0998.14036

Winn, J. C., Bester-Van der Merwe, A. E., & Maduna, S. N.2025). Annotated Bioinformatic Pipelines for Genome Assembly and Annotation of Mitochondrial Genomes. *Bio-Protocol*, *15* (5), 1–28. https://doi.org/10.21769/BioProtoc.5231

Wolf, C., Rentsch, J., & Hübner, P.1999). PCR–RFLP Analysis of Mitochondrial DNA: A Reliable Method for Species Identification. *Journal of Agricultural and Food Chemistry*, *47* (4), 1350–1355. https://doi.org/10.1021/jf9808426

Wragg, D., Eynard, S. E., Basso, B., Canale-Tabet, K., Labarthe, E., Bouchez, O., Bienefeld, K., Bieńkowska, M., Costa, C., Gregorc, A., Kryger, P., Parejo, M., Pinto, M. A., Bidanel, J.-P., Servin, B., Le Conte, Y., & Vignal, A.2022). Complex population structure and haplotype patterns in the Western European honey bee from sequencing a large panel of haploid drones. *Molecular Ecology Resources*, *22* (8), 3068–3086. https://doi.org/10.1111/1755-0998.13665

Wragg, D., Techer, M. A., Canale-Tabet, K., Basso, B., Bidanel, J.-P., Labarthe, E., Bouchez, O., Le Conte, Y., Clémencet, J., Delatte, H., & Vignal, A.2017). Autosomal and Mitochondrial Adaptation Following Admixture: A Case Study on the Honeybees of Reunion Island. *Genome Biology and Evolution*, *10* (1), 220–238. https://doi.org/10.1093/gbe/evx247

Wu, M.-C., Lu, T.-H., & Lu, K.-H.2017). PCR-RFLP of mitochondrial DNA reveals two origins of Apis mellifera in Taiwan. *Saudi Journal of Biological Sciences*, *24* (5), 1069–1074. https://doi.org/10.1016/j.sjbs.2016.12.008

Xu, K., Feng, J., Ma, X., Wang, X., Zhou, D., & Dai, Z.2016). Identification of tuna speciesThunnini tribe) by PCR-RFLP analysis of mitochondrial DNA fragments. *Food and Agricultural Immunology*, *27* (3), 301–313. https://doi.org/10.1080/09540105.2015.1086978

Yamauchi, M. M., Miya, M. U., & Nishida, M.2003). Complete mitochondrial DNA sequence of the swimming crab, Portunus trituberculatusCrustacea: Decapoda: Brachyura). *Gene*, *311*, 129–135. https://doi.org/https://doi.org/10.1016/S0378-1119 (03)00582-1

Yang, Z., & Rannala, B.2012). Molecular phylogenetics: principles and practice. *Nature Reviews Genetics*, *13* (5), 303–314. https://doi.org/10.1038/nrg3186

Zakour, M. K., & Bienefeld, K.2014). Basic considerations in the development of breeding plans for honey bees, illustrated by data on the native Syrian honey beeApis mellifera syriaca). *Journal of Apicultural Research*, *53* (2), 314–326. https://doi.org/10.3896/IBRA.1.53.2.13

Zammit-Mangion, M., Meixner, M., Mifsud, D., Sammut, S., & Camilleri, L.2017). Thorough morphological and genetic evidence confirm the existence of the endemic honey bee of the Maltese Islands Apis mellifera ruttneri: recommendations for conservation. *Journal of Apicultural Research*, *56* (5), 514–522. https://doi.org/10.1080/00218839.2017.1371522

Zayed, A.2009). Bee genetics and conservation. *Apidologie*, *40* (3), 237–262. https://doi.org/10.1051/apido/2009026

Zhang, G., Zhang, Y., Yuan, B., Tiang En, R., Li, S., Zheng, H., & Hu, F.2024). An innovative molecular approach towards the cost-effective entomological authentication of honey. *Npj Science of Food*, *8* (1), 24. https://doi.org/10.1038/s41538-024-00268-9

Zhang, H., Liu, Q., Lu, C., Deng, J., & Huang, X.2021). The First Complete Mitochondrial Genome of Lachninae Species and Comparative Genomics Provide New Insights into the Evolution of Gene Rearrangement and the Repeat Region. *Insects*, *12* (1). https://doi.org/10.3390/insects12010055

Zhang, R., Zhu, T., & Luo, Q.2023). The Complete Mitochondrial Genome of the Freshwater Fish Onychostoma ovaleCypriniformes, Cyprinidae): Genome Characterization and Phylogenetic Analysis. *Genes*, *14* (6). https://doi.org/10.3390/genes14061227

Zhang, Y. Z., Wang, S., Chen, Y. F., Wu, Y. Q., Tian, J., Si, J. J., Zhang, C. P., Zheng, H. Q., & Hu, F. L.2019). Authentication of apis cerana honey and apis mellifera honey based on major royal jelly protein 2 gene. *Molecules*, *24* (2). https://doi.org/10.3390/molecules24020289

Zhao, J., Yang, W., Cai, H., Cao, G., & Li, Z.2025). Current Progress and Future Trends of Genomics-Based Techniques for Food Adulteration Identification. *Foods*, *14* (7). https://doi.org/10.3390/foods14071116

www.ingramcontent.com/pod-product-compliance
Lightning Source LLC
LaVergne TN
LVHW052215200926
848481LV00001B/114